Practical Handbook of Warehousing

FOURTH EDITION

Chapman & Hall
Materials Management/Logistics Series
Eugene L. Magad, Series Editor
William Rainey Harper College

Total Materials Management: Achieving Maximum Profits through Materials/Logistics Operations, Second Edition by Eugene L. Magad and John M. Amos

International Logistics, by Donald Wood, Anthony Barone, Paul Murphy and Daniel Wardlow

Global Purchasing: Reaching for the World, by Victor Pooler

MRP II, by John W. Toomey

Distribution: Planning and Control, by David F. Ross

Purchasing and Supply Management: A Strategic Approach, by Victor Pooler and David Pooler

Competing through Supply Chain Management by David F. Ross

Practical Handbook of Warehousing

FOURTH EDITION

Kenneth B. Ackerman
K. B. Ackerman, Co.
Columbus, OH

 CHAPMAN & HALL

 **International Thomson Publishing**
Thomson Science

New York • Albany • Bonn • Boston • Cincinnati • Detroit • London • Madrid • Melbourne
Mexico City • Pacific Grove • Paris • San Francisco • Singapore • Tokyo • Toronto • Washington

JOIN US ON THE INTERNET   WWW: http://www.thomson.com
EMAIL: findit@kiosk.thomson.com

thomson.com is the on-line portal for the products, services and resources available from International Thomson Publishing (ITP).

This Internet kiosk gives users immediate access to more than 34 ITP publishers and over 20,000 products. Through *thomson.com* Internet users can search catalogs, examine subject-specific resource centers and subscribe to electronic discussion lists. You can purchase ITP products from your local bookseller, or directly through *thomson.com*.

Visit Chapman & Hall's Internet Resource Center for information on our new publications, links to useful sites on the World Wide Web and an opportunity to join our e-mail mailing list. Point your browser to: **http://www.chaphall.com**

A service of I(T)P

Cover Design: Andrea Meyer, emDASH inc.

Copyright © 1997 Kenneth B. Ackerman

Printed in the United States of America

For more information, contact:

Chapman & Hall
115 Fifth Avenue
New York, NY 10003

Chapman & Hall
2-6 Boundary Row
London SE1 8HN
England

Thomas Nelson Australia
102 Dodds Street
South Melbourne, 3205
Victoria, Australia

Chapman & Hall GmbH
Postfach 100 263
D-69442 Weinheim
Germany

International Thomson Editores
Campos Eliseos 385, Piso 7
Col. Polanco
11560 Mexico D.F.
Mexico

International Thomson Publishing-Japan
Hirakawacho-cho Kyowa Building, 3F
1-2-1 Hirakawacho-cho
Chiyoda-ku, 102 Tokyo
Japan

International Thomson Publishing Asia
221 Henderson Road #05-10
Henderson Building
Singapore 0315

2 3 4 5 6 7 8 9 10 XXX 01 00 99 98

Library of Congress Cataloging-in-Publication Data

Ackerman, Kenneth B.
    Practical handbook of warehousing / Kenneth B. Ackerman.—4th ed.
        p.   cm.
    Includes bibliographical references and index.
    ISBN 0-412-12511-0 (alk. paper)
    1. Warehouses. I. Title.
HF5485.A24 1997
658.7—dc20   96-43817
                                                          CIP

British Library Cataloguing in Publication Data available

To order this or any other Chapman & Hall book, please contact **International Thomson Publishing, 7625 Empire Drive, Florence, KY 41042.** Phone: (606) 525-6600 or 1-800-842-3636.
Fax: (606) 525-7778, e-mail: order@chaphall.com.

For a complete listing of Chapman & Hall's titles, send your requests to
Chapman & Hall, Dept. BC, 115 Fifth Avenue, New York, NY 10003.

Foreword

This is a fourth edition of a work first published in 1983. It contains the same number of chapters as the third edition, published in 1990. However, it has a substantial amount of new material. Major changes in warehousing in the last seven years have caused appropriate changes in the content of this text.

Nearly three decades have passed since our first published writing about warehousing. The goal of our early writing was to develop a better understanding between the third-party warehouse operator and the user of these services. Today the emphasis has changed to a work that provides the tools that every warehouse manager needs. This book intends to be a comprehensive handbook consisting of everything we know that would help the manager of warehouses. Much of the information is based upon materials previously used in *Warehousing Forum,* our monthly subscription newsletter.

While the work is designed primarily as a handbook for managers, it also serves as a guide for students. It is based upon my experience, both as a warehousing manager and executive, and later as a management advisor. The work is designed as a management reference for anyone involved in operating, using, constructing, or trading in industrial warehouses.

Acknowledgments

The dozens of people who made this work possible are credited in endnotes. Our writing could not exist without the generosity of people who have been willing to share their ideas and writings with me. We apologize in advance for names that may have been omitted in error.

For nearly a decade, Dewey Abram has been our advisor and coach in the preparation of *Warehousing Forum,* and his suggestions have greatly improved our writing.

William J. Ransom of Ransom & Associates continues to be one of our most effective advisors, and he contributed important ideas for new material in this edition.

This book is a successor to earlier books. A 1972 work titled *Understanding Today's Distribution Center* was coauthored by R. W. Gardner and Lee P. Thomas. Cathy Avenido and Jane Hill provided word processing and organization of the material. William F. Blinn provided final composition of this material. To all these people we are most grateful.

Contents

The Elements of Warehouse Management

Real Estate Aspects of Warehousing

Planning Warehouse Operations

Protecting the Warehouse Operation

The Handling of Materials

Handling of Information

Part I

BACKGROUND OF THE WAREHOUSE INDUSTRY

1

THE EVOLVING ROLE OF WAREHOUSING

History

No understanding of the changing role of warehousing can be complete without considering the unusual history of the development of warehousing as an economic function. As George Santayana said: "Those who cannot remember the past are condemned to repeat it."

Our civilization has passed through three major periods, the third of which began within the memory of most of us. The first was the Age of Agriculture, which lasted from the dawn of recorded history until two or three centuries ago. The second great period was the Industrial Era and it is linked with the development of improved transportation, steam power, and mechanical inventions that facilitated mass production. The newest period is the Age of Information. The utilization of computers and radio frequency terminals as common warehousing tools has taken place only in the last three decades.

Warehousing was initially linked to the Age of Agriculture. The description in Genesis of the use of granaries to store food and thus prevent famine in Egypt emphasizes the social benefit and commercial utility of warehousing. While a casual reading might suggest that Joseph's idea of constructing granaries was nothing more than an ancient social welfare program, the Egyptian monarch realized a commercial gain from this project.

The Renaissance saw the birth of the Industrial Era with the development of transportation systems to move spices from the Orient. The use of both transport and warehousing was also related to the Age of Agriculture. Before refrigeration, oriental spices were the only means of preserving meat and vegetables. Marco Polo and other Italian explorers went to the Orient to trade for spices and they gained a new awareness of the commercial practices of China. It was the Chinese who taught the Europeans that bank notes could replace gold and silver as a common currency. Under the same principle, a negotiable warehouse receipt could be bought and sold in place of goods stored in a warehouse. This negotiable receipt was called a Lombard, named after the northern Italian province of Lombardy. Beyond warehousing, most of the logistical activities of that time were centered on a search for new sources of spices. This was the principal motive for the financing of many 15th century explorers, the most notable of whom was Christopher Columbus.

The connection between third-party warehousing and agriculture remains strong today. At a convention of third-party warehouse operators in Cartagena, Colombia, the keynote speaker was the Minister of Agriculture. A significant portion of goods found in general deposit warehouses in Latin America is agricultural com-

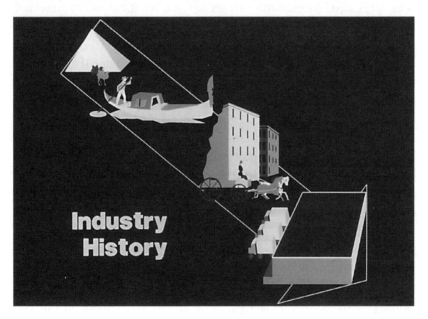

Figure 1-1. Industry History

modities, and the modern adaptation of the Lombard remains today in the deposit bonds of Latin American warehouses.

Warehousing in the Industrial Era

Warehousing was recognized as a useful reservoir for mass production as we entered the Industrial Era. When actual demand was unknown, it was prudent for the home appliance manufacturer to build a stock of finished goods so that the customer who wanted a yellow refrigerator could find it without delay. When capital was cheap and "just-in-time" (JIT) was unknown, the philosophy of marketing was to ensure that no sale was lost because of lack of inventory.

The Industrial Era was accompanied by major changes in transport. Warehousing has a "ham and eggs" relationship with transport, and warehouse design is influenced by the predominant transportation available. Storehouses constructed at piers served as a consolidation and distribution point for goods moving by ship. With railroading, warehouses constructed at rail centers were referred to as "terminal warehouses." The railroad industry developed "storage in transit" which allowed the user to move goods on a through rate from origin to destination with no extra transportation charge if the goods were held in an intermediate warehouse for up to one year. In the middle decades of the 20th century, extensive use of storage in transit created heavy use of both private and third-party warehousing in railroad junction towns in many inland states. The prime function of these transit warehouses was to provide a temporary holding point for goods at a point between origin and destination.

In the United States, the development of railroads led to monopolistic practices which caused concern for farmers and small businessmen. The Interstate Commerce Act, passed in the 1880s, was the first pro-consumer legislation in the United States. The purpose of this legislation was to prevent the railroads from using discriminatory pricing practices, and it required the carriers to publish standard tariffs and to seek the approval of the Interstate Commerce Commission for any changes in pricing. In the 1890s, a group of small businessmen who owned public warehousing firms formed an association and lobbied successfully for legislation to prevent the railroads from offering free warehousing services. The American Warehouse Association, one of the oldest trade associations in the United States, spent its early years challenging abusive practices from railroad operators.

As the Industrial Era created new modes of transportation, the need for government intervention no longer existed. The result was the deregulation that occurred in the 1980s, which was soon extended to every mode of freight transportation. With deregulation, common carriers diversified into third-party warehousing, and many of the major carriers today offer warehousing as part of a package of logistics services.

With changes in transportation, facility design changed as well. The multistory terminal warehouse became obsolete when the widespread use of forklifts after World War II permitted efficient high bay storage in sprawling single-floor buildings. Warehousing moved from the railroad yards to the suburbs as operators looked for low-priced land. Until the 1970s most warehouse operators would not consider a facility that did not have rail service. As boxcar use for packaged goods continued to decline, newer warehouses were developed that can be served only by truck.

Air freight carriers recognized that the hub and spoke concept of transportation creates an opportunity for effective warehousing at the air freight hub. Positioning of inventory at the hub allows the shipper to have a cutoff point for dispatching deliveries that is several hours later than if the warehouse were not located at the hub.

With the growing efficiency of transportation, warehouse users who formerly needed up to 100 regional warehouses to serve the marketplace could drastically reduce the number of locations needed to provide customer satisfaction. This reduction of locations is closely tied to the faster delivery times available from air and motor transport.

Warehousing in the Age of Information

The modern Age of Information continues to change the way warehousing is performed. Handheld terminals are used today to read bar codes and transmit the information by radio to a processing center. The scanning of bar codes with a handheld machine eliminates the possibility of error in reading a lengthy product code. Transmission of this information by radio allows a shipment to be deducted from stock instantaneously, thus eliminating the need to wait for paper to be processed in the warehouse office.

Until recently, the cost and size of computer equipment limited its use in warehousing. Today, personal computers have the processing capability to handle all but the largest of warehousing information systems. Very few warehouses today need a mainframe computer. "Never trust a computer that you cannot lift" was once a controversial statement, but is an accepted fact in warehousing today.

Logistics scholars such as Professor B.J. La Londe at The Ohio State University recognized that information could be substituted for materials movement. That recognition has turned to reality. A pharmaceutical manufacturer can use the computer to substitute for physical quarantine. Product moving off the assembly line must be held until quality checks are completed. At one time, the quarantine involved double handling in and out of a special storage area. Today, this is accomplished by locking the storage location on the computer so that it cannot be shipped until the testing and release process is complete.

Information systems for chain store warehouses allow goods to be moved directly from receiving to shipping docks when certain items are urgently needed in the stores. Before such systems were developed, every item at the receiving dock moved to a storage area. Effective use of information thus has eliminated some warehousing and the double handling of merchandise.

The Changing Role of Human Resources Management

Until a few years ago, a high percentage of warehousing activities in the United States were unionized. The prime emphasis in personnel management was to achieve a satisfactory contract and relationship with a union. Once this was accomplished, the union grievance procedure and the contract were the primary methods of dealing with hourly workers. Health benefits, retirement, and promotion were all specifically described and governed by a typical labor contract.

This high degree of unionization was the result of the close association of warehousing with transportation. Transportation in all modes was highly unionized by some of the most militant labor organizations, and those same unions organized most warehousing operations.

At one time the American Warehouse Association maintained a union contract file and warehouse operators who wanted to learn about labor costs in other cities were able to reference that file to compare their own costs for wages and benefits with those negotiated by operators in other cities. Today an ever smaller percentage of warehouse operations is organized by unions.

The turning point for trade unions was the unsuccessful strike against the Federal government by PATCO, the union representing air traffic controllers. This 1981–82 strike was the most important labor dispute in American history. It was the final confirmation of the proclamation made by Massachusetts Governor Calvin Coolidge during the Boston police strike of 1919: "There is no right to strike against the public safety by anybody, anywhere, anytime." It was Reagan rather than Coolidge who demonstrated that fact to organized labor and the nation by breaking the PATCO strike. The air traffic controllers learned that neither the government nor the public had sympathy for their position, and labor relations in the United States have not been the same ever since. Most of the newer corporations engaged in providing logistics services, both transportation and warehousing, are either non-union companies or firms that have labor agreements that emphasize flexibility.

Union-free status creates new challenges, since management now bears sole responsibility for the handling of grievances, compensation, and health care.

Continued emphasis on controlling costs of warehousing has caused management to monitor and improve the effectiveness of workers. In the 1990s, computer software replaced the stopwatch and clipboard as a means to measure the work accomplished.

Unlike the factory, the warehouse has no assembly line in which inspectors can be positioned to measure quality. Given the difficulty of line supervision, it was natural that warehousing would be included in those activities in which supervisors were eliminated through the use of a "self-directed work team." A flat organization with few layers between chief executive and hourly worker has been adopted in warehouses as well as other businesses. Only a few distribution centers today are managed with self-directed work teams, but the concept is being tested in more warehouses.

The different lifestyles in the past few decades have created new problems in managing people. One of these is the control of workplace substance abuse. This problem is aggravated in warehouses, because the population is younger, and more experimental in lifestyles. Second, just as close supervision is difficult in a sprawling distribution center, control of substance abuse is equally hard to accomplish. A growing number of companies have at least created a policy to handle such abuses when they are discovered. These policies usually recognize the difference in diligence between workers who may arrive at work under the influence as compared to those who consume prohibited substances on the job. The most severe punishment is reserved for those who sell such substances in the warehouse.

A major challenge in today's society is the ability to hire workers with sufficient skill to read and count the cargo handled in the warehouse. One result of our public education problems is the shortage of people with these basic skills.

Hanging over all of these personnel challenges is the threat of unions returning. Many in organized labor look for the chance to regain the members they once had in warehousing. Failure of management to handle the personnel challenges of today could result in a return to unionization. Since few managers want to see this happen, we will see increasing emphasis on successful management of the special people problems associated with the warehouse workplace.

Warehousing as Part of a Logistics System

The concept of business logistics is relatively new. The professional society for logistics managers was founded less than four decades ago. Cynics claim that the logistics manager is nothing more than yesterday's traffic manager looking for more organizational prestige. Regardless of motive, it seems clear that the concept of logistics has had a major influence on corporate organizations.

The role of logistics is mentioned in many business articles about competitive strategy. Logistics becomes the centerpiece of a new business strategy featuring time-based competition and core competencies. Using time-based competition, the winner in the logistics game will be the company that provides the fastest product delivery and the shortest order cycle time. The theory of core competencies prescribes that a manufacturer with demonstrated skills in assembling widgets should not dilute those skills by also operating warehouses.

Those who advocate core competencies would usually suggest that a third party should be retained to manage warehouses, and development of this idea is likely to extend the percentage of business that is managed by third-party warehousing companies.

As the time-based competition concept is implemented, there will be more use of cross docking and less placement of goods in long-term storage. Velocity of inventory movement has grown each year. With this increase in velocity, the warehousing function emphasizes materials handling rather than storage.

Globalization of Warehousing

Not long ago, most public warehouse operators in the United States conducted their work in only one city, frequently with just one building. Gradually the larger ones became regional operators, managing a number of facilities in several communities in one state or a group of neighboring states. A truly national third party warehousing supplier has become available only within the past few years.

The next step is international coverage of warehousing services

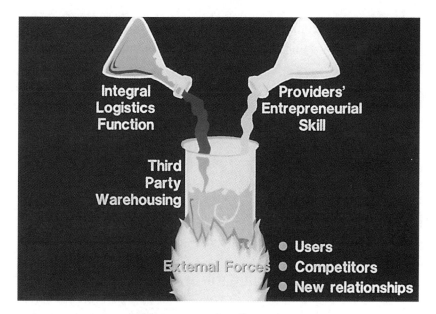

Figure 1-2. A Blending Process

from a worldwide organization. The move toward globalization is stimulated by a dropping of national barriers to trade. The European Community (EC) has effectively removed border inspections of freight vehicles. NAFTA has created the largest trading block in the world by removing tariff barriers between Canada, Mexico, and the United States. We are now seeing the beginnings of multinational warehousing organizations. Japanese third-party warehouse companies now operate in the United States as well as other countries in the Western Hemisphere. An Australian firm operates throughout the Americas. A few foreign warehousing firms now operate in Mexico. One of the largest third-party warehouses in the United States is a British organization, and several of the major refrigerated warehouses are European companies.

As warehousing goes international, United States firms may have some advantages. Third-party warehousing in the United States has had less government regulation and more entrepreneurial spirit than found in other countries. It was these characteristics that caused Mexican warehouse users to invite foreign companies into the marketplace. As our economy becomes more international, we are likely to see a growing number of warehousing organizations that offer multinational coverage.

A Profile of Tomorrow's Warehouse

The warehouse of tomorrow will emphasize fast movement rather than efficient storage. Order pick lines will be designed with an ergonomic layout, one that places the fast moving items in those locations where they can be picked with the least amount of effort or risk of personal injury. Above all, the layout will be designed to facilitate change. Warehouse managers will plan to rearrange the warehouse layout every three to six months, to adapt to new products and new demands.

The warehouse of tomorrow will have no office, since improved information transmission capabilities will have already eliminated the need for office workers to be at the same location where the cargo is handled.

Tomorrow's warehouse will be a flow-through structure with layout, equipment, and people who emphasize flexibility and capacity to accommodate quick and frequent change.

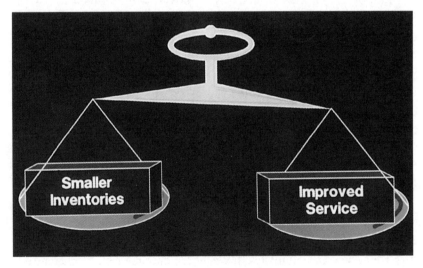

Figure 1-3. A Delicate Balance

2

THE FUNCTIONS OF WAREHOUSING

To evaluate the use of warehousing in your business, you need to understand ways in which warehousing functions to add value to products. In essence, warehousing provides time and place utility for any product, through efficient management of space and time.

Functions of Warehousing

Ultimately, the use of warehousing in commercial activity is related to its function in the business cycle. The following common functions of warehousing will be considered:

- Stockpiling
- Product mixing
- Consolidation
- Distribution
- Customer Satisfaction

Stockpiling is the use of the warehouse as a reservoir to handle production overflow. Such overflow reservoirs are needed in two situations—one involves seasonal production and level demand, and the other arises from level production and seasonal demand. For example, the canner of tomato products builds a warehouse inventory at harvest time, while customer demand for the product is fairly level through the year. For the toy manufacturer, the highest demand comes at certain seasons or holidays, but the manufacturer may need to stockpile in order to accommodate seasonal demand. In either case, the warehouse is the reservoir used to balance supply and demand.

A manufacturer who has product-oriented factories in different locations also has the opportunity to use a product mixing warehouse to combine the items in the entire line. For example, one food manufacturer has factories in several communities, with each factory producing a distinct line of products. To satisfy customers who wish to order full carloads or truckloads containing a mixture of the entire line, warehousing points are selected at locations that permit economical mixing of the product.

Some manufacturers need to stockpile semifinished products, and this function is called *production logistics*. One use of warehousing in production logistics is the principle of "just-in-time" (JIT) which requires either closely coordinated manufacturing or well-organized warehousing. It also requires precise loading of vehicles so that, for example, the blue auto seat is unloaded in precisely the right sequence to meet the designated blue automobile on the assembly line.

Consolidation is the use of warehousing to gather goods that are to be shipped to final destination. Warehousing costs are justified by savings in outbound shipping costs achieved through volume loads. In one case, a fast-food company uses consolidation warehouses to serve clusters of retail stores, thereby reducing costs and frequency of small shipments to the stores. Suppliers of the food company are instructed to place volume loads of their products in these consolidation centers. This enables the fast-food company to cut its transportation costs by moving its supplies closer to its food-serving outlets. At the same time, the food retailer reduces inventory costs by arranging for its suppliers to retain title to these inventories until they are shipped from the consolidation center.

Distribution is the reverse of consolidation. Like consolidation, it is justified primarily by the freight savings achieved in higher volume shipments. Distribution involves the push of finished products by the manufacturer to the market, whereas consolidation involves the pull of supplies by the customer.

In one case, a food manufacturer uses distribution warehouses to position products at locations convenient to customers. Fifteen distribution centers at strategic locations allow the pet food canner to achieve overnight service to major customers in the continental United States.

Both consolidation and distribution provide service improvements by positioning merchandise at a convenient location. Both involve cost tradeoffs that balance warehousing expense against transportation savings. Both provide improved time and place utility for inventories.

At times, improvement in customer satisfaction is the only motive for establishing an inventory. The five warehousing functions considered earlier all relate to production, marketing, and transportation costs. Yet, at times a warehouse stock is justified only by the demands of the customer, which may be far from frivolous.

A southern manufacturer of toilet seats persuaded a chemical supplier to place a warehouse stock of raw plastics in Cincinnati. The motive for this request was the need for backhaul freight for the manufacturer's private truck fleet. By establishing a raw material inventory in Cincinnati, the manufacturer is able to move finished goods into Ohio and then return his trucks with backhaul cargoes of plastics. The supplier positioned this inventory in Ohio to satisfy the manufacturer.

Warehousing Alternatives

The three types of distribution centers are differentiated by the extent of user control. These are the private warehouse, the public warehouse, and the contract or dedicated warehouse. Many companies use a combination of the three methods.

The private warehouse is operated by the user, and it offers the advantage of total control. If the user is distributing pharmaceutical products in which the penalty for a shipping error could be loss of life, such total control is often the only acceptable alternative. Where storage volume is large and handling volume is constant, the private warehouse is often the most economical means of handling the job. Yet, the private warehouse carries the burden of fixed costs and total exposure in the event of labor disruptions, and requires management. Though labor costs can be controlled through layoffs of unnecessary workers, many warehouse operators are reluctant to release skilled people who may be needed in the next few days or weeks. During a strike, the union can picket and impede shipments from a private warehouse. Warehousing clearly has its own required management skills, and the private warehouse operator must attract and maintain such skills in his management team.

The public warehouseman is an independent contractor who offers services to more than one user. The public operator does not own the merchandise that is stored, and usually the warehouse company is independent from the firms owning the inventory. By serving a number of customers, the warehouseman is able to balance the variations of inventory, or of handling the workload, and, therefore, develops relatively level demands for storage space and personnel. As an independent contractor, the public warehouseman is immune to involvement in the labor disputes of any clients.

Contract or dedicated warehousing is a combination of public and private services. Unlike the public warehouse, which offers a month-to-month agreement, the contract warehouse usually has a long-term arrangement. This contract may be used to govern supplemental warehousing services such as packaging, assembly, or other extraordinary activities. In such cases, the contract provides an element of stability in procuring services.

As an example, a machine manufacturer needed a parts supply depot but was unable to compile the specifications necessary for public warehouse rates. It was desirable to have the same workers handle this merchandise on a regular basis. A long-term lease was developed for the space in a public warehouse, and workers and lift trucks were hired at a fixed fee per week. In this instance, the warehousing contract included elements of both public and private warehousing.

Cost Structure

Figures 2-1 to 2-3 illustrate the cost structures found in third-party warehousing. All warehousing costs are a mixture of fixed and

Cost Structures in Third-Party Warehousing

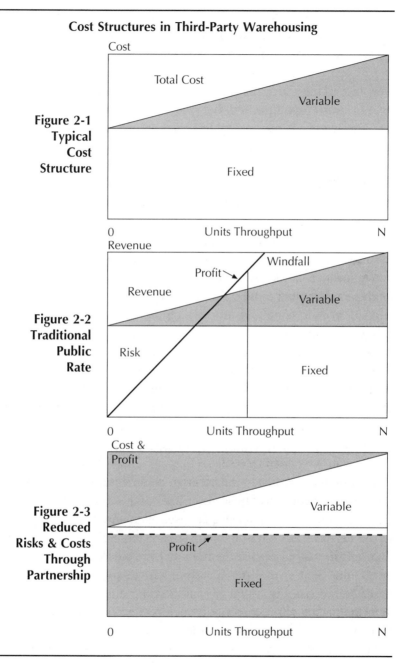

**Figure 2-1
Typical
Cost
Structure**

**Figure 2-2
Traditional
Public
Rate**

**Figure 2-3
Reduced
Risks & Costs
Through
Partnership**

Figures 2-1 to 2-3 Courtesy of W. G. Sheehan, The Griffin Group.

variable costs. The fixed costs in a warehousing operation account for more than half of the overall costs in that operation. Figure 2-1 shows how fixed and variable costs vary when volume increases.

To these costs the third-party operator must add a profit. However, in making rates, the public warehouseman must decide what level of unit throughput will be used to establish the rate.

The warehouse *user* may have negotiated a rate to pay $1 per unit to move merchandise through the warehouse. But what happens to the operator if activity is reduced? Figure 2-2 illustrates what happens. The vertical line in the middle of the chart shows the projected level of warehouse throughput. At that level, the warehouse contractor makes the profit that was planned. When throughput falls below that level, profit rapidly drops and moves to a loss. As throughput moves to the lower left corner of the chart, the warehouseman is unable to recover fixed costs. In contrast, when the throughput level moves ahead of projection, the gain turns from planned profit to windfall profit (which the operator keeps to protect against the rainy day when volume falls below projections). Much of the risk in traditional public warehousing is connected with the losses occurring when throughput is below projection, which are balanced by windfall profits when volumes are ahead of predicted levels.

The Balance Between Risk and Costs

When both parties to the contract understand the impact of the fixed and variable costs just described, and when both determine that they are going to have a long-term contractual relationship, they then have an opportunity to reduce these risks. A long-term relationship may also give them a chance to look at capital investments aimed at increasing handling efficiency, investments that might be too risky in a typical warehousing arrangement.

Figure 2-3 suggests a rate structure that controls these risks. Instead of a fixed per-unit rate, the user pays a level minimum fee that is designed to cover fixed costs and profit for the operator. That fee is represented by the solid horizontal line—a cost of $1,000 per month. In addition to this monthly fee, the user will pay a variable per-unit rate when throughput levels exceed the projection represented by the left vertical axis, in this case 1,000 units. If fewer than 1,000 units are moved, the minimum charge of $1,000 still applies. When volume doubles, fixed costs and profits stay the same, and the variable cost of moving the extra units is applied.

In choosing among the three warehousing alternatives, the user must measure the degree of risk involved in the decision. One may determine that the risk of public warehousing is unacceptable because of high penalties for service failure. However, this risk may be moderated by choosing the contract warehousing alternative with the user's own supervision to increase control over the warehousing function. A third alternative is to control the risk by having a private warehouse.

The Changing Values of Warehousing

In the current business environment, the ways in which warehousing can add value have been changed by structural changes in business logistics.

Today's logistics manager may be dealing with a system that manages tradeoffs among purchasing, manufacturing, transportation, and warehousing. For example, purchasing decisions can be altered to provide a balanced two-way trucking route. Purchasing and distribution managers may work together to position warehousing in a location that helps to create this balance. When multinational operations are developed, the warehousing job becomes much more complex.

To the extent that capital remains scarce, the logistics manager will continue to work at reducing inventory. Slow movers will be purged from the product line, and every effort will be made to speed

---

From an article by W. G. Sheehan, The Griffin Group, Volume 4, No. 3, *Warehousing Forum*, Ackerman Company.

turnover in warehouses. Getting the most out of inventory assets has become a key job for the logistics manager.

Elements of Warehousing: Space, Equipment, and People

Warehouse space is a commodity, and like any other, its price can show great volatility with changes in demand. At times, this commodity can become so scarce that the acquisition of additional space seems nearly impossible. Scarcity or high cost of space can be alleviated through changes in use or specifications of the other two warehousing components—people and equipment. On the other hand, when space is relatively cheap, people and equipment are used in a totally different fashion.

In different parts of the world, utilization of warehouse space is the product of cultural influences. In countries long accustomed to crowded living conditions, warehouses are designed to utilize space to the utmost. In contrast, part of the American frontier tradition is a general disregard for the value of both land and interior space.

Equipment consists of materials handling devices—racks, conveyors, and all of the hardware and software used to make a warehouse function. Some equipment is especially designed to save space, although it may also require greater amounts of labor. Obviously, different kinds of equipment will be used when space is cheap, when there are low ceilings, or when there is an unusual environment such as refrigeration.

The useful life of most warehousing equipment is substantially shorter than the useful life of the building. On the other hand, warehouse equipment depreciates in value, while warehouses often increase in value over time.

The most critical component in warehousing is people. The personal performance of warehouse workers often makes the difference between high- and low-quality warehousing. By comparison, the variations in quality of buildings and equipment are typically small.

It is possible to design warehousing systems that waste space and equipment, but reduce the cost of labor. An example would be a warehouse in which all freight was stacked just one pallet high and could be handled quickly, but space would be grossly wasted.

Space, equipment, and people are the three components of warehousing. At all times, consider the tradeoffs among the major components, and that judicious employment of any one of them can affect the performance or cost of the other two.

3

THE PROS AND CONS OF CONTRACT WAREHOUSING

Contract warehousing is a relatively new variation of commercial warehousing. In recent years, contract warehousing has been identified with growth and progress in the warehousing field. In the minds of many users and operators, contract warehousing is good, but the older format of public warehousing is bad.

Professor Thomas W. Speh of Miami University defines contract warehousing as "a long-term mutually beneficial arrangement which provides unique and specially tailored warehousing and logistics services exclusively to one client, where vendor and client share the risks associated with the operation." In listing principles, Speh notes a "focus on productivity, service and efficiency, not the fee and rate structure itself."

Language like this stimulates the love affair for contract warehousing which seems to promise greater stability and simplicity in the relationship between the third-party warehouse provider and the user of such services.

It has become increasingly apparent that such arguments in favor of contract warehousing miss the most important stimulus to this newer kind of service—the continuing scarcity of real estate financing. Those who advocate contract warehousing also ignore the fact that some contracts allow just as much price bargaining as the traditional public warehousing arrangement. While some contracts may build trust, others can destroy it.

The Historical Perspective

The concept of contract or dedicated warehousing reached great popularity during the past few decades. One early user of contract warehousing was a chemical manufacturer who sought to establish large third-party warehouse operations in very small towns in the southeastern United States. The manufacturer had established production plants in that region, and substantial amounts of warehousing were needed at or near the plant sites to provide an overflow storage capability and balance production with demand. The manufacturer had a strong desire to use a third party, primarily to avoid risks of unionization. As the search for a supplier proceeded, it became increasingly obvious that nobody in the public warehouse industry had the capability of financing construction of large new buildings in tiny towns in Tennessee and Kentucky. Traditional lenders were afraid of the risk, and third-party warehouse operators were equally concerned about owning a big building in a small community. The manufacturer solved the problem by offering a long-term contract designed to induce public warehouse operators to construct these specialized facilities. When the first arrangements were successful, the same manufacturer extended contract warehouse operations to many other locations, and the warehouse operators recognized a good thing when they saw it. Word spread about this new kind of warehousing, and the idea was imitated by many users and operators.

One experience with contract warehousing came about in order to solve a pricing problem. An office equipment manufacturer needed a new warehouse to handle spare parts, but no one had any specifications to allow the development of a conventional public warehousing proposal. The manufacturer did not have reliable data on product weights and dimensions, size of inventory, or average order size. Lacking such essential information, the operator suggested a contract arrangement in which the manufacturer paid a fixed rent for the space and purchased labor on an hourly rate. In effect, the risk of losing money was eliminated by making the arrangement a "time and material" contract. Many contract warehouse arrangements are still done on this basis, even when user and supplier are well able to calculate the warehousing cost on a per-unit basis. Some call this the "evergreen" contract because the supplier has a guaranteed profit. Is it any wonder that this approach is so popular with suppliers of third-party warehousing services?

Is the Contract Primarily for Lending Institutions?

From the end of World War II through the middle of the 1980s, the growth of third-party warehousing was facilitated by a very favorable real estate market. After the war, public warehouse operators needed to replace their multistory buildings with modern facilities, and private warehouse operators were doing the same thing. As a result, new suburban industrial parks grew rapidly, and commercial lenders handled many transactions to finance the real estate in the new industrial parks. A prime source of debt financing for new warehouses was the life insurance industry, accompanied by banks and S&Ls. Third-party operators with a decent track record had little difficulty in finding the money to finance construction of new space. While most bankers never really understood public warehousing, they were willing to back plant modernization. Leonard Sahling of Merrill Lynch describes the situation this way:

> *"When inflation peaked early in the '80s, the commercial property markets were all in short supply, and property values were soaring far in excess of the rate of inflation. If a property owner ran into problems in meeting the principal and interest payments on his mortgage, it was easy for him to sell the property, re-pay his debts, and probably still pocket a tidy profit. Defaults on commercial mortgages were negligible, while the fees and interest rate spreads were generously ample."*

Even though the lender did not understand third-party warehousing, this overall environment made it easy to be comfortable with warehouse expansion. Sahling observes that the Tax Act of 1981 liberalized tax treatment of commercial real estate and encouraged tax shelter driven syndications. While the resulting boom influenced warehousing much less than office properties, the influx of credit that flooded commercial property markets undoubtedly spurred warehouse construction as well. Sahling observes that commercial mortgages in the United States more than doubled in six years, moving from $408 billion in 1982 to $907 billion in 1988.

Everybody knows what happened next—the worst real estate down-turn since the Great Depression of the 1930s. In his research, Sahling describes how this influenced the three major lenders of warehouse properties. Between 1980 and 1991, the life insurance industry reduced its percentage of assets devoted to commercial mortgages from 21% to 15%. Between 1985 and 1991, the thrifts reduced the percentage from nearly 17% to a little over 13%. The banks were the last to reduce real estate lending, and that reduction is still taking place. Commercial mortgages peaked in United States commercial banks in 1989 at a little below 11%, and the percentage is still trending downward. The regulatory rules governing commercial property lending have been tightened for the thrifts, banks, and life insurance companies. More importantly, management in these lending institutions is more pessimistic about real estate loans today than it was years ago. The result is that debt refinancing for new warehouses is tougher to find today.

The "Evergreen" Contract

As contract warehousing has become more popular, users have recognized that the "evergreen" contract impedes innovation and improvement. A growing number of users have cut down the evergreen and replaced it with operating agreements that specify continuous improvement. This is accomplished first by separating the real estate from the operating contract. To finance new facilities, the user may need to make a long-term commitment to real estate to satisfy lender demands. However, the operating agreement can be separated from the real estate, thus giving the user the ability to change operators without moving the inventory. This also allows the commitment for space to be long term, but that for labor and management can be of much shorter term. Furthermore, most warehouse contracts today contain a clause that permits the user to dismiss the supplier if there is any failure in service. Service failures are not easily defined from a practical standpoint. As a result, the divorce of supplier from user is just as common today as it was twenty years ago when the traditional public warehouse arrangement existed. The repeated bargaining of the old arrangement is replaced by newer language such as "continuous improvement." The effect is the same. The ruthless customer can hammer his supplier just as effectively with a contract warehouse as he could with the older public warehouse arrangement.

The Pricing Challenge in Contracts

Pricing of warehouse services is just as complex as developing the price for building a house.[1] If you ask a builder to develop the price to build a house based on a rough sketch that you made on the back of an envelope, it would be amazing if the final cost agreed with the price developed from an inadequate set of plans. Yet users of warehouse services frequently give a similarly vague specification to the operator, and they are surprised and distressed when the price varies from the original estimate.

The Request for Proposal (RFP) sets the parameters from which the warehouse price will be prepared. If the RFP is wrong, the bid will be wrong as well. When it is wrong, what caused the error? There are six ways in which the pricing process can go wrong:

1. The RFP is not accurate because critical information about manufacturing and marketing forecasts was not included.

2. The warehouse operation is a startup with no history on which data can be based.

3. The outsourcing of warehousing is a secret.

4. The purchasing department designed an RFP with a goal of achieving simplicity.

5. The RFP contains projected growth forecasts that are not realistic.

6. Startup problems have not been considered.

There are answers to deal with each of the six pricing problems shown earlier.

1. When the buyer lacks information to create an accurate profile, the fairest way to start the operation is to manage pricing on a trial basis. The warehouse operator stipulates assumptions and then the prices themselves are based on those assumptions. Both buyer and seller sign a document that certifies that they agree on the assumptions that were used in making the prices. The contract stipulates that there will be a review and adjustment every ninety days for at least the first year of the contract. Each adjustment should be retroactive to the first day of the quarter under review. Most importantly, it should be understood that adjustments can go either way, with prices either raised or lowered depending on how actual conditions vary from the assumptions. While this approach may create some pricing uncertainty, it does ensure that the numbers are an accurate reflection of the cost of getting the job done.

2. When the operation is a startup with no data available, the same approach described earlier is the most accurate way to maintain per-unit pricing. However, buyer and seller may agree to handle the startup on a time and material basis, with a fee per square foot for space used and a handling fee based on a cost per hour. This cost plus approach removes the risk from the operator, but it is best used for a short period of time until an accurate per-unit price can be developed.

3. When the move is done in secret, the buyer may reduce the risk by a careful choice of suppliers. By considering only warehouse companies that have extensive background with the same commodity, the buyer will narrow the risk of major price variances because every bidder approaches the price development process with substantial and recent experience in handling similar products under comparable conditions.

4. When the purchasing department demands simplicity, the vendor will probably feel compelled to comply. However, the vendor should recite in his bid response a summary of the assumptions on which the prices were developed. Included in the price quotation should be this statement: *The assumptions listed above were the basis for calculating the prices shown in this proposal. Should these assumptions prove to be inaccurate, the above prices are subject*

to an adjustment that will be related to the degree of change in warehousing specifications. A quote prepared in this manner satisfies the purchasing department's desire for simplicity and still protects the warehouse operator from the possibility that the RFP does not accurately reflect operating conditions.

5. When the RFP is based upon growth projections, the operator can hedge against the future by developing multilevel rates. A certain price may be developed for startup levels, with stipulated discounts when volumes are higher. Some warehouse prices may be adjusted based on level of inventory turns, with a predetermined increase or decrease when turns are slower or faster than projected.

6. Start-up costs are a factor in every new warehouse operation, and many third-party warehouse contracts stipulate extra costs that are intended to cover these one-time expenses involved in a warehouse startup and then seek to recover that loss with a rate adjustment that is unreasonably high. Since startup costs are typically a one-time expense, they should be handled as a separate and nonrepetitive payment.

Contracts and Stability

Which third-party operator is more stable? Company A has ten major customers. Nine of the ten are public warehouse customers operating under the traditional thirty-day agreement. Of those nine customers, six have been with the company for more than twenty years, two have been there for over ten years, and one for about five. Company B has ten major customers, all of which are under contract. However, none of the contracts is more than three years old. When you look at it this way, clearly Company A has greater stability than Company B, even though contract warehousing is considered a long-term arrangement.

The real stimulus for contract warehousing has been the inability to expand third-party warehousing in any other manner. Both users and suppliers have recognized that additional space could not be constructed without a contract from the user, simply because the lender would not provide the money otherwise. If the warehouse provider could find an existing building, he might also find friendly financing from the owner of that building. Otherwise, the contract was an absolute necessity to allow the warehouse provider to obtain the space.

Enormous growth in recent years has turned *contract warehousing* into a corporate buzzword. As both users and providers focus on the simplicity and apparent attractiveness of contract warehousing, they may overlook some of the less obvious faults of the concept. The overwhelming stimulus for growth of contract warehousing has been scarcity of debt financing, and the contract in itself offers no guarantee of a stable relationship between buyer and seller.

4

WAREHOUSING AND CORPORATE STRATEGY

Strategy is a military term, and it refers to the planning and directing of military movements and operations. Like other military concepts, strategy has moved into the corporate world. In business, strategy typically includes the development of a corporate mission and objectives. Healthy companies plan to grow, and the question of how to grow is the most critical part of a corporate strategy.

Warehousing organizations deal with three primary assets: people, real estate, and equipment. The way in which these assets are managed is likely to be influenced by corporate strategy.

A Mission Statement for Warehousing

A mission statement is typically a brief description of the company's goals. One third-party warehousing company developed the following mission statement:

We define our business as providing customers with warehousing and related services designed to offer 48-hour delivery time to any significant market in the continental United States. We will build, own, and operate a chain of public distribution centers consistent with this objective. This will ensure the availability of efficient operating facilities at a real estate cost substantially below appraised value. We will aim to fill a unique role in warehousing. We will constantly seek innovative techniques in storage, handling, warehouse design, and construction. We will also seek to provide peripheral services, such as packaging and freight consolidation, to enhance our capabilities in warehousing and the service package offered to our customers.

We will become the most competent, rather than the largest, company in our industry.

Objectives are the additional detail that follows the mission statement. Here are some typical ones that could affect warehousing:

- We aim to be the largest firm in our industry.
- Our objective is to have the highest quality rather than the largest size.
- We plan to be the lowest cost provider in our industry. We will not be undersold.

If your company plans to be the largest in its industry, your warehousing job must be geared toward constant expansion. If your company plans to emphasize quality, you would naturally look for ways to both measure and improve the quality of the work done in your warehouses. When your company aims to be the low-cost provider, your prime emphasis will be to reduce warehousing costs. In each case, the mission and objectives of your company will influence the way in which warehousing is performed.

The Make or Buy Decision

If you are a logistics manager who uses warehousing services, the traditional "make or buy" decision should be periodically reviewed. Perhaps your current practice is to do all warehousing in your own facilities with your own people. As a logistics manager, your management expects you to explore the question of outsourcing. If you do the warehousing within your own shop, why was this decision made? Was it fear of being exploited by opportunistic suppliers? Was it a firm belief that you could run the warehouse for less money? Was it a fear that outsourcing would involve loss of control of a vital customer service function? Whatever the reasons, are they still valid in 1997?

"Make or buy?" has been a classic business strategy question for many years. Those who try to answer it usually start by listing priorities. If your first priority is to improve control, will you lose this capability if you buy warehousing from a third party rather than manage it yourself? Labor relations will influence the decision. If your first priority is to be union free, subcontracting warehousing may be important for you. Is ownership of real estate part of your corporate plan? This too would influence the way you approach the make or buy decision in warehousing.

Dealing with Growth

Let's assume that your company's goal is to be the biggest tree in the forest within a few years. Consider the fact that there are many ways to implement this goal. A classic strategy question is diversification versus specialization. The specialist will grow by making more of the same products produced today. The diversified firm will seek growth from new products, some of which may be radically different from anything handled today. Business strategists have debated the relative merits of specialization and diversification for decades. The implications for a warehouse are significant. If you are committed to diversification, specialized materials handling equipment that is not flexible enough to handle packages with a substantial variance in size and weight would not be a sensible investment. On the other hand, if your firm is a committed specialist, you may justify investment in highly automated equipment that is not very flexible. One of the most successful stacker crane warehouses handles a product that has not changed significantly in size and weight during the past several decades.

What about the diversification issue? A third-party warehouse company may choose to diversify into related activities such as transportation, packaging, warehouse construction, or industrial land development. Wholesalers or manufacturers might diversify into third-party warehousing by utilizing management skills that are already present in the organization.

Another strategic question regarding growth is "green field" expansion versus acquisition. Will your company grow by acquiring other corporations, or will you grow simply by buying or building new warehouses to support increased sales?

In the third-party warehousing industry, there have been a

number of acquisitions during the past decade. Yet some of our largest third-party warehousing firms have grown primarily through "green field" expansion, or the development of new warehouse buildings to meet a growing market. Why buy something that can be grown and developed? There is no one correct answer to this question, since warehouse operators clearly find good reasons to decide to grow through acquisition rather than through green field expansion.

As you look at your warehousing program, you need to understand whether your company is committed to growth, and how that growth is likely to take place. If "green field" expansion is part of the program, do you have sufficient knowledge about site development and construction of new warehouse facilities? If acquisition is the preferred strategy, how much do you know about the companies that compete with you?

Consider the possibility that the strategy is not to grow, but rather to make the company attractive enough to be profitably sold. Senior management reaching this decision will not publicize it, because common knowledge of such a strategy would impede marketing and internal morale. However, it is very important for you to understand what the strategy is, so you can adapt to it successfully.

What Business Are You in—Or, Should Form Follow Function?

We are in the business of trading information for inventory. Today, we have a real opportunity to do so. Traditional electronic data interchange (EDI) got us thinking about it and efficient consumer response (ECR) has really caught our attention. The next step will be true integration with all of the players in the supply chain—one that is transparent and merges into a true Supply Chain Management (SCM) system—a new class of software technology. SCM will let warehouse operators, suppliers, and manufacturers better integrate their operations. For example, a single purchase decision in a retail store can generate a virtually instantaneous series of domino-like responses all the way through the supply chain. Your goal is to be able to replace inventory—manufacturers', suppliers', and customers', wherever it may be—with information. To do this, all of us "touching" this inventory and its related information need to have integrated time-phased inventory management software in place.

SCM is, in effect, just one great big distributed application spanning across several firms, each with its own computer working in concert in the supply chain. It includes manufacturing and production systems; marketing and sales systems; order entry and product configuration systems; traditional logistics—including warehouse management systems; customer service; invoicing for customers; accounts payable and receivable; vendor relations; electronic commerce systems and communications; and integration, backup, security, and overall management of all of the data created by all of these systems.

Essentially, warehousing service is a commodity that is usually bought at the cheapest price. It is the relationship that you have with your customer that can change the game into a vital service that can demand a premium value. The ideal warehousing information system needs to provide the best in operations support, but it has to do that "next thing." Quality, value, and the timeliness of service have to be a given. To prosper, indeed survive, in tomorrow's business climate, you should step back, better understand what business you are in; understand what the driving forces behind the changes in your customer's industry and therefore your own industry, are; and anticipate where you need to be to meet those customer needs for tomorrow. Customers can be notoriously lacking in foresight. Twenty years ago, how many of us were asking for cellular telephones, fax machines, home copiers, mini-vans, compact disk players, or personal computers?

Motives for Contracting Out

There are many reasons for outsourcing, and these vary by industry and firm. There are several identifiable advantages to contract logistics arrangements.

First, the outsourcing of logistics functions allows the user firm to improve its return on assets. By reducing investments in warehouse facilities, materials handling, and transportation equipment, the return can be improved significantly. In the process, cash is preserved to invest in the core business.

Second, utilization of personnel can be more effective. By emphasizing the core business, the productivity of personnel is enhanced. With the intense competitive pressures in today's marketplace, streamlining and downsizing are virtual necessities; contract logistics affords a convenient and cost-effective vehicle for doing so. As market characteristics change, logistics needs change as well, and the use of an outside firm greatly reduces the risk of misplaced of outdated facilities and equipment.

One of the most important reasons for outsourcing is the change in the contract logistics companies themselves. No longer is the industry characterized by small, family-managed companies. The integrated logistic provider of the 1990s is a sophisticated firm, utilizing a combination of systems, facilities, transportation equipment, and materials handling techniques. It is managed and staffed with logistics professionals. More often than not, it is better qualified than its clients to perform the product distribution function.

Part II

THE ELEMENTS OF WAREHOUSE
MANAGEMENT

5

COMMUNICATIONS AND ELECTRONIC DATA INTERCHANGE

Because the warehouse is part of a larger logistics system, communications to and from it assume critical importance. Sometimes the warehouse serves as a buffer between manufacturer and customer, sometimes between manufacturer and supplier, and sometimes between wholesaler and retailer. It is always a place where accurate and timely communication is vital.

In essence, the most critical communications with a warehouse are those designed to answer five questions:

1. What has been or is to be received?
2. What has been or is to be shipped?
3. On what date and at what time was the customer's shipment delivered?
4. What is the correct amount of money due to each common carrier for transportation services?
5. What remains in inventory?

There are four general groups of people who will need answers to these and related questions: the operators of the warehouse, outside users of the warehouse, suppliers of transportation, and the ultimate consignees of shipments from the warehouse.

Nonroutine problems also must be handled in the communications procedure. These include quarantine holds on inventory, credit clearance, damage, and emergency shipping procedures.

The warehouse manager must frequently accept communications in a variety of forms, some of which may be easy to handle, some not. This chapter focuses on third-party warehouse communications. Because of the speed of computer development, methods of communicating are changing so rapidly that any attempt to describe them becomes obsolete even before the ink is dry. However, certain principles are more permanent.

Warehousing Forms

The warehousing industry uses serveral forms, that, while not strictly standardized, have only slight variances as they are employed throughout the country. Some forms will contain preprinted legal clauses, such as those customarily shown on a common carrier bill of lading. Figures 5-1 through 5-3 show sample public warehouse forms.

Receiving goods at the warehouse involves three different forms: the advance shipping notice, the warehouse receiving tally, and a non-negotiable warehouse receipt. The receipt normally has some clauses describing the responsibilities of the public warehouse.

The purpose of the advance notice is to give the shipper a listing of goods shipped *before* the inbound vehicle gets to the warehouse. The advance notice also identifies the trailer or box car number on which the product was shipped. Careful use of advance notice information can provide opportunities for savings in materials handling. For example, if goods contained on the inbound shipment are urgently needed to fill orders, those critical items can be kept on the dock for immediate shipment, thus avoiding a wasted round-trip to a storage slot.

A receiving tally is the warehouse receiver's independent listing of the goods that came off the inbound vehicle. This tally should be prepared "blind." The warehouseman filling it out should not know what the shipment is *supposed* to contain. The tally can then be compared with the advance notice, with an immediate recheck to resolve any discrepancies.

The purpose of the non-negotiable warehouse receipt is to

provide legal certification that the goods listed on it are now in the custody of the public warehouse.

Public warehouses usually provide a stock-status form at the

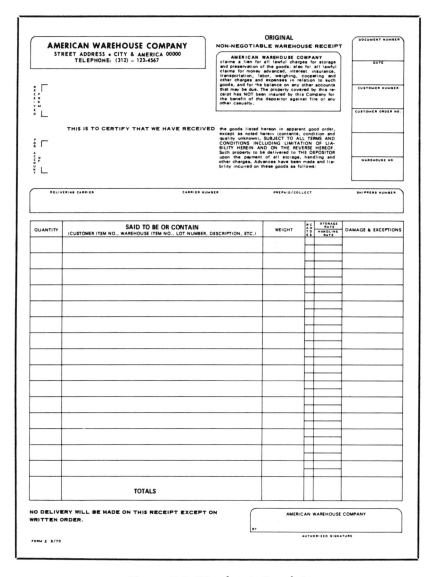

Figure 5-1. Warehouse Receipt

end of each month that also includes inventory levels. Some of these reports will show a beginning balance, a summary of receipts and shipments, and an ending balance. This provides a means of checking the public warehouse records against the user's records.

When common carriers are used, the bill of lading is a contract

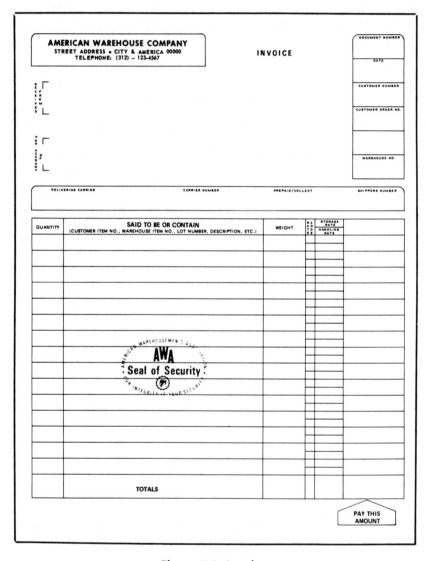

Figure 5-2. Invoice

between shipper and carrier. It provides proof that the merchandise was transferred from shipper to carrier, and that the carrier has assumed responsibility for the cargo until he can prove it was delivered.

NO. UNITS BILLED	NO. UNITS RECEIVED	DESCRIPTION		OVER	SHORT	DAMAGED	GOOD ETC OR	RE-JECTED	LOST
2000	1995	Wigit, NOIBN			5				
						14		14	

IMPORTANT - NOTICE TO CARRIER

THIS REPORT IS NOTICE THAT A FORMAL CLAIM WILL BE FILED. ALL SALVAGE ITEMS AND/OR REJECTED GOODS ARE BEING HELD FOR YOUR ACCOUNT SUBJECT TO A CHARGE FOR STORAGE IF NOT PICKED UP WITHIN 48 HOURS. WE ASSUME NO RESPONSIBILITY FOR LOSS OF WEIGHT, SHRINKAGE, LEAKAGE, OR CONDITION OF MERCHANDISE. PLEASE ARRANGE FOR PROMPT DISPOSITION.

NOTICE TO CUSTOMER

FOLLOWING PAPERS ARE ATTACHED:
[X] INBOUND FREIGHT BILL   [ ] INVOICE #
[ ] ORIGINAL BILL OF LADING   [ ] OTHERS

CARRIER CERTIFICATION OF INSPECTION

I HEREBY CERTIFY THAT I HAVE EXAMINED THE ABOVE DAMAGED SHIPMENT AND FIND AS REPORTED HEREON, AND THAT I HAVE BEEN FURNISHED A COPY OF THIS REPORT.

INSPECTOR'S SIGNATURE _____
NAME OF CARRIER ___ MTRR ___
DATE INSPECTED ___ 8/20/84 ___
BY: ___ J.M. ___

CARRIER RECEIPT

RECEIVED ___14___ UNITS AS ITEMIZED IN REJECTED COLUMN ABOVE

NAME OF CARRIER ___ MTRR ___
SIGNED _____
DATE ___ 8/30/84 ___

ORIGINAL

Figure 5-3. "OS&D" Report

Damages

Sometimes a common carrier shipment to a warehouse will either have damaged freight or a discrepancy in the count. Such exceptions are reported on a report known as the "OS&D," which stands for *over, short, and damaged.* This document, prepared by the warehouse operator, is the basis for settlement of a claim between the shipper and the common carrier.

Stock Status Information

The purpose of a stock-status report is to allow the warehouse operator and the user to compare notes on the quantity of merchandise availability in inventory. If user and operator do not agree on balances, they must find the source of the error.

The stock reporting system should have a means of reflecting errors that are identified but not yet adjusted. By adding a variance column to the stock-status report, the user is informed of the existence of these unresolved errors. Typical of such mistakes is an inventory crossover, in which a shortage of one model is balanced by an overage of another. This happens because the wrong item was shipped. The variance column provides an interim report that an error exists. When a physical inventory adjustment is made, the variance disappears.

Communications regarding warehouse releases or shipments must be precisely controlled. The warehouse user must determine which parties have the power to release merchandise, and how and when credit authorization will be handled. Some warehouse users have salesmen or brokers, and the power of these people to make warehouse releases must be carefully documented.

The bill of lading provides proof that goods moved from warehouse to common carrier. However, information about when and in what condition goods are received by the consignee can be obtained only through the carrier's freight bill.

Communications and Order Picking

The format of order pick sheets affords a labor-saving opportunity in order selection. Ability to sequence line items by location and to process orders for zone picking can greatly improve the effectiveness of the warehouse order pickers. If goods are arranged in a fixed location, the order-picking document should list merchandise in the same order in which it is located on the line.

One of the most common communications breakdowns in warehousing is confusion as to whether an order is in cases or units. Some orders may call for full cases and individual packages from broken cases. If the order is not clear as to whether cases or packages are involved, there can be substantial errors.

Another symptom of poor communication is error in transcription of orders. With today's information handling techniques, manual copying of information should be prevented, thus reducing opportunities to make errors.

Dealing with Complaints

Any busy operator will inevitably experience complaints. However, complaints should be looked upon as an opportunity for improvement. When they are handled properly, a faulty procedure may be identified. Logging complaints can be a means of comparing the relative efficiency of various distribution centers. But, to do this fairly, the number of complaints must be related to the number of transactions.

Developing a Standard Procedure

Both the user and the operator of the warehouse should provide written procedures to describe how each job is to be performed. No warehouse is too small to have written procedures, and the presence of such procedures provides a basis for training new people and maintaining quality control over existing operations. Misunderstandings about the best way to get a job done usually occur when there is no agreed upon procedure. Written procedures frequently no longer exist in mature companies where employees have been on the job for many years and are performing their jobs from memory. Unwritten procedures frequently become erratic.

For example, grocery product warehousing uses "tie and high" instructions. This is a written pallet pattern to be used when floor-loaded goods are received at the warehouse. In the absence of specific instructions, the delivering truck driver may use an ineffective storage pattern on the pallet, or even worse, he may use a pattern that makes it easy to conceal a product shortage by leaving voids in the center of loaded pallets. Precise description of stacking pallet patterns is essential when floor-loaded case goods are received.

EDI—How It Looks from the Warehouse

The concept of electronic data interchange, popularly known as EDI, has been around for decades. Most warehouse managers have few options about whether or not they will use EDI, but many have questions about when and how they will use it. As the function of the warehouse has changed from a storage depot to a distribution center to today's cross-dock facility, the owner of the merchandise has used available technology to track inventory status as it moves through the supply chain. The warehouse operator, like every other link in the chain, must cooperate in providing the visibility that the owner of the inventory demands. EDI is the mechanical way in which this visibility or control in maintained.

Justification for EDI

As long ago as 1976, The Transportation Data Coordinating Committee (TDCC) was formed to develop standard format for carriers in the different transport modes. Soon thereafter, the grocery industry started a search for a way to achieve uniformity in communications between the various links in the supply chain. Industry gurus pointed out that EDI was just around the corner, and it would soon have a major impact on warehousing.

If EDI was just around the corner, it was one of the biggest corners ever seen. One answer from a manager asked about this is "We haven't implemented EDI 100%, but we are on our way." Some cynics believe that manager will be saying the same thing 10 years from now. However, there is far more evidence that EDI is finally making its mark in business logistics and warehousing.

There are two incentives to adopt EDI. The first is to improve cycle time and accuracy. The second and more common reason why companies adopt EDI is that they are virtually forced into it by a major customer.

Is EDI a fad or a megatrend? After languishing for many years, what finally brought EDI out of the closet? Who will lead and who will lag in EDI development? What forces will spur the growth of EDI in the future?

Those who accused EDI of being a fad are clearly mistaken. While formats are not yet standardized and there may be some future changes, the concept is here to stay.

How Does it Work?

Electronic data interchange in the third-party warehouse includes at least 15 standard transactions. The table describes a typical transactions format. Additional management reports allow the tracking of productivity, measurement of product velocity, and warehouse

---

PROTOCOLS FOR EDI

Receipts
  Enroute Receipts-Warehouse Ship Advice (Inbound)
  Enroute Receipts from Principal (Inbound)
  Receipt Confirmations (Outbound)
Orders
  Shipping Schedule
  Confirmation
  Order Load (Inbound)
  Order Change (Outbound)
  Order Confirmations (Outbound)
Inventory
  Inventory Status (Outbound)
  Inventory Adjustments (Outbound)
Master File Updates
  Product Master File Add/Update (Inbound)
  Product Information Request (Outbound)
Carrier Communications
  Electronic Load Notices/Advance Load Notices
  Appointment Notification
  Delivery Notification

---

locations. While the goals from the beginning has been to develop a standard protocol, that goal is far from being reached. Very few EDI users have adopted all 15 of the protocols listed in the table, and the format used will differ between users. Third-party warehouse users report that they frequently must modify their system to meet the needs of a particular client.

The large users show the greatest savings reductions from EDI—a reduction of clerical costs to one-third of former levels. Error reduction is higher among the large users, though the smaller companies also show significant reductions. The large users are driving their trading partners into EDI. This confirms information received from third-party warehouse sources that indicates that EDI is a necessary part of the service package required by many clients.

Professor Margaret Emmelhainz at the University of North Florida cites inventory management as a contribution to growth of EDI.[1] Logistics managers today face pressure to reduce inventory without compromising service to customers. They are being asked to change inventory turns from between six and seven turns a year to ten and eleven turns within the next few years. One way to achieve improvement in inventory turns is through better matching of actual production with actual demand. EDI improves the accuracy and speed of the information needed to manage inventory more effectively.

The global economy is also a stimulus to EDI. When language is a barrier, a standard format for transmitting information is one way to ease the communication problem.

It is not easy to point to any specific circumstance that caused EDI to flourish in recent years. Presumably its growth is related to the increasing power of retail chains, and to the fact that such organizations have embraced information technology as a means of improving profits.

Today, the leaders in EDI have been large companies, primarily but not exclusively in consumer goods. It is natural for smaller companies to lag, because they lack time and money for implementation and because their customers have not asked them to adopt EDI. Eventually, many of those who lag will get on board because major customers force them to do so. EDI will continue to grow in the future as a growing number of large users press for standardization of procedure by all of their suppliers. It seems clear that EDI is here to stay because it offers significant cost reductions and because a number of large companies have pushed their customers and suppliers to use it.

The sudden and relatively recent growth of EDI could well be compared to the development of standard pallets over thirty years ago. Early in the 1960s, the conventional wisdom said that a standard pallet could never be adopted because grocery chains and other major warehouse operators each had racking systems designed for different sizes of pallets, and a standard pallet would cause many warehouses to reconfigure expensive storage rack installations. Then General Foods introduced its standard pallet and began to pressure customers and suppliers to use the same specification in order to exchange pallets at the dock. As one of the most powerful companies in the grocery industry, General Foods was successful in persuading a trade group to adopt the same standard for use by all of its members. Grocery Manufacturers of America introduced the same pallet design and with amazing speed all of the people who said that standard pallets were impossible began to adopt the standard design.

How Widely Has EDI Been Accepted?

Although a minority of users of third-party warehousing are actually implementing EDI, that minority is made up of leading consumer products manufacturers. Most of these are the largest customers in the warehouse. For example, one company reported that only 12% to 15% of its customers are using EDI, but their order volume represents over 50% of the total. Third-party warehouse operators are "pushing" EDI as a concept that will save money both for them and for their customers, and will improve speed and accuracy. EDI must therefore continue to grow.

The Future of EDI

EDI is growing rapidly because a relatively small number of highly influential companies are demanding that EDI be used in the warehouse. Because EDI can speed freight payments, major carriers have joined the pressures to implement EDI in all aspects of logistics management. EDI has significant advantages, but the most important is the reduction in total cycle time. Other advantages include reduction of clerical error rate, faster access to data, and improved customer service.

Some have said that information is power. In the warehouse of 1997, information is the price of entry. Computer programs and information are available to everyone who has a warehouse. How the information is used is the key to success in warehousing, just as it is in most other industries.

Access to information is no longer a competitive advantage. The advantage lies in how that information is integrated into management processes and management decision making. EDI is one method of transmitting and integrating that information, and success in utilization of EDI could well separate the winners from the losers.

Emmelhainz predicts that some firms will actually use the information available from EDI transactions to run decision making software. Information will be used to better manage the entire supply chain and to make your company more competitive.

Other potential new developments are a form of real time EDI, or the use of the Internet as a means of handling EDI transactions.

While adoption of EDI is not cheap, it has clearly proven its value for major users. Therefore EDI is destined to continue its rapid growth in the future. As a warehouse manager, it would be foolish for you to neglect continual review of developments in electronic data interchange. EDI is destined to continue its rapid growth in the future.

6

PACKAGING AND IDENTIFICATION

Warehousemen in the United States receive, store, handle, and ship a broad range of products. All products (with the exception of some bulk commodities) are packaged in some way. A food packaging system may consist of a primary package that touches the product, an intermediate or secondary package such as a folding carton showing the advertising message, and a distribution container that identifies and carries multiples of the secondary package. The warehouseman receives, stores, and handles the distribution container but must deal with it according to the size and strength of the total package.

Functions of a Package

A packaging system has four functions: (1) protection, (2) containment, (3) information, and (4) utility.

The package should provide protection against the common hazards of warehousing and distribution. These include stacking compression; shock and vibration in transit and handling; and protection against temperature extremes and changes, moisture, and infestation. The containment quality of a package is its ability to resist leakage and spilling, and containment methods will vary with the type of product. The information on packaging may consist of advertising messages, as well as a listing of ingredients. For the warehouseman, the most important information is that which enables the

The packaging section of this chapter was adapted from an article by Paul L. Peoples, School of Packaging, Michigan State University, East Lansing, MI.

materials handler to identify the package correctly. Utility of the package itself involves its convenience of opening and closing. In warehousing, utility is usually the ease and efficiency of handling the unit in storage.

The manufacturer naturally seeks a minimum cost for these functions, but logistics people must be concerned about the normal stresses involved in warehousing and transportation of the product. Clearly, packaging represents a compromise between the minimum cost desired by the manufacturer and the minimum strength needed to protect the product as it moves through the various channels of transportation and warehousing.

Ideally, the package designer provides sufficient packaging so that the package, when coupled with the inherent strength and characteristics of the product, is equal to its environment at every stage. Less than adequate packaging leads to damage. Excessive packaging leads to excessive costs. The variables in this equation are, of course, constantly changing and are seldom precisely known. The transportation environment varies in different regions of the country. It also varies between modes, and even between different trucks and highways. The storage environment and other potential hazards can also vary. The use of new packaging materials can cause major variations in package performance. The net effect is that even the best of packages can become a poor performer owing to unforeseen changes in its environment.

How the Package Affects Warehousing

Most warehousing operations deal with the distribution package, and the strength and dimensions of this package have a direct effect on handling and storage efficiencies. The weight of the individual unit affects handling and order picking efficiency. The compressive strength of the package and product affects the stacking heights and thereby the storage efficiency of the system. The height of the package affects the utilization of individual rack spaces; its length and width influence the stacking pattern and the effective use of space on the pallet, as well as the use of overlapping packages to make firm pallet loads.

Quite frequently, the dimensions and performance of the package–product combination are accepted without challenge by the warehouse operator. It is assumed (falsely in many cases) that changes are not possible. In reality, changes are possible and can yield large benefits in warehouse utilization and efficiency. Very minor dimensional changes of the distribution package can yield large increases in rack efficiency, pallet cube utilization, and use of trailers. If the distribution carton itself cannot be altered, the sizes of the interior package should be investigated. Minor reductions in height that do not affect product weights and volume, or changes in height with compensating increases in diameter, could permit the required changes in the distribution container. The need for such an investigation will be readily apparent if the cubic space occupied by cases of the product is compared with the cube required to store a pallet load of the product. In a similar manner, the number of units within the distribution package also can be changed to affect container dimensions and improve its storage.

The sizing of a container for interior packaging to achieve the most efficient pallet pattern, height, and good trailer utilization involves a multitude of variables. Computer programs have been developed to analyze palletization, case count and arrangement, package sizing, and truck/rail car loading.

Effect on Stacking

In recent years, the shift from metal and glass containers to plastic containers for many liquid products has resulted in significant decreases in stacking height for these products. The metal and glass containers provided the necessary strength to withstand four-pallet-high stacking. The thinner and weaker walls of the plastic container often reduce maximum stacking height to two pallets. Although the use of dividers and heavier container materials can improve plastic container stacking heights slightly, they often will not permit the stacking heights of glass and metal containers. Warehousemen can look forward to further increase in the use of plastic containers for food and liquid products.

In addition, many food products now in cans and jars will soon be marketed in packages formed from composites of plastic film, foil, and paperboard. It is quite probable that stack heights for these commodities will be reduced. The sidewalls of these containers are flexible and bulging containers will be more prevalent. The prudent warehouseman concerned with maximum utilization of warehouse space should be looking at alternative storage methods to include drive-in and drive-through racks, as well as other methods for improving warehouse cube utilization.

Most distribution containers in the United States are constructed from corrugated board. The strength of the total packaging system is a combination of the product and the container. Some products, by their nature, are inherently load sustaining. Examples are metal cans, glass jars, and industrial products with rigid cushioning, blocking, and bracing systems. Products in nonrigid containers, such as paper products and loose food products, cannot contribute to the stacking and compressive strength of the container–product system. Accordingly, the stacking capabilities of such systems are totally dependent on the strength of the fiberboard container.

By its nature, fiberboard is a hygroscopic (water-absorbent) material. As moisture is absorbed by the fibers, they swell and the bond between them begins to weaken. At a temperature of 73°F and relative humidity of 50%, fiberboard has a moisture content of about 8%. If the temperature remains constant and the relative humidity is increased to 90%, the moisture content of the fiberboard increases to 20%. This increase in moisture content lowers the compression strength of fiberboard box by nearly 50%. Another characteristic of corrugated fiberboard is its tendency to creep and lose strength when subjected to a steady load, as in a storage stack. The rate at which this creep and strength loss occurs is based on many factors, including the support provided by the product, and stacking misalignment.

For best use of space with minimum risk of stack failure, the warehouseman should work closely with available packaging specialists to verify stacking capabilities under various temperature, humidity, and storage duration situations.

Packaging for Unitization

Packaging can save labor at the distribution center through unitization, which is simply the consolidating of smaller boxes into a larger single unit in order to reduce the number of pieces handled. There are several ways to unitize packages, the most common being palletization. A pallet load of small packages may be glued together with nonslip adhesive. Such an adhesive has a high shear (side-to-side) strength and low tensile strength. This prevents the packages from sliding off the pallet, while enabling them to be easily removed by lifting.

Pallet loads also can be unitized with shrink film. This is a plastic film that is wrapped around the load and then heated to shrink and form a tight outer shell around the pallet, offering protection as well as load integrity. Strapping or tape also may be used to stabilize a unit load. A disadvantage of unitization occurs when one of the cartons in the unit is damaged, and the entire load must be rehandled.

Identification

There are three hidden costs resulting from poor package identification:

1. Labor cost involved in extra searching during order selection.

2. Cost of correcting shipping errors resulting from identification problems.

3. Customer dissatisfaction created by shipping errors.

One cause of poor identification is package advertising that obscures the identification code. Another is complicated marking systems that are difficult to read, understand, or recall. Some numbering systems are conducive to numerical transposition. An example is quantity-oriented coding, such as 12/24 ounce and 24/12 ounce. These numbers describe the quantity and size of the inner packages, but are easily reversed or transposed.

The packaging department, in conjunction with marketing, determines the system used for identifying products—often without considering the problems of order selection in the warehouse. Small numbers, combinations of alphanumeric identifiers, poor lighting, and poor color contrast will inevitably lead to mistakes in order picking and shipping errors. Warehouse operators should insist on product identification systems that are legible in the warehouse storage environment and use only the minimum number of letter and number combinations necessary for identification.

When the product moves in fewer than truckload quantities, the carriers normally require that each container be stenciled with the appropriate address. Container printing should allow space for the stencil or label.

Effective identification has certain characteristics. One is the use of large and bold marks with adequate background contrast. Another is marking on two or more sides of the container. A third is the use of color.

Increasingly, package identification systems are equipped with bar-code printing designed for reading by scanners. More about scanner technology will be found in Chapter 50.

Container Handling

Many products are picked and shipped in fewer than pallet quantities. Order picking and handling costs are typically developed around the cost-per-container. Each of the units within the container carries a portion of that handling cost. By increasing the number of units within each container, the handling cost per unit can be dramatically reduced as can the material costs for containers. By tradition, many products are sold at the retail and wholesale level in some multiple of six, twelve, and ten. This is often perpetuated by sales order blanks that identify only those multiples. But when the wholesaler or retailer is purchasing in quantities of 96, 100, or several hundred, there is no need to package the product in smaller multiples. The use of master cartons that approximate the quantity typically purchased, or the most economic order quantity, can reduce a number of case handlings and thereby the handling cost per unit.

Package Design

Product packaging should be designed with consideration for all of the transportation and handling operations involved. If the package is designed to save on handling and transportation costs as well as provide protection, it may be different from one designed solely to minimize costs. A tradeoff calculation will demonstrate whether additional packaging expense is justified.

Increased packaging costs may be offset by savings in handling or transportation. Unitized handling with a carton-clamp truck can provide substantial labor savings in the warehouse, but the strength of the package must be sufficient to protect the product during carton-clamp handling. The extra protection built into the package to allow clamp handling also will provide a margin of product protection throughout the distribution cycle.

There are also cost tradeoffs in improved storage capability. Most newer warehouses have a clear pile height of 20 feet or more. Unfortunately, there are many products that cannot withstand a stack height of 20 feet so the overhead space is likely to be wasted. While the addition of steel storage racks will improve space utilization, it is costly to equip a warehouse with racks. Moreover, the storage height gained by racking is frequently offset by the lateral space that is lost for aisles. For this reason, storage racks are not always an economical investment. Furthermore, insurance underwriters have expressed concern as to whether normal fire protection systems can control fires in storage racks.

Any distribution manager with products now limited to less than 20 feet of free standing pile height would be wise to evaluate packaging that would allow increased stacking height.

Package design must also afford protection from damage, so the designer must understand the causes of damage in physical distribution. In storage, carton compression is a potential problem in the stacking of merchandise. When a stack falls in a warehouse, the causes of the collapse should be studied. In many geographic areas, humidity is the most serious environmental problem because most packages will be weakened by high moisture absorption.

The packaging engineer also must consider risks of mishandling of his product throughout the distribution system. He must know what happens to the package and to the product when it is abused. For example, one appliance manufacturer discovered that a cartoned washing machine could be dropped from a height of 15 feet; if it landed squarely on the floor, the package would show no evidence of damage. Yet, inside the package, the machine itself was a total loss.

The responsibility for such concealed damage usually cannot be determined. Frequently the damage is not discovered until the product reaches the consumer. Sometimes it is not discovered until the product is actually put to use. An ideal package is one designed to reveal *all* damage, thus eliminating the problem of assigning responsibility for concealed damage.

In planning a package for distribution, the designer must consider the value of the product to be packaged and the length of time the product will be in the package. "Overpackaging" is the popular term for a carton stronger than it needs to be. Overpackaging may be an economical investment for a product of critical value. Overpackaging also is desirable for an item likely to be in the distribution channels for a long period of time, or subjected to an unusual environment or to extensive transportation or rehandling.

Choosing a Contract Packager

Each year, manufacturers spend an increasing amount of money on value packs, displays, couponing, and other forms of specialized packaging. Increasingly, for a variety of reasons, they choose an outside contractor to perform these services. As a result, contract packaging has become one of the newly popular third-party logistics specialties.

There are several product/customer/facility/equipment constraints that can be eased by the use of contractors. A few of these are:

- Packaging for military and export markets (materials, methods, marking, and testing requirements often differ markedly from the domestic)
- Packaging of hazardous materials
- Short production runs and regional marketing tests
- Seasonal products (antifreeze, suntan lotion, etc.)

Packaging can add substantially to product weight and cube and thus escalate shipping and warehousing costs. Therefore, some firms defer the packaging operation by shipping products in bulk to a downstream distribution center. Final packaging and labeling is then performed as a part of the warehousing/order picking function. A similar procedure is used when the same product is sold under different labels and markings.

At least one United States computer manufacturer ships units in bulk to an European distribution center where country/language specific packaging and markings are applied.

Similarly, United States firms import flashlights, batteries, auto parts, and bike parts loose, in large boxes. They are then packaged singly or in kits with United States logos for sale in the United States market.

Consider these guidelines for choosing a vendor:

- *Precise Specifications:* Develop a complete description of the service or package desired—materials, color, graphics, marking, performance, etc. This could be performed by your own people, by consultants, or with advice from potential vendors. Once agreement is reached, this specification becomes your primary means of measuring vendor capability and subsequent performance.

- *Location:* A contract packager should be conveniently located relative to your manufacturing or distribution facilities. This will save valuable time in delivery and lower freight charges that might impact the total cost of the project.

- *Experience:* No two packaging projects are exactly alike. Go with a company that has performed contract packaging services for other companies with similar product lines and packaging needs.

- *High Standards:* Does your prospective supplier have a clean facility? Are there training safety programs? Is there a good inventory control and planning system? And is it up to your standards?

- *Organizational Capability:* Many contract packagers are capable of producing the package but may not have the control systems and records necessary to satisfy your internal quality control requirements. This is particularly true with foods and pharmaceuticals. If your packaging includes detailed paperwork and controls, make sure the vendor you select understands your requirements and has the capability to perform.

- *Quality:* No matter how good a contract packager is, the group should always be striving to get better. Find out if your candidate has a quality improvement program in place and if it's functioning.

- *Experienced Personnel:* A key to successful contract packaging is understanding good manufacturing practices and having an experienced staff capable of solving problems and implementing solutions. Check their credentials.

- *Other Capabilities:* Although some companies focus solely on contract packaging, there are many that offer additional services such as transportation and warehousing. Using a vendor like this is a smart move, because it can help you avoid double handling, reduce costs of freight and administration, and cut delivery times.

- *Your Own Needs:* Even if your candidate passes all of these tests with flying colors, he still may not be the right contract packager for you. Don't go with any vendor unless you feel comfortable and confident he can do the job. If something tells you to keep looking, trust that instinct. Exercise your right to choose until you're confident that your packaging needs are all wrapped up.

A Series of Compromises

The design of packaging and packaging identification always represents compromises between the desire of the manufacturer, packaging engineer, warehousemen, and customer. The package must be designed for various available handling methods in the distribution cycle. Warehouse operators naturally hope that the packages they handle will be as strong as possible. Users of those warehouses must suffer with the economic limitations of packaging. The fire underwriter wants all packaging to be perfectly safe. The final result is always a compromise between the desires of the various people handling the package.

7

TRANSPORTATION

In some companies, warehouse managers are responsible only for operations within the warehouse.[2] Perhaps the most important thing those in this position need to know is the right person to talk to in their transportation department when they need help!

If there is no corporate transportation department, or you don't have confidence in that department, use these seven steps in managing transportation from your warehouse:

1. Carrier selection
2. The bill of lading
3. Shipper's load and count
4. Released valuation
5. Terms of sale—F.O.B.
6. Claims
7. Hazardous materials regulations

Carrier Selection

How do you find a reliable motor carrier? If you do the selection job, your sources may come from directories, advertisements, or personal contacts. Directories are found in many places. The most common is the Yellow Pages of the telephone directory. Trade associations have member lists, and you may find carriers through these lists. The trade magazines have advertisements and some have directories. There are also directories designed for carriers of each mode. Other sources of contact are direct mail advertising or personal contact from vendors who call you or who may be known from other sources.

As you consider whether or not to use a carrier, there are six points to be checked out:

- Operating authority
- Financial stability
- Insurance certificate
- Claims ratio history
- References
- Service requirements

It would be extremely rare for any carrier to solicit or handle freight without the legal authority to make the delivery you need, but it has happened. You should ask the carrier to prove that operating authority exists to cover the move involved. Common carriers, like other suppliers, include companies that are financially stable and others that are not. Therefore, your request for evidence of financial stability is a reasonable safeguard. If the carrier publishes an annual report, you should have a current copy in your files. Common carriers must usually publish an operating ratio. The lower the ratio, the more profitable the carrier. There are also standard reference sources on financial stability, including Carrier Report and Dun and Bradstreet.

Unlike public warehouses, common carriers have common law responsibility for all merchandise in their custody. Ample insurance coverage is therefore of great importance to you as a shipper. Ask for evidence of such coverage, and if you are in doubt get the details.

The relationship between the number of claims filed and the total number of shipments handled is called claims ratio. The lower the ratio number, the better the carrier. Every carrier has a salvage policy for damaged goods. Depending on the product you ship, you may or may not want damaged merchandise to be left in the hands of the carrier.

Ask about salvage policy, and get it in writing. If you intend to have a salvage allowance, establish this before you make the first shipment.

Reference checking is no different for carriers than for any other kind of supplier. Comparing notes with other shippers in your industry or in related industries may help you learn more about the reputation of carriers.

Be sure you define your service requirements in detail, then be sure that the carrier is able to meet them.

In outlining the six points above, why didn't we say anything about prices? Obviously a favorable price is important, but in an industry noted for price competition and "low ball" rates, checking on the quality features mentioned earlier becomes even more important. When you ask for a rate quotation, get it in writing and be sure that the price has been filed with the ICC or any other appropriate regulatory authorities. Then look out for the fine print, including accessorial charges, liability limits, and other limitations listed in the carrier's "rules" tariff.

The Bill of Lading

The legal contract governing your relationship with a common carrier is a bill of lading. The short form is most commonly used, and you want to check to be certain that there are no unusual exceptions. Bills of lading are covered by a law that dates back to 1916, and one section of that law gives the carrier the right to limit liability for damage caused by improper loading or non-receipt or mis-description of the goods in the bill of lading. If there are any variations from the standard short form, be sure you understand why they are there and how they will affect your operation.

Shipper's Load and Count

In motor freight, the standard practice is for a truck driver to count and sign for the freight receipt. This act of signing the bill of lading makes the carrier responsible for any irregularities that occur after the bill is signed. However, sometimes carrier and shipper agree to change this responsibility by permitting "shippers load and count" (commonly abbreviated as SL&C). This change transfers responsibility for count and condition of freight from the carrier to the shipper. However, SL&C does not mean that the shipper loses the right to make a claim against the carrier. However, to do so the shipper must satisfy three burdens of proof.

- Prove count and condition at origin.
- Prove count and condition at destination.
- Measure of damage.

There have been cases when a carrier tried to impose SL&C unilaterally. Be sure that you are aware of your rights in this area.

Released Valuation

Either party to the transportation contract can negotiate a maximum value for the goods being shipped, based on a declared "cents per pound." Sometimes shipper and carrier agree to a reasonably low value in exchange for a favorable freight rate.

Terms of Sale—F.O.B.

In any shipping situation, one party is responsible for the merchandise at any given time. F.O.B. is an abbreviation for "free on board" which indicates the point at which title to the goods transfers from seller to buyer. A starting point on any claims situation is to determine who is responsible, and sales and shipping terms are part of that determination.

There are several variations in sales and shipping terms. There are nine questions that are answered by the terms governing shipments from your warehouse:

1. When and where is title and control of the merchandise to be transferred?

2. How is the total cost to be determined?

3. Who pays the freight?

4. When is payment due?

5. Who is to bear the risk of transportation?

6. Who is to select and make any necessary special arrangements with the carrier?

7. Who absorbs cost of loading into the carrier's equipment?

8. Where is the point of shipment?

9. Where is delivery to be made?

Be sure you know what terms govern every shipment made from your warehouse.

Dealing with Claims

A freight claim is a shipper's legal recourse against the carrier to recover the value of losses and damages to goods occurring while in the carrier's possession.

There are five exceptions to the carrier's liability and they are the following:

- Act of God

- Public enemy

- Act of the shipper

- Public authority

- Inherent nature of the goods

There are three types of freight claims. An apparent shortage or damage is something that appears at the time of delivery. Concealed shortage or damage is a situation that was not apparent at the time of delivery, either because unitizing of freight or packaging concealed the problem. Non-delivery refers to shortage or disappearance of an entire shipment.

A most important point about claims is the fact that the clock is ticking. You have only nine months from date of delivery to file a claim in nearly every shipping situation.

There are three documents required to file a claim against a common carrier. The first is a delivery receipt which contains accurate notation of the problem. Second, you should request and receive an inspection report signed by the carrier, or a letter that waives such inspection. Finally, be sure you produce copies of the bill of lading, invoice, photos of damage, and any appropriate correspondence covering the situation. Documentation requirements are slightly different when you are dealing with United Parcel Service (UPS) or other parcel carriers.

Figure 7-1 shows the format for a loss and damage claim.

For more detail on freight claims, you should review National Motor Freight Classification 100-R, referred to as Principles and Practices for the Investigation and Disposition of Freight Claims.

```
Claim No.                                          Claim No.

------------------------------------------------------------------------
                   PRESENTATION OF LOSS AND DAMAGE CLAIM
------------------------------------------------------------------------
Claim Presented To:  CARRIER CODE      ! Claim Presented By:
                                       !
         CARRIER NAME                   !    COMPANY NAME
         ADDRESS                        !    DEPARTMENT
         CITY, STATE, ZIP              !    ADDRESS
                                       !    CITY, STATE, ZIP
                                       !
         Acknowledgement #             !    Forward All Correspondence and
         Action Date                   !    Payment To Above Address
------------------------------------------------------------------------
Consignor Location:                    ! Consignee:
                                       !
         COMPANY NAME                   !
         CITY, STATE, ZIP              !
                                       !
                                       !
------------------------------------------------------------------------
General Claim Information:
                                                Max Liability      0.00
Claim #              File Date          Pieces     0 Claimed @     0.00
Invoice              Ship Date   /  /   Credit Amount       (      0.00)
Pro No.              Ship Pieces    0   Additional Charges         0.00
Shipper              Ship Weight    0   Freight Charges           0.00
Comments:
                               Total Claim Amount Due =====>       0.00

------------------------------------------------------------------------
!L!      ! Product !   !      Product      ! Unit ! Full  !Release!  Line    !
!D! Qty. !  Code   !UM!    Description      !Weight! Value ! Value !Extension !
------------------------------------------------------------------------




Note: L=Loss. D=Damage
------------------------------------------------------------------------
INDEMNITY AGREEMENT: The claimant agrees to protect the carrier and its
connections against any loss from non_surrender of the Original Freight Bill.
Original Bill of Lading. or Both.
------------------------------------------------------------------------
ie following documents are submitted __ Bill of Lading  __ Freight Bill
n support of this claim:  __ Invoice/Certified Copy  __ Other _____
------------------------------------------------------------------------
The foregoing statement of facts is hereby certified to be correct:

                              _____
                              Claim Prepared By
```

SOURCE: Penny Weber, V.P., Genco, Warehousing Forum, Vol. 8, No. 7, © Ackerman Co.; Columbus, Ohio.

Figure 7-1. Format for a Loss and Damage Claim

This publication includes eight pages of details concerning freight claims.

The Special Problem of Hazardous Materials

The definition of what materials are hazardous changes almost daily. Be sure you know whether any of the materials that you ship from your warehouse could be considered as hazardous materials, and then be sure you follow all of the required steps for identifying and controlling those shipments. Because the rules are changing so fast, familiarity with the latest regulations is of critical importance.

In the old days of regulation, transportation management was considered to be a mysterious and highly specialized practice. Today, common sense and good business practices are at least as important as any aspect of the technical side of transportation management. The seven points we have explored cover the technical side of the job. Decisions involving common sense or judgment are no different than any other purchasing decisions in industry.

Integrated Logistics

Integrated logistics is the effective combination of two or more of the logistics functions that are typically included in moving goods from sources to users.[3] The two that are most frequently integrated are warehousing and trucking. When a warehouse operator moves into integrated logistics, customer service is extended from merely doing things right at the warehouse and shipping the product in a timely manner. The integrated logistics supplier assumes responsibility for the product until it has been delivered at the customer's door. Therefore the monitoring of service quality is extended to the delivery function as well as the warehousing function. In some cases, the integrated supplier also monitors and controls transportation from the manufacturing plant to the warehouse, so that in-bound as well as out-bound transportation becomes the warehouse operator's responsibility. In this situation, it is critical for the operator to control the in-bound moves in a way that will eliminate, or at least minimize, out-of-stock conditions in the warehouse.

Effective consolidation requires teamwork with the consignee. For example, if your warehouse is shipping over a million pounds a week to a grocery chain, you must work closely with that company to move that freight in a manner that saves money for them and for you. One of the ways to do this is to consolidate many small shipments on the same vehicle. But this concept is not limited to local delivery—it also works over fairly long distances. Such consolidation can be done with outside truckers, but usually it works more effectively when you do it yourself. Freight consolidation means savings—savings for the shipper and for the receiver. It also requires very good planning and control.

8

ACCOUNTABILITY

Under common business law and practice, someone is accountable for goods at all times. In distribution, goods change hands from manufacturer to carrier, to distribution center, then on to another carrier, and finally to the customer. In these exchanges, the responsibility limits of each handler of the goods must be clearly defined.

Accountability in distribution covers the following: title to goods and inventory responsibility. Inventory responsibility includes overages, shortages, damages, product liability, insurance, and claims.

An "identifiable party," in the legal term, must be held accountable for the care of merchandise at each point in the distribution channel. Accountability does not become a problem until there are claims or losses. However, since claims are inevitable in distribution, accountability must be carefully defined.

The definition of accountability can be less rigid for a private warehouse, since the warehouse inventory is the property of the operator. In contrast, the third-party warehouseman is handling merchandise that belongs to others, and must deal with freight carriers who also handle goods belonging to others.

Title

As the goods move through the distribution system, transfers of title occur. The precise point at which a transfer occurs is very important. Traditional selling terms for freight shipments define the owner of in-transit goods at different points in its movement. Various selling terms are used. When sales terms are "F.O.B.," the buyer assumes title to the goods from the time they leave the seller's premises. The seller retains title to the goods while in transit if the transaction is "F.O.B. Destination."

The responsibility of the public warehouse is based on the Uniform Commercial Code (UCC) which is law in every state except Louisiana. Most third-party warehouses use standard contracts developed by the AWA based on that code. Under the UCC, carriers are legally liable for nearly all losses to their cargo, but warehousemen are not. Therefore, a loss that takes place at the warehouse is frequently handled differently from a loss of cargo held by a common carrier. The warehousemen's liability for goods is limited to "reasonable care," defined in Section 7-204-1 of the UCC as follows:

A warehouseman is liable for damages or loss or injury to the goods caused by his failure to exercise such care in

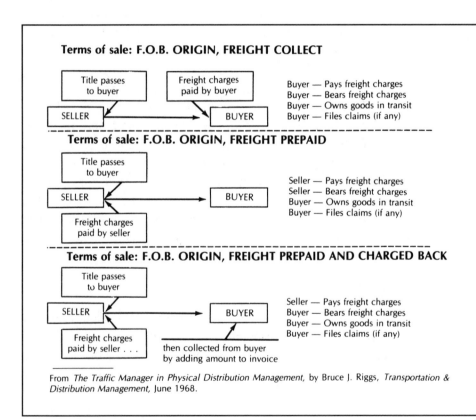

From *The Traffic Manager in Physical Distribution Management,* by Bruce J. Riggs, *Transportation & Distribution Management,* June 1968.

regard to them as a reasonably careful man would exercise under like circumstances, but unless otherwise agreed he is not liable for damages that could not have been avoided by the exercise of such care.

Where and when the public warehouseman's responsibility for goods begins and ends is not always clear. The point of transfer of responsibility may vary with different operations. For example, if the warehouseman operates delivery trucks, he retains custody of the cargo until it is delivered. If goods are received by rail, responsibility may begin when the rail car enters the warehouseman's property, even though he may not have broken the seal to open the car. A question may arise concerning freight that is still on common carrier trucks when those trucks are parked on the warehouse company's property.

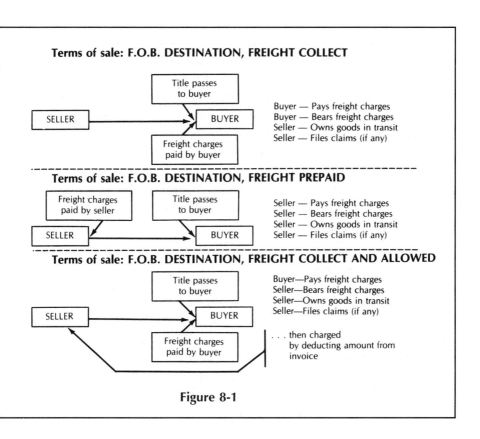

Figure 8-1

The creation of a public warehouse receipt is a legal acknowledgment of responsibility for care of those goods. The signing of a bill of lading transfers responsibility from the public warehouse to the carrier.

Goods consigned to a public warehouse should always be addressed to the owner of the goods in care of the warehouse company. This is the legal way to show that title to the goods remains with the owner, and that the public warehouse is acting as his agent.

The "reasonably careful man" theory is as old as English common law, though standards of what is "reasonably careful" are changing.

For example, standards of fire protection have been upgraded in recent years. Advertising a "fireproof warehouse" is almost never seen today, though it was once common practice. Because contents are nearly always flammable, there is no such thing as a fireproof warehouse. Claims to the contrary are misleading and could be considered an assumption of additional liability.

Protection standards against theft and vandalism also have become more stringent. Increases in national crime rates have affected warehousing practices as much as they have other aspects of life. Burglary protection systems believed adequate in the past must be constantly reevaluated by warehouse management.

Effective security against fire and theft is partially a matter of attitude. The finest systems in the world are worthless if a conscientious interest in loss prevention has not been instilled in all distribution center personnel.

Inventory Responsibility

Accurate maintenance of records is the keystone of any inventory responsibility system. The correctness of the book inventory must be checked by physical counts. A complete physical inventory should be made at least once a year. Cyclical inventory checks, the counting of a few line items each period until the entire stock has been counted, should be used as a backup.

Since book inventory records often have clerical errors, one way to ensure accuracy of warehouse inventories is to maintain two book inventory systems—one posted by the warehouse user and one by the warehouse operator. When the two book inventories are not in balance, they can be compared for omitted, duplicate, or incorrect entries.

Inventory discrepancies also result from errors in shipping and receiving. If the wrong item is shipped, but the correct item is posted to the book inventory, a compensating overage and shortage (a crossover) will show up when a physical count is taken.

The warehouse user should develop a policy for handling overages and shortages or the warehouse operator will be constantly penalized for declaring overages. To hold the warehouseman is responsible for all shortages, without giving credit for overages is considered a violation of the "reasonable care" standard outlined in the Uniform Commercial Code.

An occasional shipping error is normal in any warehousing operation. If the warehouseman must pay for the shortage without receiving credit for the overage, the owner of the merchandise "realizes a monetary benefit that ordinarily would not be gained in the conduct of business," as the lawyers would say. Most public warehouse users today recognize that inventory responsibility is best maintained through policies that deal reasonably with a predictable percentage of errors.

Product Liability

Some kinds of losses may result from the nature of the product itself or damage that occurs through use of the wrong product, resulting from an identification error in shipment. An extreme example would be two similarly marked products in an inventory—one a food ingredient, the other a highly poisonous compound. If the items are confused in shipment, the results could be disastrous.

The manufacturer must specify the conditions necessary for protecting such products. He should determine whether his product would be damaged by temperature and humidity variations, or by cross-contamination through odor transmission or leakage. Infesta-

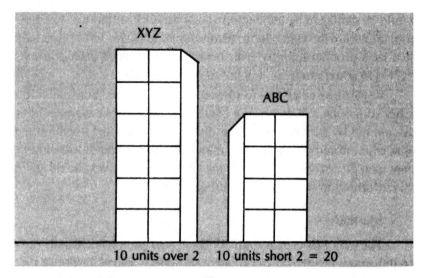

Figure 8-2. The "Crossover"

tion may also be a product liability risk, because rodents and insects are attracted to some products, but not to others.

Regardless of efforts to foresee them, unexpected product liability situations will occur. For example, an appliance manufacturer discovered by accident that his product's white enamel finish turned pink when stored near tires. Any manufacturer, carrier, or warehouseman who discovers a risk due to the nature of a product should share this information with others handling the product.

Insurance

The major risks in accountability in distribution are usually covered through insurance. In private warehouses, broad insurance policies cover factory and warehouse alike.

When third-party warehousing is used, the insurance contract must be more precisely defined. Fire and extended coverage insurance for goods in storage normally is carried by the warehouse user, since he owns the goods.

The public warehouse operator insures only his building, not the goods stored within it. However, the user's cost of coverage depends on the insurance rate established for the warehouse. The building rate represents an assessment of risks made by the insurer. Obviously, a warehouse with a relatively high insurance rate is one considered unsafe for one reason or another.

Unlike common carriers the public warehouse operator is not legally liable for cargo casualty losses, except for losses resulting from failure to exercise ordinary care.

The Uniform Commercial Code allows the public warehouse operator to limit liability for loss or damage. For example, a public warehouse operator may limit claims to an amount not to exceed 200 times the monthly storage rate for each item. Thus, if the storage rate is 50 cents per case, claims are limited to $100 per case.

Most public warehouse operators carry insurance to protect themselves against legal liability claims. However, such insurance policies include a clause that may cancel the policy if the warehouse operator signs any contract extending responsibility beyond the standard set forth in the Uniform Commercial Code (UCC). For this reason, most public warehousemen will not sign any agreement drafted by their customers without first securing approval from their own insurance company. A customer's contract that cancels the warehouseman's insurance coverage would obviously remove legal liability insurance protection for the warehouse user in the event of a major loss.

Insurance protection through fidelity bonds covers the risk of dishonesty of warehouse employees. It is carried by most public warehouses, and the user should request proof of coverage.

Summary of Claims Procedures

The distribution center user should set up procedures for handling claims. Claims will inevitably occur, and without procedures, the necessary supporting documents may be lost.

Procedures for handling overages, shortages, and warehouse damage are good business practice and the distribution center user must determine whether he will provide an allowance for this damage.

A procedure to appraise damage claims must be developed. If

an insurance company is required to pay the claim, there must be proof that the claim is fair and reasonable.

Prompt handling of carrier damage claims is important. Some carriers are notoriously poor in handling damage claims, and delay in submitting the claim will only aggravate the problem of prompt settlement. In today's deregulated environment, negotiation may be as important as documentation in settling carrier damage claims.

Repackaging is often possible when the product is not totally destroyed. A high percentage of some types of damaged products can be salvaged by repackaging. But this action, known as recooperage, could result in contamination. If recooperage is not handled carefully, a damaged product may reach a customer in a new carton. Training and inspection will help prevent improper recooperage at the warehouse.

Other methods of handling carrier damage include scrapping all damaged merchandise, with costs of the product and its scrapping paid by the responsible party. Damaged merchandise may be disposed of through a sale handled by the owner, the warehouseman, or the carrier.

Accountability is essentially the assigning of responsibility. When a disaster occurs, emotions frequently cloud the process. Both the user and the operator of a distribution center must clearly understand the assignment of responsibility for property stored. The points at which responsibility passes from one party to another must be defined. Accountability covers title to goods, inventory responsibility, product liability, and claims. Insurance is a means of controlling the risks. Since someone *must be* responsible for goods in distribution at all times, these responsibilities should be clearly defined *before* something goes wrong.

9

STARTING-UP OR MOVING A WAREHOUSE OPERATION

Reasons for Finding Another Warehouse

When people who say they want to replace their present warehouse, they almost always list one of these eight reasons:

- Wrong size
- Operating problems
- Policy change
- Market changes
- Transportation changes
- Taxes
- Technical changes
- As a symbol of progress

The most common reason most people give for replacing a warehouse is that the present one is the wrong size. When it is too small, expansion may be expensive or impractical. When it is too large, renting excess space to other users may be difficult or impractical.

The second most mentioned change is operating problems. A frequently mentioned problem is labor—either an expensive labor market or union problems.

Policy change sometimes stimulates the search for another warehouse. This could either be a decision to move from third-party warehousing to private warehousing, or the reverse.

With the passage of time, market changes with either suppliers or customers can create the need to change the warehouse location.

If your customers or your suppliers are no longer in the same locations, moving the warehouse could cut costs or improve service. A shoe company designed a distribution center in the right place to receive products from its resources in New England. When its sources of supply moved from New England to Asia, the company moved its warehousing from Ohio to east and west coast ports of entry.

Transportation changes also make an existing facility obsolete. Many warehouses constructed before the 1980s were equipped to handle high-volume boxcar shipments. These buildings were convenient to a rail switching yard and contained extensive indoor loading platforms. Today's operator usually will not need rail services, and the design of the older facility may no longer serve current needs.

Changes in *taxes* can also diminish the value of an existing building. A newly developing community may create tax incentives that did not exist when your present warehouse was established.

The search for a new warehouse is sometimes stimulated by *technical changes.* An older building may not be high enough to handle the stack heights that are easily available with today's materials handling equipment. Other older buildings may have inadequate sprinkler systems and therefore create an unnecessary fire risk. In either case, the move to a new building may allow a warehouse to operate more economically or more effectively.

Another reason for moving is strictly qualitative, and that is the hope that a better warehouse will be seen as a symbol of progress. While a move for this reason may be the least tangible motive to find another warehouse, there is ample evidence that this is a powerful drive for change in some organizations.

Options for Your Next Warehouse

Finding a new warehouse begins with the review of several options. Consider whether to own or lease the real estate, with a third option of using public warehousing.

Another choice is whether to construct a new building or to buy or lease an existing one. There are times when a building can be purchased with the assumption of a mortgage at a favorable interest rate. Attractive financing is often a prime reason for buying an existing property. For many corporate warehouse users, the question of whether to own or lease is quickly answered by a company policy that tends to discourage capital investment in industrial buildings in order to preserve capital for necessary investments in inventory and production equipment.

Requirements Definition

The search for another building should begin with a management meeting designed to reach consensus on a definition of what the next warehouse must be. This requirements definition should state the maximum and minimum size warehouse that would be considered. It should define any limits that exist on location. For example, if your management has a strong feeling that the next warehouse must be east of the Mississippi, this would be part of the definition. It should define any other priorities that exist with respect to technical features, transportation, or any other aspect of warehouse operations. Before beginning the search, a careful definition of your requirements should be negotiated with every manager who will be involved in the final decision. When agreement is reached, the requirements definition should be put in writing and formally approved by those who will be making a decision on the project.

Finding the Best Location

Once you know that you will have a new warehouse, you need to decide where it should be located. If your only reason for making the change is that the existing warehouse is too big or too small, you will want to find a new facility that is as close as possible to the existing location. This would allow you to retain a maximum number of the present employees who understand the operation and have been doing a good job. On the other hand, if the move is dictated by problems with the labor force, the user may require that the new location be far enough from the present plant to reduce the possibility of a labor disturbance from the laid off workers in the existing warehouse.

Just as changes in transportation have made some older warehouses obsolescent, present and future changes in this field must be considered in finding a location. If your company is increasing its use of air cargo, you may want to look at the proximity of major airports. Consider each available mode of transportation and calculate the possibility of switching modes. Then look at distances to appropriate terminals for rail, motor freight, or marine transportation.

Taxes and tariffs continue to be a critical element in the choice of warehouse location. Competition among states has tended to diminish differences in state inventory taxation policy, but a newer battleground has been the development of free trade zones that offer shelter not only from customs tariffs but also from state taxation. While the free trade zone legislation was originally designed to facilitate imports, many users have now recognized that it offers a shelter from state inventory taxes as well. Other tax shelters are found in tax abatement zones.

Traditionally, warehousing has been a labor-intensive business. Increasing mechanization has reduced the labor demand for some operations, but for warehouses dedicated to packaging, less than full case shipments, or fulfillment, the labor content remains high. If your operation is one in which labor content is high, the quality of the labor market can be a defining factor in selecting a location. Look beyond the mere availability of workers and consider crime rates, educational systems, and the experience of other users with local problems involving theft or substance abuse. In earlier years, industries tended to move south to exploit the availability of cheap labor. In more recent years, labor cost differentials have substantially narrowed. Furthermore, in some situations, productivity differences will allow a higher priced labor area to produce lower unit handling costs than a cheap labor area. The warehouse user with high labor content should emphasize quality rather than cost and availability of warehouse workers.

Utilities can be a key factor in selecting the location. The most important of these is the water necessary to power a sprinkler system.

New sprinkler designs require substantially more water at higher pressure than the older sprinkler systems. While the pressure can be increased through pumps, the supply cannot. Therefore the availability of ample amounts of water may be a determining factor in choosing between several available warehouses or construction sites. In considering other utilities, the next most important consideration is telephone communications. Growing use of EDI (electronic data interchange) and fax machines has required substantial increase in the number of available telephone lines.

As warehouse users have expanded beyond national borders into third world countries, the quality as well as quantity of telephone lines must be considered. In some parts of the world, the telephone infrastructure is seriously deficient in quality, and in some regions there are significant delays in providing additional phone lines. While electricity use is low in most warehouse operations, the dependability of the source is critical both for computerized offices and for electric materials handling equipment. While quantity and quality of electrical sources may be taken for granted in your own country or region, overseas operations may require far more attention to the dependability of electricity. In some cases, backup sources in the form of an internal combustion engine generator may be a necessity.

Architectural Features

Whether you are buying or building your next warehouse, always remember that the most important feature is the floor. Other construction deficiencies can usually be corrected, but it is sometimes not practical to repair an unstable floor. Occasionally warehouses are actually abandoned for this reason. For an existing building, quality problems with the floor are usually apparent when the building is inspected. With a new building, test borings are the only way to determine the strength and stability of the subsoil. If you discover an existing building with floor problems, both structural engineering experts and test borings of adjacent land should be employed to determine whether the floor can be repaired economically.

Next in importance is the quality of the roof. A warehouse that is chronically subject to leaks is an obvious problem, and some roof systems are more easily repaired than others. For this reason, the best time to inspect existing buildings is on a very rainy day or shortly after a hard rain. Stains on the interior or exterior of walls are a good sign that a building has had leak problems in the past.

Whether the building is new or used, measure the cost and feasibility of adding additional dock areas in the future. Construction of truck docks will be enhanced or limited by availability of land and the depth of the foundations on each warehouse wall. Many warehouse buildings will design one long wall of the warehouse with a deep foundation to facilitate addition of more dock doors at any position on that wall.

Estimating the Cost of Moving

After setting the target date, make an assessment of the inventory to be on hand at that date. The simplest approach to this is to forecast the future date's inventory levels as a percentage of the present inventory levels.

Remember that it's always possible to plan the moving date to coincide with a lower inventory level period or a period of lesser shipping activity—as long as the new building is ready!

After estimating the level of on-hand inventory, determine the number of transfer loads currently on hand. You can do this by counting the full and partial pallet loads and pallet equivalents on hand, and then dividing the number of anticipated pallets per transfer load into the total currently on hand.

For example, assume the current inventory level consists of 3,500 full and partially loaded pallets and pallet equivalents. (Pallet equivalents are items, not palletized, that equal a pallet in space and cube requirements.) Anticipate 26 pallets per transfer load. The number of loads of inventory to be transferred may be calculated as:

3,500 pallets/26 pallets per load = 134.6 loads

A total of 134.6 current inventory loads adjusted by a 92% factor to compensate for reduced inventory levels at the move date yields:

$$134.6 \times 92\% = 123.8 \text{ loads of inventory to be transferred}$$
on the targeted move date

Experience has shown that, for a short-distance move, floor loads (goods stacked just one pallet high) allow for quicker turn-around of transfer vehicles and are more efficient; for longer moves, fully loaded vehicles are more economical. When the trucks used to move the inventory can be turned around quickly by using a floor load, you will reduce driver waiting time.

Cost of Transferring Each Load

The cost of each load to be transferred can be calculated by totaling the following costs:

- Transfer vehicle operating cost, or vehicle rental cost plus driver labor and fuel, per round-trip
- Outloading manpower and machine cost per load
- Unloading and "put away" manpower and machine cost per load
- A small additional allowance for damage, extra clerical labor, and contingency costs

The sum of these costs is an indicator of the cost per load and, when applied to the number of loads to be transferred, will yield the total cost of transferring the inventory.

Example:
Truck cost: 20 miles/round trip @ $1.80/mile 36.00
Outloading cost: 0.3 hours/load @ $22.00/hr........... 6.60
Unloading/putaway cost: 0.25 hours @ $22.00/hr... 5.50

Subtotal
Overtime allowance @ 25% × labor cost................. 3.03
Damage/clerical/contingency allowance 1.44

Total cost per transfer load $52.57

In addition to relocating the inventory, the warehouse "business" must itself be moved. So the following costs should be added to our example:

- Cost of relocating the office operation
- Cost of relocating the warehouse maintenance shop and recouping operation
- Cost of transferring the materials handling equipment
- Cost of disassembling, transferring, reassembling and lagging down storage racks and other equipment

These can be costed by estimating the number of transfer loads involved, multiplied by the cost per load previously calculated, and then adding additional estimated labor costs such as rack relocation and lag-down labor.

How Long Will It Take?

The following is an example of how to calculate the number of days that will be required to make the move:

Assumptions:
Inventory loads to be transferred .... 124 loads
Additional loads to move the office equipment
    and files .... 5 loads
        Total loads to be transferred .... 129 loads

Further assumptions:
Truck travel time, one way loaded .... 25 minutes
Return trip time, one way not loaded .... 25 minutes
Outloading time, one lift operator................... 18 minutes
Unloading and putaway time, one lift
    operator .... 15 minutes
Total round trip cycle time............................... 83 minutes

Time available, one shift operation
Available truck time, 7.5 hours..................... 450 minutes

Minus allowance for delays/interruptions,
   10% ....45 minutes
Minutes available per truck per shift..............405 minutes

405 minutes available per truck per shift @ 83 minutes per round trip cycle = 4.9 loads/truck/shift.

With moderate overtime or extra efficiency, five loads may be achieved per shift per truckload with one lift operator loading and another unloading.

The 129 total loads to be moved will require:

129 loads/5 per day = 25.8 days with one truck, two lift operators on one shift

129/10 = 12.9 days with two trucks

129/15 = 8.6 days with three trucks[4]

Communications

Customer communications have top priority in a warehouse move. Customers need to be sold on the benefits of the proposed move. All customers must be informed well in advance of the move, and then provided with progress reports. An overinformed customer seldom complains. Points that must be covered in communications with customers include:

- Out of service dates, if the warehouse is to suspend operations during the move. Also provide for emergency service requirements, which will surely arise.

- Date and time of transfer of each class of inventory. This is needed to coordinate delivery and pick-up carriers. It requires daily, even hourly, contact.

- Phone number and address changes and their effective dates.

- Dates and times when data communications will be out of service. These breaks in service should be held to an absolute minimum.

- Directions to the new location for users to provide to their carriers and customers for pick-up and delivery.

- New hours of service, if these are to be changed temporarily or permanently.
- Changes in service charges or standards of service should be announced and explained.

Other Communications

Customer communications have top priority, but carriers, employees, suppliers, the phone company, and letter carriers need to have timely notice of the move.

Employees should hear about the relocation well in advance; if you don't inform them, the grapevine will. Employees should be informed—in the early stages of planning—what is happening, why, and what to expect at the new site. Employees should be given regular information updates. It is especially important that they tour the future facility while it is in the later stages of preparation. A complete walk-around, explanation of the location system, and any needed training in the use of new or different equipment will pay off in higher employee morale and a more favorable employee learning curve at the new site.

Bear in mind that union organizers will take advantage of any uncertainty—such as a move—to play on employee concerns. Good communications will help you avoid this.

Carriers also need to be informed, even though they may already know about the move from industry contacts or your warehouse employees. This is a good time to look at any special arrangements with carriers.

Opening the Relocated Warehouse

Relocating a warehouse inventory or opening a new warehouse is a major undertaking. Before making the "go" decision, anticipate all the costs and be sure there are ample benefits to offset these costs. Moving or opening a new center involves planning, communications, planning, and more communications. You cannot possibly have too much of either. Checklists are always an important part of planning. Shown as Figure 9-1 is a checklist to find your next warehouse.

A CHECKLIST TO FIND YOUR NEXT WAREHOUSE

Editor's note: most of this checklist is designed to use whether you plan to build a new facility or acquire an existing structure. However, a final section contains items used only for new construction. KBA

GOVERNMENTAL RESTRICTIONS:

1. Is the property zoned for warehousing and related uses?
2. Is the zoning description likely to impede conversion of the building to other uses?
3. Are there any variances or special exceptions in the zoning?
4. What are the limits or restrictions on employee parking?
5. What is the maximum expansion capability of the facility?
6. Are there any easements or protective covenants on the property?
7. Are there any restrictive load limits on roads leading to this facility?

GEOGRAPHIC RESTRICTIONS:

1. Are there grade problems on the land?
2. Are there any problems with drainage?
3. Are there streams, lakes, or wetlands near this site?
4. If so, what is the 100 year flood plan?
5. What is the level of the ground water table?
6. Is there any history of earthquakes in the area?
7. If so, have you received the US Geodetic Survey of fault lines?
8. Does the site have any fill?
9. If so, what material was used?
10. What is the load bearing capacity of nearby soil?
11. Is any part of the site wooded?
12. Are there any restrictions on tree removal?

TRANSPORTATION:

1. What is best access to the nearest freeway interchange?
2. What is the distance from primary motor freight terminals?
3. Have you interviewed motor freight managers?
4. Is the facility inside the commercial delivery zone of the city?
5. If not, what is the extra delivery cost compared to deliveries within the commercial zone?
6. Is rail access available?
7. If there is no rail siding now, what is the cost of adding one?
8. What is the cartage cost to a TOFC (piggyback) terminal?
9. What is the cartage cost to nearest marine terminal?
10. What is the cartage cost to nearest air freight terminal?
11. Will those customers who pick up at your warehouse have any transportation problem with this facility?

SECURITY CONSIDERATIONS:

1. Does the sprinkler system meet preferred risk standards?
2. If not, what is the cost of upgrading the system?
3. Is the sprinkler capable of conversion to ESFR?
4. If so, what is the conversion cost?
5. Is there a secondary water source for the sprinkler system?
6. Have you seen available insurance inspection reports?
7. Have you interviewed law enforcement officials?
8. Have you interviewed neighboring managers?

UTILITIES:

1. What is the size of the water main serving the site?
2. What is the water pressure and when was it last tested?
3. Are there water meters on hydrants and sprinkler systems?
4. What is the source of the water supply?
5. What is the capacity of electric power at the site?
6. Is submetering of electricity permitted?
7. Have you received copies of recent utility bills?
8. Are incentives available for reducing peak electric demand or installing HID lighting?
9. Is submetering of gas permitted?
10. Are there limitations on capacity or demand?
11. If fuel oil is used for heat, what is the source of supply?

LABOR MARKET:

1. Are neighboring businesses unionized?
2. Have you measured community attitudes toward unions?
3. Is there a labor shortage or high unemployment?
4. How are nearby schools rated in comparison to other sites?

COMMUNITY ATTITUDES:

1. Do community leaders welcome industrial expansion?
2. Is there any discernable opposition to warehouse operations?
3. Is there any history of difficulty in rezoning?

TAXATION:

1. What is the date of the most recent property appraisal?
2. What is the real estate tax history over the last ten years?
3. Have there been any recent tax assessments?
4. Are there any tax abatement programs in effect or available?
5. Is the facility in a duty free or enterprise zone?
6. Are there planned public improvements?
7. How do inventory tax levels compare with other sites?
8. Is this a free port state?
9. What are tax rates for:
   - Corporate income tax
   - Payroll tax
   - Sales tax
   - Franchise tax
   - Other taxes
10. What are the costs of workers compensation?
11. What are the recent trends in taxation?
12. Are industrial revenue bonds available, can you qualify?

NEW CONSTRUCTION CONSIDERATIONS:

1. Have test borings been made at all key points on the site?
2. Are there landscaping or buffer zone requirements?
3. How will roof drainage be discharged?
4. What are estimated costs and time requirements for all construction and occupancy permits?

SOURCE: Warehousing Forum, Vol. 9, No. 9, copyright Ackerman Company, Columbus, Ohio.

Figure 9-1

10

AUDITING WAREHOUSE PERFORMANCE

How well am I doing in managing my warehouse? How well are each of the managers who report to me doing in managing their warehouse operations? Which of our warehouses is doing the best job, and what changes would bring the others up to that standard? How well are each of my foremen doing? How effective is my warehouse crew, and what can I do to upgrade the performance of these people?

All of these questions are asked regularly by people at various levels in warehouse management. Senior managers are concerned about the entire warehouse network. Distribution center managers are concerned about the facility that they manage. Warehouse supervisors want to evaluate their own performance and that of their crew. It is easier to ask the above questions than to answer them. Absence of standards or benchmarks is a common complaint in the industry, and the difficulty of developing an audit causes many to throw up their hands and abandon any effort to measure.

There is no single set of benchmarks that would apply to every warehouse operation. A framework for auditing performance will enable you to design your own procedure. Based on this framework, you can create a detailed audit format based on history, projection, and reasonable goals.

A first step is to separate those performance items that are quantitative from those that are qualitative. On the quantitative side, every warehouse manager has control over three prime elements of warehousing: space, equipment, and people. Simple ratios can be used to compare the actual versus maximum utilization of warehouse space and materials handling equipment. The quantitative output of people who work both in the warehouse and in the warehouse office can be compared to established benchmarks.

Quality measures may be just as important as the quantitative items. First on the list is customer satisfaction—how well has management met the expectations of the customers who are served from that warehouse? Another qualitative measure is human relations. How successful is management in recruiting and maintaining a harmonious workforce? Overages, shortages, damage, and errors are referred to as "claims." They are a barometer of management's success in preserving the inventory. Safety is another measure of quality in warehousing.

With an organized effort, you can develop an audit format for all of these areas. Consider the steps involved in developing the audit.

Quantifying Warehouse Space

Every warehouse has a measurable capacity, typically measured either in square feet, cubic feet, or their metric equivalents. From the theoretical capacity we arrive at a practical capacity by deducting space dedicated to aisles, staging, and support activities. Since the amount of such lost space is controllable by management, the simplest way to measure capacity is to compare actual utilization to theoretical capacity.

Suppose you are in charge of a warehouse containing 100,000 square feet of high ceiling space, and it is your job to store pallets containing grocery products. The pallets are all standard size 48″ × 40″ pallets, each of which occupies 13.33 square feet, which we round to 14 square feet to allow for overhang on the pallets. The product you are storing can be stacked 3 pallets high. By dividing 100,000 square feet by 14 square feet per pallet, we find that a little over 7,000 pallets could be stacked on the floor, and at three pallets high the theoretical capacity of the warehouse is 21,428 pallets. Obviously the theoretical capacity can never be reached, but a measure of your success as a space manager is to compare actual capacity to theoretical capacity. If there were actually 10,000 pallets in the warehouse on the first day of January, your utilization would be calculated as 47% of theoretical capacity (10,000 ÷ 21,428 = 47), and 47 would be your score for storage. By maintaining such ratings over time, you can develop a history and a comparison between a chain of warehouses.

But what if the inventory is more complex? Not every warehouse is filled only with products that go on standard sized pallets. Where different kinds of cargo are involved, divide the operation into departments. One department will be palletized freight on standard pallets, and another might be spare parts stored in bins, major appliances, bulk containers, or other categories of products that have different storage characteristics. Once the categories are divided into departments, you can develop an individual rating for each department and then convert these into a composite rating for the entire warehouse.

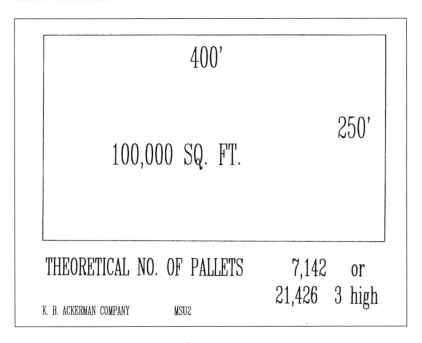

Figure 10-1

Quantifying Equipment Utilization

Every powered lift truck can and should be equipped with an hour meter that shows the time that the engine was running. For a one-shift operation, the theoretical maximum is 2,080 hours per year, based on 52 five-day weeks at 8 hours per shift. Therefore a utilization ratio for each truck might be calculated by dividing the hours actually shown on the meter by the theoretical number of hours available for use of the truck. Then an overall fleet utilization can be determined by adding the individual results. Consider lift truck number 103 which was used 1,000 hours in 1996. Divide 1,000 by 2,080 to calculate a utilization of 48%. The same calculation is done for each truck. Admittedly, a score of 100 is practically impossible, but the important thing is relative progress rather than actual scoring. Management has the ability to move actual closer to theoretical capacity and to identify underutilized vehicles in the fleet.

Quantifying Productivity of People

Measuring the productivity of people is more difficult than calculating space or equipment utilization. Benchmarks are rarer and less reliable. One way to start might be from history. In 1997, our 100,000 square foot warehouse handling palletized products moved 72,800 pallets through the facility during the year. Five full-time people were employed, which means that each worker handled 14,560 pallets. Dividing this by the number of work hours in the year (2,080) we find that in 1997 the average productivity was 7 pallets per hour. If this number 7 is used as a benchmark, performance in 1998 can be compared with 7 pallets per hour and at least management will know whether productivity is improving or slipping backwards. Figure 10-2 shows a way to track productivity. For warehouses with different kinds of cargo, separate benchmarks will be established for palletized products, single case order picking, binned parts, and other specialized cargo. Just as storage was departmentalized, materials handling hours are separated by the various departments.

By utilizing the scoring system described above for space,

Productivity Record in Pallets per Hour

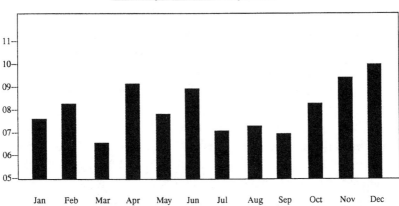

Figure 10-2. People Productivity Summary

materials handling equipment, and people, each warehouse manager can be given a score for progress in controlling the three essential elements of warehousing. Improvements in productivity can be evaluated and estimated before the change takes place, and then appraised after the change is implemented. By comparing results to theoretical capacity rather than practical capacity, the guesswork involved in rules of thumb for aisle loss, equipment maintenance, and lost time can be eliminated.

Qualitative Measures

Since most warehouses exist to provide dependable delivery for the end user or customer, the most important qualitative measure for any warehouse is the degree to which customer satisfaction was delivered.

The best measure of customer satisfaction is presence or absence of complaints. Therefore one quality audit approach would be to measure the number of customer complaints received for each warehouse during each month and compare the results over time. The accuracy of such an approach must be questioned because not every customer will issue a complaint when bad service is received.

Another way to measure customer satisfaction is to log the

number of times that orders are shipped behind schedule, or the number of times they are delivered at a later time than was promised. If detailed records of shipping and delivery are maintained, this measure will at least be reasonably accurate. However, some customers are not as concerned about on-time delivery as we think they are. Therefore, this measure of on-time performance may not accurately reflect customer satisfaction.

If both measurements are considered, management will have two ways of detecting improvements or slippage in customer service.

Warehouse claims are made up of known shipping errors, damage caused both by transportation and warehousing, and inventory discrepancies which include both overages and shortages. By logging the number of claims discovered each month, and by comparing these incidents with the number of actual shipments made during the month, you can develop a barometer of inventory accuracy. If there were 23 claims last month out of a total of 7,000 shipments, inventory accuracy is calculated at 0.9967.

Safety is a measurable qualitative goal. Most companies know the number of work days that pass without a lost time accident.

Since the task of a warehouse operator is to protect the products as well as the people, another measurement might be a ratio between the number of products damaged and those that move through the facility without damage. The result is a ratio of damage-free movement that is over 99%, which means that far less than 1% of the product is lost or damaged.

The last qualitative measurement is in the human side of warehousing. How can you measure a manager's success in human relations? In a union environment, logging the number of grievances filed might be considered. Both union and non-union operations should measure the longevity of service of their workers. A happy person stays on the job, and high turnover is typically seen as a sign of weak management. Therefore, perhaps the most universal method of measuring success in human relations would be to calculate the longevity of service throughout the workforce, possibly separating the categories of warehouse worker, warehouse supervisor, and office worker.

Consider a warehouse that has fourteen employees—seven day shift warehouse workers, three on the night shift, and four people in the office. During 1997, two of these people were replaced, which provides a retention rate of 86%. However, when we look at departments, we find that the two who were replaced were both in the office, which means that the office has a retention rate of only 50%, and each of the two shifts in the warehouse has a rate of 100%. Clearly the quality problem in human relations is in the office, which is where management's attention should be centered.

Figure 10-3 shows a measurement of retention (turnover) and health claims.

Tracking Accuracy

A multicity warehouse operator should try to compare the service performance of the different cities where warehouses are located. The most obvious way to track service performance is to count the number of reported errors. However, those errors should be compared to the volume of orders, since a warehouse that is doing very low volume should also have a low number of errors. A service rating is arrived at by dividing the number of errors by

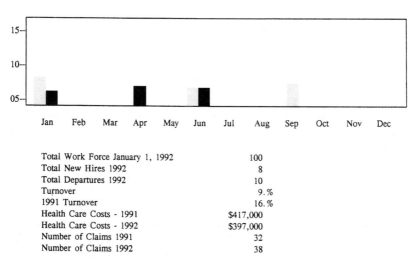

Total Work Force January 1, 1992	100
Total New Hires 1992	8
Total Departures 1992	10
Turnover	9.%
1991 Turnover	16.%
Health Care Costs - 1991	$417,000
Health Care Costs - 1992	$397,000
Number of Claims 1991	32
Number of Claims 1992	38

Figure 10-3. New Hires and Departures and Health Claims

the total number of orders, but one should consider the order volume compared to a six-month rolling average. If order volume is above normal, the overload could be the source of difficulty. Tonnage throughput should also be considered. If the tonnage is above normal, this could create a different type of overload. An overcrowded warehouse frequently causes excess damage and shipping errors, and for this reason space utilization should be examined. By using a service performance report, the multicity warehouse operator can track service rating and compare it with two measures of volume, throughput and order count. Space utilization must be considered not only in connection with service performance but as an indicator that the warehouse has become too full (or too empty).[5] Figure 10-4 shows the report.

Account Profitability

For the third-party warehouse operator, the most difficult cost tracking is that which measures profit contribution for each customer. This is done by establishing targets for earnings per man-hour in the warehouse and earnings per-square-foot per-year. When total earnings for time and space exceed the targets, there is a contribution to profit which is recorded. In any warehouse, there is some time or space that cannot be allocated, and this is identified and tracked.

Service Performance Report

Warehouse	Orders	% of 6 Mo. Avg.	Errors	Service Rating	Thruput	% of 5 Mo. Avg.	Space Utilz.	Space
Boston	1,870	105%	6	99.68%	67,521	101%	90%	78,000
Atlanta	1,746	85%	11	99.37%	105,879	87%	84%	105,000
Baltimore	1,571	98%	13	99.17%	95,072	102%	75%	98,000
Los Angeles	1,412	107%	21	98.51%	140,238	103%	85%	140,000
Chicago	1,268	97%	18	98.58%	117,816	99%	95%	125,000
Houston	900	95%	30	96.67%	130,965	89%	77%	127,000
TOTAL	8,767	96%	99	98.87%	657,491	100%	83%	673,000

SOURCE: John T. Menzies for "Tracking Warehousing Costs and Performance," Warehousing Forum, Vol. 3, No. 8 copyright Ackerman Company, Columbus, Ohio.

Figure 10-4

Account Performance and Contribution
$30.00 Hourly Target
$3.84 Square Foot Target per Year

Date	Square Feet	Hours	Storage	H$Earn	Total Revenue	$/Ft	$/Hr	Contribution	Contribution %	Contribution /Ft
Total*****	120,000	1,559	36,533	47,562	84,095	3.35	32.08	(1,085)	-1	-0.01
Total May-87	120,000	1,645	37,500	53,918	91,418	3.75	32.78	3,668	4	0.03
Total Apr-87	120,000	1,605	36,930	46,930	83,530	3.66	29.24	(3,020)	-4	-0.03
Total Mar-87	120,000	1,550	38,399	48,991	87,390	3.84	31.61	2,490	3	0.02
Unalloc*****	9,291	66	0	0	0	0.00	0.00	(4,943)	0	-0.53
Unalloc May-87	6,530	65	0	0	0	0.00	0.00	(4,040)	0	-0.62
Unalloc Apr-87	8,765	32	0	0	0	0.00	0.00	(3,765)	0	-0.43
Unalloc Mar-87	9,152	65	0	0	0	0.00	0.00	(4,879)	0	-0.53
Cust A*****	28,835	461	9,224	16,523	25,747	4.11	35.60	2,685	10	0.09
Cust A May-87	24,428	475	8,196	18,021	26,217	4.03	37.94	4,150	16	0.17
Cust A Apr-87	23,498	458	8,123	15,540	23,663	4.15	33.93	2,404	10	0.10
Cust A Mar-87	24,171	489	8,863	16,636	25,498	4.40	34.02	3,094	12	0.13
Cust B*****	20,390	122	5,422	3,415	8,836	3.19	28.43	(1,358)	-15	-0.07
Cust B May-87	22,466	128	5,355	3,812	9,167	2.86	29.78	(1,862)	-20	-0.08
Cust B Apr-87	21,160	117	5,601	3,470	9,071	3.18	29.66	(1,210)	-13	-0.06
Cust B Mar-87	19,941	108	5,302	3,558	8,860	3.19	32.94	(761)	-9	-0.04
Cust C*****	8,841	71	2,634	1,927	4,562	3.50	30.01	(382)	-8	-0.04
Cust C May-87	8,582	73	4,034	1,947	5,981	5.64	26.67	1,044	17	0.12
Cust C Apr-87	9,383	61	3,435	1,579	5,014	4.39	25.89	182	4	0.02
Cust C Mar-87	9,897	84	3,421	1,906	5,327	4.15	22.69	(360)	-7	-0.04

SOURCE: John T. Menzies for "Tracking Warehousing Costs and Performance," Warehousing Forum, Vol. 3, No. 8 copyright Ackerman Company, Columbus, Ohio.

Figure 10-5

In the illustration shown above, Customer A has a healthy contribution to profit, Customer B is a loser, and Customer C has a mixed result. The lines with asterisks are a 6-month rolling average. This shows that Customer C performed considerably better in the last two months, which shows either an improvement in rates or productivity.

Then that time and space that can be allocated to individual customers is separated, and the contribution is calculated for each customer. This allows the operator to know whether or not each customer is making a reasonable contribution to profit. If the unallocated portion is significant, then an improved method of assigning costs must be found. If a given customer is not producing a satisfactory contribution, an upward adjustment in warehousing charges should be negotiated. Figure 10-5 shows this report.

The private warehouse operator could adapt this report to measure costs according to different products rather than different accounts. By measuring in this manner, the warehouse operator is able to measure not only overall results, but those that pertain to specific accounts. Note that there is always a small amount of space and labor that cannot be assigned to any particular account, and this is shown as unallocated.[6]

The warehouse performance audit may be no better than the auditor who performs it. If the audit is not done frequently, its effectiveness will be diminished. Performance audits can do a great deal to ensure consistency of service in multiple locations. When used as a basis for recognition or reward, they can be a powerful motivator.

Part III

REAL ESTATE ASPECTS OF
WAREHOUSING

11

FINDING THE RIGHT LOCATION

Site selection is an art as well as a science. Decisions usually involve weighing priorities, determining which features are critical, and a process of elimination. Since every location has both advantages and disadvantages, the final selection of a site will likely involve some compromises.

Location Theory

Location theory is a useful exercise in picking the site for a new warehouse. Whether you accept or reject any of the theories that follow, at least consider them as helpful guidelines for the location decision.

First, move from a macro to a micro decision. We might first determine that the new warehouse must be somewhere within the continental United States. Let's say that the facility should be in the southwestern region of the country, in the state of Texas, Dallas–Ft. Worth metropolitan area; and that it be in the southwest area of the city of Dallas; and finally on a specific site on Duncanville Road.

Over a century has passed since J. H. Von Thunen wrote about the advantages of growing the bulkiest and cheapest crops on the land closest to the city needing that commodity. The same principle can certainly be applied to all kinds of manufactured products—location of inventory is most critical for those that have the least value added. An inventory of gold or diamonds, for example, can be economically kept in one spot and moved by air freight all over the world, but the producer of bagged salt may find it necessary to place inventories very close to the market.

The theory of postponement also has an impact on site selection. Some products can be manufactured to a semifinished state, or finished but not given the final brand or identification. Inventory of postponed goods can be held until specific customer orders are received, at which time the goods are finished or branded to the customer's specifications. The postponement warehouse then becomes an assembly center, as well as a distribution center. The need for these assembly capabilities could affect the warehouse location.

Outside Advisers

It is difficult to pick a location for a new warehouse without eventually turning to people outside your organization for advice. For this reason, we should consider some commonly used sources of such advice.

Real estate brokers are often involved in any search for new property. Remember that the broker's compensation for this activity is based on successful completion of a transaction, so the broker is motivated to complete a commissionable sale. When the value of the broker's time invested begins to approach the value of the commission, the broker would be foolish not to abandon the search. Not all brokers have multiple-listing services. Furthermore, commission arrangements that may not be known to the prospective buyer could make it difficult for the broker to maintain objectivity as a source of outside advice. Not all real estate brokers are biased, of course, but the user should be aware that the commission system for brokerage tends to create pressures that may influence the results of the site search.

If existing warehouse space is sought, a warehouse sales representative could be useful. The public warehouse industry has several national marketing chains, each with representation in the Chicago and New York metro areas. Typically, the chain sales representative is not paid a sales commission, though a few may work on a commission as well as a salary basis. Since the chain sales agent normally cannot represent competing warehouses, each chain usually has one outlet in each major city.

Major railroad companies are a source of advice on sites for new buildings, as well as existing warehouses. Railroad people do not work on a commission basis, but they naturally will want to have the new warehouse located on their line rather than on a competing one. If railroad competition is not a factor in the decision, the railroad representative can be a reliable source of advice. If only two railroads are under consideration, the use of two competing railroad development agents may be a valuable exercise.

The area development departments of electric and gas utilities are an excellent source of outside advice, once the general area for the facility has been determined. With a few notable exceptions, there is usually just one electric or gas utility to service a city and its suburbs. Therefore, the representatives of that utility should be unbiased in recommending sites within the entire metro area. Utility people try to work with all local real estate brokers to become a clearinghouse of information on available buildings and land within their service area.

State and local government development agencies, as well as local chambers of commerce, are also reliable information sources, once the choice of community has been made. Some large metro areas have several chambers or government development officers. Since these people are motivated to attract new industry into their own area, make sure their area is where you really want to be before relying on this source of advice.

Finally, the use of a management consultant may provide an unbiased source of advice. In selecting a consultant, one important consideration is to be sure that the individual is truly independent and objective in approaching the site search. Some individuals who use the term "consultant" also are involved in other activities, such as real estate brokerage, which may affect their objectivity in handling the site search.

There are three reasons to bring in an outside consultant: The most important of these is objectivity. Second, the consultant may have extensive past experience in site seeking, and this experience will save time. Finally, the consultant represents a source of executive time with no long-term commitments involved. A detailed site search takes many hours of time, and hiring managers to handle it would be foolish if their job will end once the site has been selected.

The most objective sources of outside advice are either a public utility or a consultant. Whatever source you use, be sure that the outside adviser is truly independent and objective.

Requirements Definition

Your search for a location should be modified and directed by your specific requirements as a warehouse user. For a successful site search, you need to define these requirements carefully.

Since transportation is so closely linked with warehousing, the transportation requirements for the new warehouse are of utmost importance. Which modes of transportation are required? All warehouses need trucks and some depend heavily on rail, water, air, or a successful combination of all three. While delivery zones are less important in these deregulated times, the user also must consider zones and transportation costs to specific sites.

To the extent that a warehouse is a marketing tool, it is important to consider the relationship of the proposed site to the locations of the planned users. Beyond just looking at locations, you should consider transportation costs and order cycle times to best serve various potential users. Remember that the warehouse user should not really care *where* the inventory is held, as long as product can be delivered in a timely fashion and at a reasonable price.

The labor market is often less important for most warehouses than for the typical manufacturing plant. If the proposed operation has a high degree of automation and relatively little hands-on labor, the availability of well-motivated workers may be of minor importance. However, labor can be of critical importance in warehousing operations that require more package handling.

There can be wide variations in taxes, particularly inventory taxes, within some metropolitan areas. Taxes can be a critical competitive factor when the warehouse inventory has high value.

Community attitudes toward the new warehouse should be carefully studied by every site seeker. In most communities today, a clean and quiet warehouse development is considered preferable to the many manufacturing operations that cause pollution, congestion, or other conditions perceived as detrimental to the quality of life of a community. Yet, some communities are opposed to *any* new industrial development, even warehousing. If such opposition exists, it's best to recognize it and look elsewhere.

Requirements should certainly consider whether a new building is necessary, or if an older one could be adapted for this use. Sound, older buildings are not easily found, particularly in newer communities.

Financial questions are an important part of defining requirements. How important is the availability of mortgage financing at favorable rates?

If space needs are not fixed, the availability of overflow space either in public warehouses or short-term lease space is an important consideration. It is seldom possible to construct or purchase a building that is the right size for year-round space needs. Therefore, the warehouse user should acquire a relatively small permanent space, utilizing overflow space within the community for seasonal requirements.

Location Models

Computer modeling techniques to help in selecting a location have been available since the earliest days of computer use in business. Typically these models use the combination of in-bound and out-bound freight costs to indicate that spot on the map where total freight costs will be the lowest. In other situations, they may work on delivery times to show that location which will be most convenient to the greatest number of customers.

With deregulation, freight costs became subject to negotiation and thus far less predictable. Negotiating ability became more important then geography in establishing favorable freight costs. For this reason, modeling has become less popular in recent years.

One computer model is based on research data rather than strictly quantitative items. The computer tool known as Location

Site Analysis Checklist

General Information
1. Site location (city, county, state):
2. Legal description of site:
3. Total Acreage: Approximate cost per acre:
   Approximate dimensions of site:
4. Owner(s) of site (give names and addresses):

Zoning
1. Current: Proposed: Master plan: Anticipated:
   Is proposed use allowed? ____ yes ____ no
2. Check which, if any, is required:
   __ rezoning __ variance __ special exception
   Indicate: Approximate cost   Time required:
   Probability of success:
   __ __ excellent __ good __ fair __ poor
3. Applicable zoning regulations (attach copy):
   Parking/loading regulations: Open space requirements:
   Office portion: Maximum building allowed:
   Whse./DC portion: Percent of lot occupancy allowed
   Other: Setbacks, if required:
   Height restrictions: Noise limits: Odor limits:
   Are neighboring uses compatible with proposed use?
   __ yes   __ no
4. Can a clear title be secured? __ yes __ no
   Describe easements, protective covenants, or mineral rights, if
   any:

Topography
1. Grade of slope: Lowest elevation: Highest elevation:
2. Is site: __ level __ mostly level __ uneven __ steep
3. Drainage __ excellent __ good __ fair __ poor
   Is regrading necessary? __ yes __ no
   cost of regrading:
4. Are there any __ streams __ brooks __ ditches __ lakes
   __. ponds __ marshlands __ on site __ bordering site
   __ adjacent to site?
   Are there seasonal variations? __ yes __ no
5. What is the 100-year flood plan?
6. Is any part of site subject to flooding? __ yes __ no
7. What is the ground water table?
8. Describe surface soil:
9. Does site have any fill? __ yes __ no

Figure 11-1

10. Soil percolation rate __ excellent __ good __ fair __ poor
11. Load-bearing capacity of soil: __ PSF
12. How much of site is wooded: How much to be cleared?
    Restrictions on tree removal: cost of clearing site:

Existing Improvements
1. Describe existing improvements.
2. Indicate whether to be __ left as is __ remodeled
   __ renovated __ moved __ demolished
   Cost $ __

Landscaping Requirements
1. Describe landscaping requirements for building parking lots, access road, loading zones, and buffer if necessary.

Access to Site
1. Describe existing highways and access roads, including distance to site. (Include height and weight limits of bridges and tunnels, if any).
2. Is site visible from highway? __ yes __ no
3. Describe access including distance from site to a) interstate highways b) major local roads c) central business district d) rail e) water f) airport. Describe availability of public transport.
4. Will an access road be built? __ yes __ no
   If yes, who will build?   Who will maintain?   Cost?
   Indicate a) curb cuts   b) median cuts   c) traffic signals
   d) turn limitations
5. Is rail extended to site? __ yes __ no Name of railroad(s):
   If not, how far?   Cost of extension to site:
   Who will maintain?   Is abandonment anticipated? __ yes __ no

Storm Drainage
1. Location and size of existing storm sewers:
2. Is connection to them possible? __ yes __ no
   Tap charges:
3. Where can storm waters be discharged?
4. Where can roof drainage be discharged?
5. Describe anticipated or possible long-range plans for permanent disposal of storm waters, including projected cost to company.

Sanitary Sewage
1. Is public treatment available? __ yes __ no
   If not, what are the alternatives?
2. Is sanitary sewage to site? __ yes __ no
   Location of sewer mains?
3. Cost of materials (from building to main)—include surface restoration if necessary:

4. Tap charges:
5. Special requirements (describe fully):
6. Describe possible or anticipated long-range plans for permanent disposal of sewage, including projected cost to company.

Water

1. Is there a water line to site? __ yes __ no
2. Location of main: Size of main:
3. Water pressure: Pressure variation:
4. Hardness of water:
5. Source of water supply: Is supply adequate? __ yes __ no
6. Capacity of water plant: Peak demand:
7. Who furnishes water meters?   Is master meter required?
     __ yes   __ no
     Preferred location of meters:
     __ outside   __ inside
8. Are fire hydrants metered? __ yes __ no
   If yes, who pays for meter installation?
9. Attach copy of water rates, including sample bill for anticipated demand if possible.

Sprinklers

1. What type of sprinkler system does code permit?
2. Is there sufficient water pressure for sprinkler system?
   __ yes __ no
3. Is water for sprinkler system metered? __ yes __ no
4. Is separate water supply required for sprinkler system?
   __ yes __ no
5. Where can sprinkler drainage be discharged?

Electric Power

1. Is adequate electric power available to site? __ yes __ no
   Capacity available at site:
2. Describe high voltage lines at site:
3. Type of service available:
4. Service is __ underground __ overhead
5. Reliability of system __ excellent __ good __ fair __ poor
6. Metering is __ indoor __ outdoor
7. Is submetering permitted? __ yes __ no
9. Indicate if reduced rates are available for a) heat pumps __ yes __ no b) electric heating __ yes __ no
10. Attach copy of rates, including sample bill for anticipated demand, if possible.

Fuel

Gas:

1. Type of gas available:
2. Capacity   Present: Planned:

Figure 11-1. Continued

110

 3. Peak Demand   Present: Projected:
 4. Location of existing gas lines in relation to site:
 5. Pressure of gas:
 6. Metering is __ indoor __ outdoor
 7. Is submetering permitted? __ yes __ no
 8. Is meter recess required?
 9. Who furnishes gas meters?
10. Indicate limitations, if any, on new installation capacity requirements.:
11. Attach copy of rates, including sample bill for anticipated demand, if possible.

Coal
 1. Source of supply: Reserves:
 2. Quality of coal available: Cost (per million BTU) delivered:
 3. Method of delivery:

Oil:
 1. Source of supply: Volume available:
 2. Quality of oil available: Cost (per million BTU) delivered:
 3. Method of delivery:

Taxes
 1. Date of most recent appraisal:
 2. Real estate tax rate history, last five years:
 3. History of tax assessments, last five years:
 4. Proposed increases, assessments and tax rates:
 5. Are any abatement programs in effect? __ yes __ no   If yes, describe.
 6. Is site in an Enterprise Zone? __ yes __ no
    Duty free zone? __ yes __ no
 7. Have any special taxes been assessed? __ yes __ no
    If yes, describe:
 8. Indicate anticipated or possible major public improvements:
 9. Services provided for taxes paid: Local: County:
    State:
10. What is state policy on inventory tax, floor tax, etc.?
    Is it a free port state? __ yes __ no   If no, describe assessment dates, procedures and tax rates.
11. Indicate rates for: Personal income tax:
    Corporate income tax: Payroll tax:
    Unemployment compensation: Personal property tax:
    Sales and use tax: Workmen's compensation:
    Franchise tax: Other taxes:
12. Indicate taxation trends:
13. Are industrial revenue bonds available? __ yes __ no

Quest has loaded into it data on every county in the United States. The user can select those criteria that are most important, or the main points of the requirement's definition. Using a prescribed definition, the model will make comparisons from one county to the next, or from one metropolitan area to the next. The model is a compilation of research material in a computer format that makes it easy for the site seeker to establish criteria and then make comparisons. In effect, this computer model is a compilation of research designed to ease the process of site selection.

The Final Selection Process

With all requirements defined, the user should now move through the selection process.

As sources of information are checked in the move from "macro" region to "micro" street location, be very skeptical. If one party advises that a specific site has never had a flood history, for example, get a second opinion from another party who cannot possibly be influenced by the earlier informant. If multisource checking produces the same answers to critical questions, one can then have some confidence in the information.

As you move through the selection process, and particularly as you zero in on a specific site, it is essential to have a contingency plan. If site A is the preferred one, it is wise to have a site B that is almost as good and equally available. It is well to let the seller discover that there is a fallback site in the event that bargaining for site A should break down.

Because site selection is one of the most important decisions the average distribution executive ever makes, the process of making the decision cannot be discussed too often. Other distribution decisions are correctable, but a poor choice for a warehouse site is a decision that, while not irreversible, is very costly to correct.

The checklist in Figure 11-1 can be a useful aid in reviewing the final decision process.

12

BUILDING OR REHABILITATING YOUR WAREHOUSE

There has been a quiet revolution in the construction of warehouse buildings. While the price of other products and services has steadily escalated, the cost of warehouse construction has remained fairly static, and in some cases has gone down. As you look at new construction, consider where the priorities for controlling quality should be placed.

The most important component in any warehouse building is the floor. If extremely bad, a defective floor can result in the need to tear down the building. In less extreme cases, broken floors typically stay bad for the entire life of the building.

Second in priority is the framing and roof system. If the structure and the roof are done badly, they can be repaired, although such repairs are difficult and expensive.

Next in importance are docks and dock doors, which are the key points for flow of materials in and out of the building. If the doors are poorly placed, the material flow will be difficult and expensive.

Least in importance is that part of the building that gets the most attention from the uninitiated—the walls. Many warehouse buildings have walls that do not bear any load, with the entire roof supported by the upright columns. When the walls are not load bearing, it matters little whether they are made of granite or cardboard, as long as they are accompanied by a perimeter alarm system and insulation to protect stored products.

To illustrate the point, one distribution center for building materials was designed with no exterior doors or walls at all. The buildings are used for storage of materials that are loaded and unloaded from flat bed trailers at grade level.

The Floor

The goal for the end product is a floor that requires a minimum of maintenance and has durable wear surfaces. The floor cannot be stronger than the earth beneath it. If the ground is not properly prepared and compacted, a void in the earth beneath the floor will inevitably cause the floor itself to fail.

To reduce maintenance, the floor should be poured in a manner that minimizes curling, which is the development of concave patterns in the floor as the concrete cures. The joints must be created in a way that minimizes spalling or failure of the floor surface at the joint. To create maximum durability, it is important that the concrete itself be dense and nonporous.

In one warehouse, the floor was poured with a series of butt joints. A strip of concrete is poured, with an adjacent area of similar size left vacant, followed by another strip of concrete. The voids between each strip are poured after the first strips have already cured. The builder uses this method to create a higher quality floor.

Another builder places prime reliance on controlling the installation environment. The floor is not poured until the walls and roof are nearly complete in order to reduce temperature variance and wind flow over the curing floor. Direct exposure to the sun can cause concrete to dry too quickly. In the controlled installation environment, relative humidity remains as high as possible to slow the drying process.

Since the subbase is of critical importance, a laser bulldozer and vibrator roller are used to ensure a solid and level base. That base is then surfaced with 46D, a gravel compound that possesses concrete properties. The concrete is poured in the same direction in which traffic will move in the warehouse. The wet cure process should take four weeks in order to achieve maximum strength.

A flatness test is utilized as the concrete is finished to ensure that the floor will be as level as possible. A floor process called *FACE* uses statistical analysis to define variances in flatness.

A laser screed allows larger amounts of concrete to be finished at one time, thus reducing the number of construction joints that are needed. The laser screeded floor reduces flatness variance from 0.75 inches down to 0.25 inches over a length of 50 feet. Most damage to concrete appears at the joints, and the best floors are the ones with the least number of joints.

Strength of the concrete is controlled by reducing the amount of water mixed with cement and carefully managing the amount of water used in the mixture. Dowel joints in the concrete are used at high traffic areas, and the thickness of the floor in high traffic areas may also be increased. A surface hardener is always installed, and when budget permits, a plasticizer may also be added to produce a mirror finish.

Structural System and Roof

The expectation for the end product of the roof and framing system is to achieve the longest possible life with minimum maintenance and no leaks. Outside of the sunbelt, it is also important for the roof to be able to handle significant temperature changes during its installation and useful life and through weather hazards.

For most of the United States, the best option is a preengineered metal roofing system. In some parts of the western United States, some or all of the structural support systems use wood rather than steel beams. Wood systems are used primarily when they are acquired at a more favorable cost. For most structures in the United States, metal is more economical.

The metal roof is built with expansion and contraction capability in the design. A 25-year-old building in Baltimore required a completely new roof. The roof itself was not worn out, but the fasteners holding roof panels had failed to the point where leaks around the screw holes could no longer be repaired. Built with state of the art technology in 1970, this roof had developed more leaks than could be reasonably repaired. The new roof for the building is expected to last for 40 years.

The typical metal roof technology of the 1970s had a slope of 1 inch in 12 inches. Current technology allows a slope that is so gentle (1 in 48) that the building does not appear to have a pitched roof. The gentle slope also makes it safer to walk on the roof when maintenance is performed. The fastener holes in the roof system are prepunched to ensure that the roof is properly aligned and to minimize the stress of expansion and contraction. A predrilled splice plate with a free-moving backer plate allows the fasteners to move without leaks. The length of each metal sheet is staggered to control the transfer of thermal stresses from one sheet to the other. An important feature is the sealing system of the metal sheet which uses the Pittsburgh double-rolled seam. The fastening is similar to that seen in the caps of soft drink cans. The two pieces of metal are crimped and turned to form a very tight seal. A mastic adhesive is used to splice the seals between each lap of the roof.

Drainage is also a critical factor, and the gutter is outside of the building envelope with no parapet walls and no internal gutters. In contrast to the heavier and more complicated roof systems which use several plies of felt roofing and asphalt, the metal system is simpler, lighter, and easier to maintain and repair. Because the framing and roof system is significantly lighter than the older systems used, there is less risk of settlement as the building ages.

Docks and Drive Areas

Because docks are the heart of the material flow in and out of the building, they must be designed for low maintenance and maximum life.

The exterior drive area is one that is frequently neglected in warehouse construction. It should be constructed to eliminate any ponding by maintaining positive drainage both above and below the wear surface. Proper preparation of the subbase soil is just as critical for the drive areas as it is for the warehouse floor. Yet this process is neglected in the construction of some warehouses.

The dock apron is prepared to specifications that exceed those of the warehouse floor. It should be of 8 inch thick concrete with reinforcing steel mesh. The steel is not required for the warehouse floor, but it should be used for the dock to handle the stress of truck trailers. That apron extends at least 50 feet from the face of the dock doors.

As the drive areas are installed, at least 8 inches of heavy gravel is placed very early in the construction cycle to allow cleaner and faster work progress. An additional 2 inches of 46D compound is used for sealing the surface before asphalt or concrete is poured.

The dock doors should be designed to minimize maintenance. Metal and plastic have replaced wood for many of the overhead doors, and the best of them today have a channel that brings the door straight up rather than to make a 90° turn like those found in most residential garages. Mechanical dock levelers are an investment that pays back in reduced operating costs for any busy truck dock. Dock lights are another feature that improves productivity and safety.

Illumination and Heating

Approaches to warehouse lighting have made revolutionary changes in the past two decades. Fluorescent lights have proven to be relatively ineffective and uneconomical in any warehouse where the temperature falls below 60°. Even where the warehouse is heated above that temperature, they are less effective than the newer HID (high-intensity discharge) lighting.

Skylights have been used to provide available sunlight and reduce electrical costs. However, as more warehouses are used on a second and third shift, the importance of sunlight is reduced. Furthermore, skylights have poor insulating value and are a source of heat loss. They can also be a source of roof leaks. Design improvements in lighting systems provide maximum flexibility at an economical cost.

The best way to heat a warehouse today is by the use of an air rotation system. This type of heating allows a single unit to heat over 100,000 square feet of space. However, if the building is designed for partitioning into several rooms, then several smaller air rotation units will be necessary to provide heat in all areas.

At one warehouse building with 112,000 square feet of storage area and 4,000 square feet of office, the floor-mounted air rotation system maintains 60 degrees in the warehouse. In the last five years, costs were $907 in the lowest year and $4,022 in the most expensive year. Costs during the other three years ranged from $1,131 to $1,880. The heating system maintains a variation of less than 5 degrees throughout the warehouse. The floor-mounted furnace has two 7-horsepower fans that circulate air throughout the building.

Another builder prefers to install several smaller heating units that are punched through the exterior walls just below the eave of the building. The units are designed to allow the building to be heated even if interior dividing walls are installed later. The air rotation units are designed to be 100% efficient in combustion of gas, and the movement of air allows a cooler temperature at the ceiling. The circulating fans are frequently run in hot weather in order to prevent heat buildups inside the building. Where propane fork lifts are used, maintaining ample air circulation is critically important, but it also raises the heating costs. Temperature sensor controls are equipped with carbon monoxide sensors which will start the fans if there is a buildup of exhaust fumes.

Foundations and Building Heights

Foundation walls are precast or cast in place. One builder uses precast wall panels that are load bearing, and they are erected with the foundation to reduce costs and increase construction speed. Building height is dictated by the specifications of ESFR sprinkler systems, which are the most effective fire protection available. Current specifications for ESFR call for a minimum height of 28 feet and a maximum of 33 feet.

Wall Panels

One builder uses hollow core prestressed concrete wall panels that have foam insulation in the cores. Similar wall construction is used for both foundations and dock walls. Other builders argue that a metal wall is less expensive. However, the insulated concrete wall is durable, abuse resistant, and it has more eye appeal to the bankers who finance the building and the brokers who lease it.

Wall panels should be designed to allow additional dock doors to be installed later, since changing uses of the building will frequently require new doors.

Interior and Exterior Finish

Bright white paint is the most economical way to make a building attractive to workers and visitors. Columns that contain fire extinguishers usually are indicated with red paint, and other color coding may be used within the warehouse. Use of a maximum amount of white finish will improve light levels and general appearance.

While the warehouse purist would maintain that the exterior of the building has little to do with operating efficiency, it has a great deal to do with the impressions left with visitors, sources of financing, and customers. Landscaping is one of the most cost-effective ways to improve the appearance of a warehouse exterior. Just as good housekeeping is so important as a symbol of warehouse management, similar attention to the appearance of exterior and grounds will greatly influence the impression of those who observe the warehouse from the street. Perhaps the best argument for the use of concrete wall panels is that they appear to be more durable and more attractive than concrete block or metal walls. While the materials used in walls may be relatively less important for physical function of the warehouse, they do impress the casual observer. Although a trashy or poorly kept exterior may not necessarily signify a bad warehouse operation, many observers will immediately downgrade management if the exterior of the building presents an eyesore.

The Rehabilitation Alternative

The preservation, conservation, and adaptive reuse of older buildings has gained increasing attention during the past two decades. Some of this interest was spurred by tax legislation that provided significant tax credits for rehabilitating both historic structures and buildings of some age.

Between 1978 and 1986, such tax credits provided a meaningful incentive for the warehouse user who would recondition an older building rather than construct a new one. By 1986, changes in the tax law had reduced the credit for nonhistoric structures to only 10%, and this applies only to buildings placed in service before 1936. In effect, the rehabilitation tax credit no longer makes it worthwhile for the warehouse operator to rejuvenate an old building. This does not mean that rehabilitation may not sometimes have economic incentives regardless of taxation. Sometimes the best location for future growth is an available building in the center of the city, and it may be more economical to rehabilitate that structure than to build a new one. At times, the community in which an older building is located may offer tax abatement, favorable financing, or other financial incentives to persuade the user to purchase an older building. Obviously these incentives must be balanced against the cost of rejuvenating the building and the cost of operating that building versus operations costs for a new structure.

Planning for Future Uses

Given reasonable care, a good warehouse building should last for many years. If abused, the building will start to deteriorate relatively early in its life.

New construction techniques have caused warehouses to be more economical and of higher quality than ever before. With a flexible design and proper maintenance, such a building should have a useful life of many decades.

13

THE 21ST CENTURY LOGISTICS FACILITY

As we look at probable changes in the 21st century, consider the relationship between warehousing and business logistics. Take a more detailed look at the building design itself and how that design can be changed. Understand the major logistics megatrends and how they will influence warehousing. Consider financial factors and in particular the ways in which financial institutions have influenced the design and construction of logistics facilities.

Some Definitions and Concepts

A logistics facility is defined as a high-cube warehouse or materials handling facility of 100,000 square feet or greater, designed with a clear pile height of at least 20 feet, and intended for the storage and handling of packaged materials and products. This kind of building is defined as institutional grade, which means that its location and configuration would make it suitable for the institutional investor. Such a building is not a single-purpose unit, but one that is convertible to the handling of other commodities or to light manufacturing. The logistics facility may be a multitenanted building, or it could be occupied by a single user.

Not included in this definition are facilities that are designed for storage and handling of dry or liquid bulk materials (silos, tanks, grain elevators) or other special-purpose structures. Also excluded are smaller storage buildings (less than 100,000 square feet) or others that are so specialized in their design that they could not be considered as institutional grade.

Having indicated that these buildings are logistics facilities, we need to have an understanding of what business logistics is. The Council of Logistics Management, the leading professional society in the field, has defined business logistics in this way:

The process of planning, implementing, and controlling the efficient, effective flow and storage of goods, services, and related information from point of origin to point of consumption for the purpose of conforming to customer requirements.

Within the field of business logistics, there are two kinds of operations, a distinct one being third-party logistics. Third-party logistics is defined as any operation that is managed by a party that does not own the merchandise being handled. In warehousing, the traditional understanding was that the third-party warehouse was one that issued a warehouse receipt for goods placed in its care. In transportation, the third-party issues a bill-of-lading, which is a contract to carry merchandise from one place to another. In both cases, the operator does not take title to the goods that are placed in the vehicle or the warehouse. The legal responsibilities of the third-party are strictly defined, with a legal tradition going back to English common law. In warehousing, fewer than 20% of goods stored in the United States are in the hands of third-party operators. However, the percentage is growing rapidly and in some places overseas it has reached as high as 50%.

Who Uses Logistics Facilities?

The largest users of logistics facilities are producers or manufacturers. Producers would include mining and agriculture, but the most significant producer group is manufacturing. Producers place goods in logistics facilities that are well suited to the product line being warehoused. In some cases, this dictates the use of a specialized storage facility rather than an institutional grade building.

A second major user group is merchandisers, both wholesalers and retailers. This group may use some of the same types of warehousing just described, but their primary motive is to facilitate the flow of goods from sources to retail stores. English and many other languages use the word "store" to refer to both warehousing and retailing.

The third major user group is the third-party operators. Third-party logistics operators can also be divided into two groups, the traditional public warehouse or common carrier, and the contract or dedicated logistics operator. The fastest growing part of the industry is the latter, and in a contract operation a building is typically dedicated to the warehousing needs of a single client. In contrast, the public warehouse operator functions on a month-to-month agreement, with substantial fluctuations in the size of inventory and a payment system that relates revenue directly to the amount of inventory handled. Dozens or even hundreds of different clients may be stored within the same building, with inventory sizes ranging from one pallet to thousands of pallets. In transportation, the public warehouse counterpart is the common carrier who moves and distributes freight with no long-term contract and gains or loses clients with a telephone call. A shipper who grows unhappy with a common carrier simply phones a different trucker for the next shipment. Similarly, the public warehouse can be dismissed with a telephone call and the goods can be transferred to a competitive facility or the owner's warehouse.

Of the three user segments—manufacturers, merchandisers, and third-parties—the third-party segment has the fastest growth, and the manufacturing segment has the slowest.

Regional and National Differences

There are distinct regional differences in logistics facilities within the United States, though these differences are not as great today as they were a few decades ago. The differences have been dictated partially by climate, partially by geography and seismic conditions, and partially by the tastes of institutional investors.

For example, preengineered metal structures were widely accepted in the south and the midwest before they were considered acceptable in the east and the far west. In the far west, cost and availability of steel encouraged construction of buildings with wood roof systems. In recent years, a change in those cost relationships has resulted in the nearly universal use of steel as a framing system.

Dock designs are influenced by climate. Interior truck courts are commonly used in northern cities with severe weather to allow loading and unloading with total protection from the elements. In most of the country, truck dock doors are on the exterior walls of the building, and the users rely on plastic and canvas dock shelters to provide a weather seal.

The Changing Players in Logistics

There have been five key participants in the logistics industry: manufacturers, merchandisers, material suppliers, consumers, and common carriers. The role of each has changed. The oldest and the largest has been the manufacturer or producer. For the manufacturer, the warehouse was a reservoir to guard against changes in supply and demand. Traditionally, the manufacturers pushed their finished goods through the market to the consumer.

The merchandising industry has grown in scope and sophistication. Until recently, the merchandisers adapted to the logistics program of each manufacturing facility. More recently, with the rise of more aggressive and sophisticated merchants such as Wal-Mart, the merchandiser has dictated the logistics program to the manufacturing sources. The result is a logistics facility that is under the control of the merchandising organization rather than the production company. Furthermore, the flow changes from a *push* by the manufacturers to a *pull* by the merchandisers. This change in the balance of power between manufacturer and merchandiser has been one of the major logistics changes of recent years.

A third key participant is the raw material supplier, and some suppliers utilize logistics facilities to ensure prompt delivery of the materials they supply to manufacturers.

The fourth key participant is the consumer, and some warehouses are dedicated to the delivery of goods directly to the ultimate user. The technical term for this logistics operation is fulfillment, also known as mail order or catalog operations. There is nothing new about mail orders, since the first major catalog and mail order warehouses were established in the early years of this century. However, the business was revolutionized with the development of credit card sales and more creative methods of advertising such products to the general public. As a result, many more goods are purchased from fulfillment centers today than ever before, and the business is rapidly growing. The result is the rise of many new fulfillment warehouses, some operated by third parties and some by the merchandising organization.

The last key participant group is common carriers. With deregulation, carriers sought to diversify as the profits from their core business were squeezed by competitive pressures. Many common carriers diversified into third-party warehousing and now operate logistics facilities for some of their clients.

Seven Megatrends

There are seven megatrends that will influence logistics facilities in the 21st century. The first is the age of information. More than ever before, the logistics facility is designed to handle an ever growing need for communications and computing equipment. For example, one new warehouse has fiber optics outlets located every 50 feet around the walls of the building.

The growth of third-party logistics is the second megatrend which will influence logistics facilities in the next century. Robert V. Delaney of Cass Logistics made a prediction in 1993 that the contract logistics market would grow from a $10 billion level in 1992 to a $25 billion level in 1996 and would reach a level of nearly $50 billion by the turn of the century. His prediction from 1992 to 1996 appears to be on target. This means that third-party logistics may be one of the fastest growing industries in the United States, and therefore an increasing number of logistics facilities will be dedicated to the needs of third-party operators.

Overnight package delivery is the third megatrend. The business was virtually invented by Fred Smith, founder of Federal Express. As the concept was proven, it attracted competitors and became an indispensable part of marketing and operations plans for many other businesses. It has also created new markets for logistics facilities. For example, a major industrial park has been developed at the airport that Airborne Express owns and operates in Wilmington, Ohio. Because Wilmington is the hub for Airborne overnight delivery operations, a shipper located at the hub can release orders as late as 2:00 a.m. and still achieve delivery later the same morning. Hub warehousing is a new and rapidly growing use of logistics facilities.

The concept of "just-in-time," the fourth megatrend, was popularized by Toyota and later by other Japanese auto companies. The initial theory was that JIT would eliminate warehousing, and in the compact Japanese islands, this frequently happened. In the United States, JIT has not eliminated warehouses, but rather it has created the need for new ones in different locations. The JIT warehouse must be close to the assembly plants using the JIT process, and new logistics facilities have been constructed near assembly plants and other users of JIT. The growth of cross-docking is part of the JIT megatrend. Sometimes referred to as flow-through facilities, a growing number of warehouses are dedicated to the movement of products *through* a facility rather than storage within it.

The change of control from producer to merchandiser is the fifth megatrend. Retail organizations can now dictate the location of new logistics facilities. In many cases, the merchant seeks an arrangement in which title to goods remains with the source until it is withdrawn from the warehouse. The logistics facility may not be operated by the merchant, but its operation is controlled by the buyer rather than by the vendor. This power shift has caused changes in location and scope of some logistics facilities.

The sixth major megatrend is globalization. With the implementation of NAFTA and their treaties for international trade, logistics managers are often responsible for operations in at least three different countries. Mexico and Canada are considered part of the same market, even though logistics procedures in those countries are quite different. Until recently, the third-party warehousing business in Mexico was dominated by the banks and utilized primarily by growers and manufacturers who needed to borrow money and use inventory as collateral. These facilities were storehouses rather than distribution centers, with extremely slow turns. One Latin American observer called bank warehousing "a large-scale pawnshop." Logistics managers today are dealing with export and import operations to a greater degree than ever before, and they are asked to warehouse in overseas markets where operations are far different from those at home.

The seventh megatrend has been the de-unionization of the logistics industries.

The turning point for trade unions was the unsuccessful strike against the Federal government by PATCO, the union representing air traffic controllers. This 1981–82 strike was the most important labor dispute in American history. The union representing the air traffic controllers learned that neither the government nor the public had sympathy for their position, and labor relations in the United States have not been the same ever since. Most of the newer corporations engaged in providing logistics services, both transportation and warehousing, are either non-union companies or firms that have labor agreements that emphasize flexibility.

The Location Decision

Where logistics facilities are placed depends on consideration of these factors: geography, locations of sources of inbound, locations of consignees, carrier terminal locations, JIT, cost and availability of labor, and taxes or tax incentives.

Geographical considerations come first. It is obvious that you cannot build a logistics facility at a site with significant grade problems or poor soil conditions. Water is also a necessary feature, either through piped sources or lakes or ponds. Fire protection in the United States is based on the utilization of sprinkler systems, and significant amounts of water are needed to power these. Less recognized is the fact that this need not be supplied by water mains. Underwriters will allow the use of ponds or lakes as a water source if there is ample pumping capacity. Sprinklers are not used in parts of Europe, Asia, and Latin America.

Since a logistics facility exists to facilitate materials flow, it has to be at a location that is convenient to the sources of inbound shipments, the destinations of outbound ones, or both. Logistics planners typically determine the locations of sources and consignees as an early step in finding the best spot to locate the building.

Because of the close relationship between warehousing and transportation, the location of carrier terminals must be considered. The hub warehousing concept has already been described, but other warehouse operations will require proximity to motor freight terminals or TOFC (trailer-on-flat-car) ramps. Others may need to be close to a marine port or an airport. Proximity does not always allow ease of access, and therefore the location should be measured on accessibility rather than distance.

Since the JIT concept is based on ability to make precisely timed deliveries, the JIT logistics facility must be located at the best site to serve its customers.

While the differences in labor costs throughout the United States have narrowed, labor supply is a major issue. Success in warehousing is based on the availability of productive and well-motivated workers.

Unlike the factory, the warehouse has no assembly line in which inspectors can be positioned to measure quality. Given the difficulty of line supervision, warehouses have been included in those activities where supervisors were eliminated through the use of a "self-directed work team." Only a few distribution centers today are managed with this concept, but it is likely to be tested in more warehouses.

Because high-quality labor is needed, availability rather than cost is a key factor in finding a location for a new logistics facility. Sometimes the availability includes nontraditional sources, such as part-time workers who may be housewives, farmers, or military personnel.

Finally, taxes and tax incentives will influence the location decision. In some situations, occupants of the new industrial park achieve tax savings that are almost equal to the rent cost of the building.

In considering any location, it is essential to measure the eventual effect of obsolescence. What conditions would make a location that is attractive in 1997 supremely unattractive ten years later? Abolition of tax incentives would be one example, but another could be a change in the location of sources or consignees which destroys the strategic value of the facility.

The Building Design

A logistics facility is typically measured more by operational efficiency than by the attractiveness of the real estate. Therefore the building design should emphasize operations. Building columns should be of dimensions that are designed to facilitate storage rack and aisle layouts. Therefore, the efficient warehouse is designed from the inside out. On the other hand, the majority of logistics facilities will outlive their initial user, and therefore the design must contain flexibility for change. An institutional grade building is not a white elephant, and every building must be designed so that it can be used by future occupants as well as the present ones.

Clear height should be just as great as possible. While there is always a tradeoff between floor space costs and equipment costs, a building with a high cube configuration will be attractive to far more users than one that has a low ceiling. Therefore, even if the initial occupant does not need high cube space, the next occupant probably will.

The most important criterion influencing building design and clear height is the development of the ESFR (Early Suppression Fast Response) sprinkler system. This sprinkler technology is the most important advancement in warehouse fire protection of the past half century, and will remain important in the 21st century.

Two developments of the past few decades have made sprinkler system protection more complicated: these are increased use of storage racks and a far greater volume of products containing chemicals that are dangerous in a fire. The insurance underwriters' solution to the problem was to specify intermediate sprinkler heads installed within the rack. Storage rack equipped with intermediate sprinklers can never be adjusted or moved, because the sprinkler system is not portable.

Plastic content in products is greater than it was a few years ago. However, some plastics emit toxic fumes when they burn. Aerosol containers can be hazardous but because of their convenience, they are unlikely to disappear.

The ability to eliminate in-rack sprinkler systems is the primary virtue of the ESFR sprinkler system. ESFR has two unique features: a quicker response time and a heavier sprinkler discharge. When a sufficient quantity of water is dropped onto a fire at an early stage, the fire will be suppressed before it presents a severe challenge to the building itself. Furthermore, the ESFR system typically opens fewer sprinkler heads than a conventional system, and therefore reduces water damage. Reducing the size and duration of the fire will also reduce the extent of smoke damage.

The clear height of the building should be determined after consulting with fire underwriters and developing a configuration that will facilitate the installation of this sprinkler system. Even if the initial tenant does not need the new system, the next tenant is likely to want it.

Velocity of product turn will influence building design. The warehouse of tomorrow will emphasize fast movement rather than efficient storage. It will be longer and narrower than the one used today, simply because a narrow building facilitates quick flow from receipt to shipment. While a building planner today might suggest one truck dock for every 5,000 square feet of warehouse, tomorrow's warehouse could have one dock for every 1,000 square feet of space.

Heating and lighting systems will be replaced by new equipment that is far more efficient in its use of energy. Floor-mounted heating systems can heat a warehouse more efficiently and more economically than the older gas fired heaters hung from the ceiling. High-intensity discharge (HID) lighting provides distortion-free illumination at lower costs than fluorescent fixtures.

Operational Changes

A growing proportion of lift trucks will be operated by electric motors because of the need to avoid pollution, though internal combustion engines will still be used because of lower cost and superior multishift capabilities. However, those using internal combustion will be fueled with compressed natural gas (CNG) rather than propane or gasoline. CNG offers significant reductions in exhaust gas pollution as well as improved safety.

Order pick lines will be designed with an ergonomic layout, one that places the fast moving items in those locations where they can be picked with the least amount of effort or risk of personal injury. Above all, the layout will be designed to facilitate change. Warehouse managers will plan to reset the warehouse every three to six months, to adapt to new products and new demands.

Adapting to quick change and using electronic guidance systems for order picking vehicles will allow warehouse managers to eliminate the painting of aisle stripes. Since the aisles could be in a different place a few months from now, it will not be practical to mark these aisles with stripes.

The warehouse of tomorrow will have no office, since improved information transmission capabilities have already eliminated the need for office workers to be at the same location where the cargo is handled.

Relatively few warehouses will operate on the traditional 40-hour work week. As multishift operations become widely used to improve velocity and reduce asset investment, the work schedule may be a multishift system with a four day work week at 10 hours per day. Three of these "4/10" shifts will allow the warehouse to be open 20 hours per day, six days per week.

Tomorrow's warehouse will be a flow-through structure with layout, equipment, and people who emphasize flexibility and capacity to accommodate quick and frequent change.

Financial Considerations

Developers of logistics facilities must find mortgages, and they must also deal with insurance and other aspects of risk management. The people they deal with in the financial and risk management fields may have minimal understanding of business logistics. Therefore, much of the process is one of educating the financing source.

A key first step in this education process is to point out that logistics facilities are a relatively attractive real estate investment risk. A 1994/95 study commissioned by NAIOP (National Association of Industrial and Office Parks) is titled *Industrial Income and Expense Report*. This national survey of industrial buildings has some valuable support for the attractiveness of logistics facilities.

The survey demonstrates that of those buildings designated as warehouses, the average percentage of occupancy was 90.7. Those classified as warehouse/bulk distribution have an occupancy rate of 89.3. Multitenant industrial/commercial buildings have an occupancy of 91.7%. These occupancy rates compare favorably with those found in office, retail, or other commercial real estate. Furthermore, there is relatively little difference in occupancy rates as different regions in the country are compared. As compared to other types of real estate, the logistics facility is relatively simple to construct, with minimal opportunities to have difficulties in completing a new construction project. Furthermore, the risk that the building will be empty is quite low.

Third-party warehousing represents a special challenge for the investment community. Financial people have seldom been comfortable with the risks involved in traditional public warehousing, since the operator has no leases and no guarantees of occupancy. Yet the same financial sources will fund the construction of a hotel, which has exactly the same occupancy risks as a public warehouse. In effect, the public warehouse is a hotel for merchandise, but it is difficult to explain this to many financial sources.

The fast growth sector of third-party warehousing is contract or dedicated warehousing. To a great extent, this growth is a monument to the ignorance of our lending institutions. As contract warehousing became more popular, users began to realize that the old "evergreen" contract impedes innovation and improvement. A growing number of users have cut down the evergreen and replaced it with operating agreements that specify continuous improvement. This is accomplished first by separating the real estate from the operating contract, thus giving the user the ability to change operators without moving the inventory. This also allows the commitment for space to be long term, but the commitment for labor and management can be of much shorter term. Furthermore, many warehouse contracts today contain a clause that permits the user to dismiss the supplier if there is any failure in service. Service failures are not easily defined from a practical standpoint. As a result, the divorce of supplier from user is just as common today as it was twenty years ago when the traditional public warehouse arrangement existed. The ruthless customer can hammer his supplier just as effectively with a contract arrangement as he could with the older public warehouse arrangement.

The real stimulus for contract warehousing has been the inability to expand third-party warehousing in any other manner. Both users and suppliers have recognized that additional space could not be constructed without a contract from the user, simply because the lender would not provide the money otherwise. If the warehouse provider could find an existing building, he might also find friendly financing from the owner of that building. Otherwise, the contract was an absolute necessity to allow the warehouse provider to obtain the space.

Conclusion and Recommendations

If you are planning the development of new logistics facilities for the immediate future, the most important consideration is to design features that will cause your building to remain superior to most of the buildings that exist today. Fire protection is perhaps the most significantly changed feature, and any developer who constructs new space without the ESFR sprinkler systems could be making a mistake that will become more painful in the future. Since the great majority of logistics facilities will be used by more than one corporation during their economic life, the design should emphasize flexibility with allowances for changes in truck docks and the potential for dividing the building or adding to it in later years. Do not assume that lenders have an adequate understanding of the function of logistics facilities. An important task for the developer is to educate the lender on business logistics.

A well-designed and well-located logistics facility is an investment that compares very favorably to the typical commercial real estate property. This belief can be supported by the presence of 40-year-old warehouse structures that still return a profit to their owners. While the design of such facilities will change in the 21st century the changes will be evolutionary rather than revolutionary.

Part IV

PLANNING WAREHOUSE OPERATIONS

14

PLANNING FOR
FUTURE USES

Whenever construction of a new building is planned, you must face the probability that the building will become inadequate before the structure is actually worn out. The building could become obsolescent for several reasons. First, the volume and storage pattern needed when it was planned has changed. Second, the design of the building may become outdated by the introduction of new warehousing technology, or by a change in the product line to be stored. Third, the function and design may still be good, but the location may no longer be effective for the use originally planned. In some cases, faulty construction or poor soil conditions may cause the building to be simply worn out. Finally, substantial expansion or contraction of warehousing volume may cause the structure to be either much too large for current purposes, or too small with no practical means of expansion.

Design Structure for Versatility

How does future obsolescence affect building design? Consider the fact that a building designed for your use alone may have very limited appeal if it must be put on the market for a quick sale. In general, there is a somewhat smaller market for very large warehouses than for those of a more common size. If your space needs are quite large, running from 500,000 to several million square feet, consider a cluster of separate buildings, rather than one huge structure. Separate buildings can be leased or sold to different users when no longer needed. Furthermore, a cluster of separate buildings reduces fire risk, since it is relatively easy to control the spread of fire from one building to another. While some warehousing operations require a single building for effective shipping and receiving, many others can be managed in a way that does not require all merchandise to be under the same roof. Don't assume that one building is essential just because one manager thinks so. Carefully simulate an operation that separates different product lines or inactive reserve stock into adjacent and separate buildings.

Planning the construction of a new building for easy expansion is a valuable means of extending the lifetime of the facility for its current use. Preengineered construction is frequently selected just because it is particularly easy to expand a preengineered building. On such buildings, the end walls can be removed and the structure readily extended. However, if the side walls (eaves) are to be expanded, either the roof must become lower than the present eave height, or the new roof must slope upwards, which leaves an interior drain in the warehouse.

Many warehouses are designed for each conversion to a future manufacturing plant. When this is done, certain construction specifications for heavy-duty power, lighting, or drainage may be added to the building for manufacturing that are not needed while the structure is still used for warehousing. The number of parking spaces needed for manufacturing workers is usually far greater than those needed for warehousing. At the same time, bear in mind that construction to manufacturing specifications could greatly escalate the cost per square foot.

In deciding between a low-cost square building and a higher cost, long and narrow building, the planner should also consider future uses and resale, as well as the current use. The long building may cost more to build, but it is easier to subdivide into smaller spaces.

Changes in Basic Function

At one time warehouses were storehouses, providing long-term storage of products that might not be needed for years. In South

America, there are inventories of green coffee that have been in warehouses for years, with banks financing the inventory and growers waiting until the price is right. Fortunately, the product does not deteriorate even if stored for several years.

The hallmark of warehousing today is that goods are moving at a greater velocity than ever before. Cross-docking is the term used to describe the movement of merchandise that comes in on one truck and leaves on another within a few hours. With growing emphasis on just-in-time and its various industry equivalents, more of the goods received at warehouses will move across the dock rather than into a stack for storage.

Internal and External Changes

Every warehouse represents a combination of raw space, labor, and equipment. An inventory that turns 30 times per year represents a highly labor-intensive operation, and effective use of people may be more important than using nearly every foot of space. But under changing conditions, the space may become a great deal more valuable than the labor. These changes, because they affect the balance among the three prime elements of warehousing—space, labor, and equipment—would be considered as internal changes.

A more common cause of building obsolescence is external changes, those that take place outside the facility and may well be beyond the control of the warehouse operator. These include a changing neighborhood, new developments in transportation, new taxes, legislative changes, or changes in fire regulations.

At times, the future of a neighborhood can be predicted with reasonable accuracy by local development people. Consider what the area is projected to be like 20 years from now, as well as its current character. For example, a sudden acceleration in crime rates in a neighborhood could make a facility undesirable for you and for most future buyers.

Taxation changes can make a building more or less attractive in the future. If the tax rates in your community were to be increased substantially compared to prevailing rates in nearby towns, your building might be less attractive to the user who could find a better tax cost in a comparable location. States and communities have changed their policy on inventory taxation, and when a community makes such a change, warehouse users often seek to take advantage of favorable inventory taxation. Inventory taxes are an example of changes caused by legislation. An increasing number of states and communities have recognized that offering "free port" status to warehouse users will create a substantial increase in business in the community.

When buildings are planned for future use, the user frequently finds that the useful life of the facility will be greatly extended. However, if there is no way to extend its use, resale of the facility may be the best course. Proper planning can make your building more attractive for resale than a "white elephant" that once was a precise fit to your requirements but now has little value for any other user.

Managing Change

The most important requirement for today's warehouse manager is the ability to deal with change. Warehousing has changed more in the last 10 years than it did in the previous century. The desktop computer and information technology have changed warehouse operations more than anything since the introduction of the industrial lift truck. Many warehouse managers today are not using readily available technology to implement bar coding and other features found in many warehouse management systems. We suspect that resistance to change is the underlying cause of this neglect.

What changes will warehouse people need to survive in the 21st century? If all of your people are not accustomed to working with a computerized system today, they should be acquiring this skill soon. As systems develop, the majority of warehouse workers will be using a warehouse information system, and in some warehouses this is happening today. Computer inquiries and entries are made by material handlers as well as office workers.

Demands for improved customer service will change jobs and

Adequate room for expansion and additional parking are important considerations in site selection.

Figure 14-1

job scheduling. The five-day 40-hour week may eventually be remembered as a relic of the 20th century. More workers will have different kinds of work weeks.

Nearly one out of every three workers in the United States had been with their present employer for less than a year, and nearly two-thirds had worked at the same place for less than five years. With this kind of turnover, your training capability is not a luxury, but a necessity.

If you cannot reverse today's trend to fast employee turnover, your success will be limited by your ability to train effectively. Most importantly, the warehouse layout may change several times per year to accommodate seasonal inventory requirements.

Storage rack systems that contain in-rack sprinklers make warehouse layout expensive. Developments in sprinkler technology (ESFR) can eliminate the need for in-rack sprinklers, and the most important reason to adopt this technology is the fact that racking systems must be flexible and capable of frequent change. A warehouse that does not lend itself to easy layout changes will become an impractical warehouse in the 21st century.

The age of information continues to change the nature of the tasks performed by a warehousing team. The job of warehouse manager is no longer that of a work engineer who is concerned about how to stay on schedule and get the job done. Tomorrow's manager will spend far more time as a coach—teaching, training, and developing people. That manager will also need to be a leader, one who motivates and creates an environment where people work together to get the job done. Three major changes—outsourcing, increased velocity, and ability to manage change—are the hallmarks of success for the 21st century warehouse manager. Do you and your people measure up to this task today? If not, what must you do to survive in the future?

15

SPACE PLANNING

In essence, the warehousing is nothing more than supplying adequate protection for stored goods in a minimal amount of space. Therefore, your goal should be to protect the inventory in your care and at the same time use as little space as possible. Achieving this goal depends primarily on careful planning.

Sizing the Warehouse

How much space is needed to hold a planned inventory?

Assume that a warehouse operator needs to develop a total space plan for toilet paper. A carton measures $20 \times 24 \times 10$ inches, and is received in box cars containing 1,500 cases, with just one line-item in each car. The product can be stacked 15 feet high. The operator determines that the product can be stacked in tiers measuring 2-by-2 units, five tiers high on a 48×40 inch pallet. With a stacking limitation of 15 units, this converts to a height limitation of three pallets. A pallet load, allowing for normal overhang, will occupy 15 square feet.

Layout plans show that 40% of the building will be used for aisles, docks, and staging areas, leaving 60% available for storage. From this 60% net storage, an additional 20% is lost by "honeycombing." ("Honeycombing" is the space lost in front of partial stacks.) After making the space calculation shown in the exhibit below, we find that 31.25 square feet are needed for each stack, or 0.52 square feet per case of product. With this information and with inventory projections, you can plan your total space requirement. Similar calculations should be made for each item planned in the inventory.

Figure 15-2 shows how a storage analysis is done for this or any other item in inventory.

Storage Requirements

The environmental requirements for different products vary, and this must be considered in planning for space. For example, some merchandise may require temperature or humidity control. Other products may be hazardous, and still others represent abnormal security risks. Some products must be stored in racks because packaging limits the ability to stack them. Some pallets must be stored in racks because of volume considerations, while others have both the quantity and packaging to allow a 20-foot freestanding stack.

Interaction of Storage and Handling Systems

The storage layout and handling system in a warehouse are inseparable. The handling system must fit the building and the storage job to be done. The storage plan must be adapted to the handling system and aisle width requirements. It is impossible to plan one without considering the other.

While the storage method aims at maximum cube use, the warehouse operator must recognize the ever-increasing cost of labor. There must be a balance between an effective storage system and an effective handling system. This balance changes if labor costs increase faster than storage costs. The goal is to plan a distribution center that operates at maximum cost efficiency today, but can be modified to keep in step with changes in the cost relationship between storage and handling.

Stock Location Systems—Fixed versus Floating Slots

Public warehouse operators are frequently required by their customers to keep all of one customer's merchandise in the same area, even though it might be more efficient to share storage areas among several customers. A warehouse in which all storage locations are fixed will lose space because specific slots must be dedicated

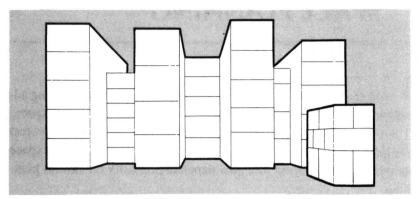

Lost storage space in front of partial stacks is called honeycombing. It exists in all warehouses to some degree.

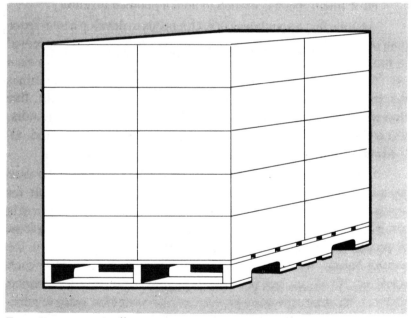

Twenty cases per pallet.

Figure 15-1

to items that fluctuate substantially in volume. When this loss is multiplied by the number of items stored in the warehouse, a totally fixed slot system will waste enormous amounts of space.

A floating slot system is one in which stock locations are

Assumptions:

1. A portion of gross space must be dedicated to aisles and staging, leaving 60% net space.
2. Honeycombing losses further reduce net space, leaving only 80% of the net space (48% of gross) available.

Calculations:

A. Each pallet of product contains 5 tiers with 4 cases per tier, or 20 cases total.
B. Each stack of 3 pallets contains 60 cases.
C. Each stack is the size of one pallet plus one inch overhang on each of the 4 sides: 42 times 50 = 2,100 square inches or 14.6 square feet. Round this to 15.
D. Each stack consumes 25 gross square feet, or 15 divided by 60%.
E. Each stack, after honeycombing, requires 31.25 square feet, or 25 divided by 80%.
F. Each case takes up .52083 square feet, or 31.25 divided by 60 cases.

Figure 15-2. Storage Space Calculation

assigned on a random basis. But such a system could result in the storage of products having incompatible storage characteristics. For example, a floating slot system that mingles grocery products and poisonous chemicals would cause needless danger. A floating slot system is effective only when inventory records are precisely maintained with location identified. When a floating slot is filled, the stock picker must be able to find another location in which the inventory can be stored. The floating system works best when inventory records are updated on a real-time basis.

Family Groupings

A compromise between fixed and floating slots is the grouping of items by family. One product family may be located in a fixed area, but members of that family will be in floating locations. The family grouping method allows the storer to handle products that have different storage characteristics. For example, a grocery wholesaler will find it necessary to store all tobacco products in the same area. Soaps and detergents might also be stored as a family, and since these frequently have an odor, they are stored far from products that would absorb that odor.

A Locator Address System

Every location system needs an address designation system, preferably one having a sensible logic that anyone can readily understand. Figure 15-3 is a basic address system. Warehouse bays are assigned a number to indicate their relative east–west position, No. 1 being the westernmost side of the warehouse. Letters are used to indicate relative north–south position, with "A" being the southernmost and "D" the northernmost. A storage address with a lower number is always reached by going west, while one with a lower alphabetic designation is reached by going south. A worker who must move from B5 to D2 can quickly calculate that he or she must go to the north and west (see Figure 15-3).

A locator system can also help rotate the stock. If the date of receipt is noted at the same time as the location is recorded, outbound

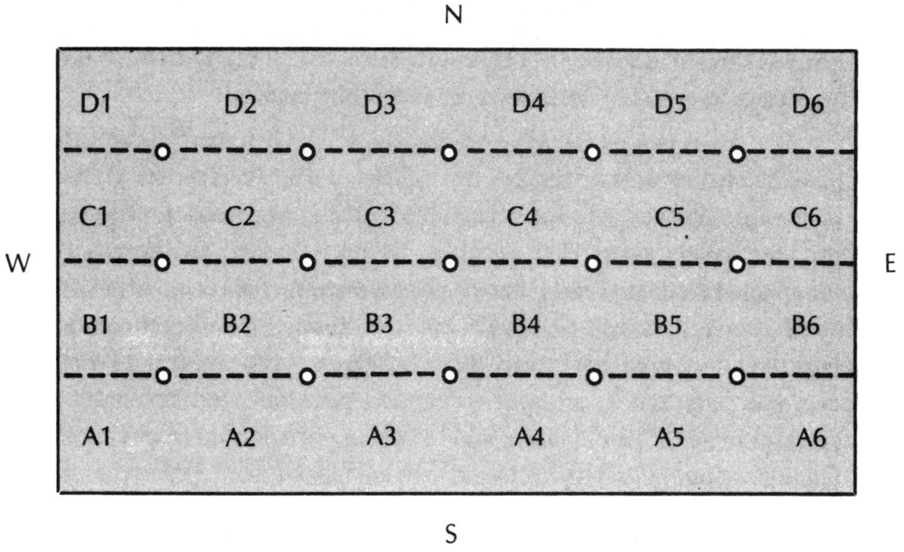

SOURCE: The ABC's of Warehousing, © 1978, Marketing Publications, Inc.

Figure 15-3. Basic Address System

shipping instructions can send the order picker to the location with the oldest stock.

Cube Utilization

There are three ways to make better use of the cubic space available in your warehouse. The first is to increase the stacking height, the second is to reduce aisle width, and the third is to reduce the number of aisles.

It is easy to overlook the lowest cost space in any warehouse—that which is close to the roof. Most contemporary buildings are high enough to allow at least a 20-foot stack height, and some allow substantially more. Fire regulations typically require that high stacks be at least 18 inches below sprinkler heads, but with this knowledge you can and should calculate the highest feasible stack height in the building. Then determine whether or not it is being used.

For some warehouses, a mezzanine is a way to use the cube by allowing two levels of storage, or in a few cases even three. Mezzanine components are produced and sold by the same vendors who supply steel pallet racks. Like racking, the mezzanine can be readily moved and relocated, and it can be depreciated on the relatively short life allowed for warehouse equipment. Mezzanines have other advantages over conventional multistory buildings. If the floor is of metal grid, you may avoid an underwriter's requirement for intermediate sprinkler heads. The elevating and lowering of materials from a mezzanine is usually done by lift truck, which is faster and more economical than a freight elevator.

Consider the use of higher lift trucks, racking designed to use the cube, and mezzanines. One or more of these options is likely to enable you to use the height of the building more effectively than it has been used in the past.

Lot Size

The typical quantity of each line item to be stored can critically affect space utilization in any warehouse. If storage spaces are planned to accommodate stacks four units deep, and the average inventory is two units deep, the result is either 50% wasted space or double-handling to store one line item behind the other.

Foods, pharmaceuticals, and other controlled items must be completely segregated. Unless these lots are the same size as available storage slots, more warehouse space is wasted.

Reducing Aisle Losses

The width of aisles is dictated by the turning radius a lift truck needs to make a right angle turn and place goods in storage. For a 4,000-pound counterbalance fork lift, this is typically a 12-foot aisle. Increased lift capacities, carton clamps, and slip-sheet attachments will further increase aisle width because of the space buffers needed to use the attachments as well as the extra weight caused by the attachment itself.

Aisle width can be narrowed if special types of lift truck are used. There are two types available. The more economical of these is the stand-up lift truck or reach truck which differs from a conventional fork lift, with the driver standing rather than seated. The stand-up lift truck makes a right-angle turn in about two-thirds of the space needed by a counterbalance truck.

Much narrower aisles can be planned by using a turret lift truck, which makes a right-angle turn by swiveling the lift mechanism rather than the entire truck. Because the truck itself remains stationary, the turret lift truck can operate in an isle that allows only a few inches of clearance on either side. Such trucks are typically wire-guided so that steering in a tight space is automated. But the turret truck is far more costly than a conventional fork lift truck, and it is a more complex machine which can be difficult to maintain. Two types of turret trucks are available, the man-aboard design and a conventional lift with the driver sitting down in the main truck body. The man-aboard type allows the operator to see exactly where the fork carriage is, which is not easy at extreme lifting heights and even lets the operator ride up and select less-than-pallet quantities without removing the pallet from the stack. These very narrow aisle trucks will permit aisle widths to be reduced from 12 feet to a little over 5 feet.

Reducing Number of Aisles

One way to reduce the number of aisles is a rack system designed for two-pallet-deep storage. The only disadvantage is the possibility that overcrowding and insufficient volume may cause one item to be blocked behind another.

Another means of improving space use is movable storage racks. These are roller-mounted rack installations that can be shifted from side to side to create an aisle whenever needed. This equipment is costly, but may be justified by the space it saves. It may also be impractical for a very fast-moving warehouse, because access to some facings must be blocked while the rack is rolled to create an aisle on the opposite facing. Mobile racks can be moved to close and open different aisles. A warehouse that is busy enough to require all aisles to be open at once cannot use the mobile track.

Managing Your Space

While equipment is useful, the most critical element in improving use of space is good management and planning. How many cases of product will your facility hold? What variables in the operation will change its capacity? Finally, how close to capacity is your warehouse operation today?

Space planning starts as goods are received. When a truckload of new merchandise arrives at your receiving dock, who makes the decision about where that product will be stored? In altogether too many warehouses, the inbound location decision is delegated to the fork lift driver. It is only natural that the driver will make an expedient decision which saves time without saving space. Your warehouse should have at least one person designated as the space planner. The planner should know where currently empty storage locations are, where additional space can be created by re-warehousing, and what inbound loads are expected as well as the identity of the items that will be on each of these loads. From this the planner develops specific storage instructions for every arriving load of freight.

The storage layout and handling system in a warehouse are inseparable. The handling system must fit the building and the storage job and the storage plan must fit the handling system and aisle width requirements. It is impossible to plan one without considering the other.

While the storage method aims at maximum cube use, the warehouse operator must recognize the ever-increasing cost of labor. Balance is essential between an effective storage system and an effective handling system. This balance changes if labor costs increase faster than storage costs. The goal is to plan a distribution center that operates at maximum cost efficiency today, but can be modified to keep in step with changes in the cost relationship between storage and handling.

16

PLANNING FOR PEOPLE AND EQUIPMENT

Before the warehouse doors ever open, management should simulate the personnel and equipment requirements for the new operation. The number of people involved will affect parking requirements, as well as numbers of restrooms, locker, and lunch room facilities. A projection of volume and nature of work flow will dictate equipment requirements, and spaces needed for staging, shipping, and receiving.

Personnel planning includes determining and quantifying the work to be performed, measuring the number of man-hours needed to perform that job, and translating this calculation into numbers of people required. In addition to the touch labor associated with conventional warehousing tasks, the ability to handle office work and supervision should be included.

Productivity measurements are described in Part VII, although some rough productivity measurements must take place in the planning stages.

The Process of Making Flow Charts

Flow-charting helps the planner learn more about a job by forcing you to describe that job, step by step.[7] While such charts are now computerized, an understanding of the process is essential.

Flow-charting leads to work simplification, which in turn increases productivity. But first you must go back to the what, where, when, who, and how questions, such as:

Purpose:
What is done?
Why is it done?

Place:
Where is it done?
Why is it done there?
Where else might it be done?
Where should it be done?

Sequence:
When is it done?
Why is it done then?
When might it be done?
When should it be done?

Person:
Who does it?
Why does that person do it?
Who else might do it?
Who should do it?

Means:
How is it done?
Why is it done that way?
How else might it be done?
How should it be done?

Flow-charting will help you identify questions you may have overlooked. It imposes a discipline, makes you plan ahead, and will provide a clue as to what a worker might do in a given situation when faced with a decision you may have failed to anticipate.

The flow chart in Figure 16-1 describes a simple materials handling job. Three different symbols are used. The first and the last are the same. This is the terminal symbol and shows where a job begins and where it ends. It is never used for any other purpose.

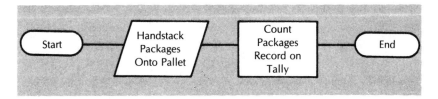

Figure 16-1

The second symbol is the physical process symbol, which identifies an element involving the use of energy or power. The third symbol, the clerical process symbol, identifies an element involving clerical work; in this instance, counting and recording the count.

This flow chart briefly and correctly describes the main elements of work involved in hand-stacking packages onto a pallet. It describes the basic job, the work elements, and the sequence in which the elements are performed. However, it does not give a complete description of the job. We might want to know how the worker got to the job site, where and how he got a pallet, and what happens after the packages are counted.

Figure 16-2 gives us a more detailed picture of a job. It tells us what must happen before our worker can hand stack the packages onto a cart. It tells us what happens after the cart is loaded and makes us think about the sequence of the elements. It also may

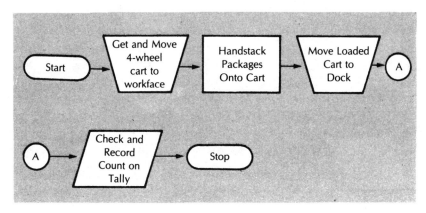

Figure 16-2

make us ask whether the elements are being performed in the most efficient sequence.

We have introduced three new symbols. Notice the arrowheads. As our charts become more complex, it will be necessary to use arrowheads to show the direction in which the work flows. The arrowhead is an important symbol as will become apparent later in this discussion. The second and fourth symbols denote input/output or transport. The fifth and sixth symbols are connectors.

The Best Sequence

Refer to Figure 16-2 and prepare to use a little imagination. Where is the check and record count element being performed? In the carrier vehicle or on the dock? To find out, we locate the check and record-count element, refer to the transport element immediately preceding the check-and-record-count element, and then establish that the work is being done on the dock. Is this the best place to

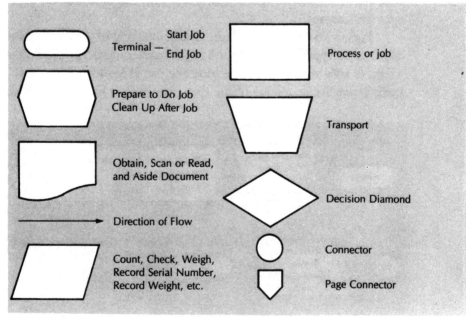

Figure 16-3

do the work? If the dock is well lighted and the inside of the vehicle is not, the dock is the place to get the most accurate count-and-check. But if the carrier vehicle is well lighted and the check and count can be made in the vehicle, then the transport move can be made directly into the warehouse and bypass a stop on the dock. Is this a good idea?

Shown in Figure 16-3 is a summary of all of the symbols commonly used in flow charts. Note the addition of new symbols to indicate documents and decisions.

Figure 16-4 shows flow charts used in some typical warehousing tasks.*

---

* The flow charts and text section are taken from Vols. 7, 18, and 19 of *Distribution/ Warehouse Cost Digest,* © Alexander Research and Communications Inc., New York.

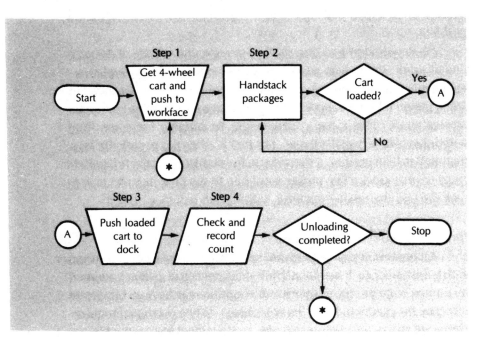

Figure 16-4

"What If" Questions

Flow-charting will force you to think and ask "what if" questions. For example, what will the order picker do if he or she arrives at the pick line stock location called for by the order and finds no stock? Remember, we made the simplifying assumption that stock would always be available, although this is clearly not realistic. Sooner or later, the picker will find the pick line location empty or with insufficient stock to satisfy the order. What will the worker do if he or she has an order pallet on the forks with a portion of the order already picked? How will the worker know where the reserve stock is and whether to take the order pallet to the reserve stock area, or set the pallet aside and proceed to the reserve stock to restock the pick line so as to continue picking? Should the picker be able to communicate with the office in the event that stock can't be located? What kind of communication system should be used? Should it be mobile or fixed? When you begin to cost out steps in terms of machine and labor time you may find you need to look into the communication problem.

One method of locating stock is to maintain records at the pick line location. When new stock is received, the receiving warehouseman posts its location at a designated point on the pick line. Thus, when the pick line is empty, the picker knows where to go to obtain reserve stock. This same system might be suitable for controlling withdrawals of full-pallet loads. The picker often has to pick the less-than-pallet-load portion of the order in the pick line and the full pallets from reserve stock. The picker would go to the pick line location to pick the less-than-pallet quantity, leaving full pallets until last.

An important part of personnel planning is the exact description and definition of each warehouse job. Particularly in a plant governed by a union contract, the written job descriptions can have an important effect on the productivity of the work force. When personnel requirements are still in the planning stages, management has a valuable opportunity to ensure that these job descriptions are written to provide maximum flexibility to get the job done. If such descriptions are written narrowly, labor contracts may restrict the assignment of workers to secondary jobs when the primary job is completed.

Once jobs have been described, the number of man-hours needed to do the work is determined. Figure 16-5 shows a table on work standards for a commonly performed warehouse task.

The Critical Role of the Supervisor

Planning is absolutely essential for supervisors. It starts with time management. Better time management means planning, prioritizing, and scheduling. You should determine how your time is actually spent and then compare it with how it should be spent. One way to do this is to keep a daily time log to track your own day. Every half hour, list what you have done, including problems that required your attention during that time. One warehouse supervisor was surprised to learn from such a log that he had spent 60% of his time one week on grievances. He thought that limited time was his major headache, but his real problem was too many grievances.

There are two types of lost time: obvious and hidden. Obvious lost time is a fork lift driver leaning against his machine with nothing to do. Hidden lost time is more subtle, such as problems within operations. For example, one warehouse loses time at the beginning of each shift because work is not scheduled and waiting for the crew. Each morning, 15 workers gather while the first line supervisor sorts and distributes orders. Then they have to find equipment and pallets before actually getting down to work. A good supervisor solves the problem by having someone from maintenance arrive before the shift begins and prepare all equipment for the first orders of the day. Then the supervisor schedules the day in advance so there are orders waiting for each worker when the shift begins.

Though the pace in many warehouses is hectic, the good supervisor uses planning to complete odd jobs that might otherwise fall by the wayside. Carry a notebook—a traveling "job jar"—in your pocket to jot down one-time tasks. Use these for assignments when certain workers are idle. Here are some ideas for your job jar:

Labor required, per occurence, for receiving elements when dead skids and low-lift platform trucks are used

Element description	Workers in crew	Base labor	Allowances	Productive labor
	Number	Man-hours	Man-hours	Man-hours
Position skid for loading: Begins when worker starts toward empty skid. Includes walking to empty skid, lifting, transporting an average distance of 24 feet, and placing skid in position for loading. Ends when skid has been released.				
1. Manually	1	0.0078	0.0020	0.0098
2. Manually	2	.0085	.0013	.0098
3. Manual platform truck (ends when skid is in position on floor)	1	.0069	.0010	.0079
4. Electric platform truck (ends when skid is in position on floor)	1	.0070	.0007	.0077
Enter carrier: Begins when transporter contacts bridge plate. Includes crossing bridge plate, transporting to pickup point, and aligning truck to move under loaded skid. Ends when truck is backed against loaded skid.				
1. Manual platform truck	1	.0028	.0004	.0032
2. Electric platform truck	1	.0023	.0002	.0025
Pick up loaded skid: Begins when truck is backed against loaded skid. Includes elevating skid for movement. Ends when loaded skid is clear of floor and ready to move.				
1. Manual platform truck, mechanical	1	.0021	.0004	.0025
2. Manual platform truck, hydraulic	1	.0067	.0013	.0080
3. Electric platform truck	1	.0027	.0003	.0030

Leave carrier: Begins when skid is clear of floor and ready to move. Includes moving skid clear of original position, maneuvering it when necessary in order to cross threshold of carrier and bridge plate. Ends when skid clears bridge plate.

1. Manual platform truck	1	.0030	.0006	.0036
2. Electric platform truck	1	.0025	.0002	.0027

Position loaded skid for setdown: Begins when transporter starts maneuvering skid for setdown. Includes turning and pushing or pulling truck to align skid for setdown. Ends when transporter touches lowering control.

1. Manual platform truck	1	.0072	.0018	.0090
2. Electric platform truck	1	.0038	.0003	.0042

Set down loaded skid: Begins when transporter touches lowering control. Includes lowering skid to floor, and moving truck out from under skid. Ends when truck clears skid.

1. Manual platform truck, mechanical	1	.0014	.0003	.0017
2. Manual platform truck, hydraulic	1	.0032	.0008	.0040
3. Electric platform truck	1	.0011	.0001	.0012

Position truck to pick up load: Begins when transporter starts maneuvering truck for pickup. Includes turning and pushing or pulling truck, and positioning truck under skid. Ends when truck is backed against skid.

1. Manual platform truck	1	.0043	.0006	.0049
2. Electric platform truck	1	.0015	.0002	.0017

Figure 16-5

SOURCE: U.S. Department of Agriculture.

161

looking for empty slots, counting products, sorting or repairing pallets, housekeeping, renumbering aisles or slots, and repacking damaged products.*

Planning for Equipment Use

In most warehouse tasks, as well as many office tasks, the number of man-hours needed to finish the job will depend greatly on the equipment or tools available to the worker. The planner begins by considering the kinds of equipment available.

For example, the lift-truck is available with a wide array of attachments besides the conventional forks. A warehouse operation designed for carton-clamp handling will require significantly different planning for space and people than one designed for the receiving of loose-piled cases that will be palletized at the receiving dock. Use of slip-sheet handling devices requires still different planning. The best time to do this planning is before the warehouse is opened.

At the planning stage, the warehouse manager should select the kinds of equipment that can best be used in the operation, taking into account lifting capacities and storage heights, and any environmental considerations that would restrict the design of the lift-trucks. For example, internal combustion engine trucks may not be acceptable in freezers or other confined areas.

Six basic items of information should be gathered before planning for manpower and equipment is completed:

1. The average order-size provides a good indication of the amount of time needed to ship each order.

2. Physical characteristics of the merchandise also may define the amount of effort needed to handle it.

3. The ratio of receiving units to shipping units, and the question of whether goods are received in bulk or as individual units will greatly affect the effort involved.

4. Seasonal variations and the intensity of these seasonal swings must be measured.

---

* This section was adapted from *Supervising on the Line*, © 1987 by Gene Gagnon, published by Margo, Minnetonka, MN.

Figure 16-6
Illustration courtesy of Gene Gagnon, from "Supervising on the Line."

5. Shipping and receiving requirements should be defined, including the times of day when goods will be received and shipped.

6. Picking requirements must be defined. A strict first-in first-out system is more costly from a labor, as well as a space, standpoint.

When this information is collected, a useful definition of manpower and equipment needs can be reliably projected.

17

CONTINGENCY PLANNING

Contingency planning is preparing in advance for the emergencies not considered in the regular planning process.* The best approach is to ask the question, "What if?" For example, what if our major suppliers are on strike? What if we are unable to obtain sufficient fuel to run our truck fleets? What if we are hit with an earthquake or a tornado at our biggest distribution center? What if we simply cannot find enough secretaries, order pickers, and industrial engineers to staff our distribution facilities? What if we have to recall one of our major products?

Managers who fail to anticipate certain events may not act as quickly as they should, and in critical situations this will exacerbate the damage. Contingency planning should eliminate fumbling, uncertainty, and delays in responding to an emergency. It also should bring more rational responses.

A contingency plan should have certain characteristics. First of all, the probability of an adverse occurrence should be lower than for events covered by the regular planning process. Second, if such a critical event occurs, it would cause serious damage. Finally, the contingency plan applies to occurrences the company can deal with swiftly.

Both "How critical is it?" and "How probable?" must be considered. Some events have a low probability. For instance, the sudden loss of a distribution center could be disastrous if no plans existed

---

* Based on an article by Bernard J. Hale, DSC Logistics, Warehousing & Physical Distribution Productivity Report, Vol. 14, No. 3, Alexander Research and Communications Inc. New York.

to meet that contingency. Even if such a distribution center is considered relatively fireproof and well-protected, possibly it could burn down. Contingency plans should be selected commensurate with the impact on the organization of the particular occurrence. The estimate of potential impact can be made in financial terms, competitive position, employee availability, or a combination of such considerations.

Strategies

Strategies in contingency plans should be as specific as possible. For instance, an organization may develop a contingency plan to be implemented in the event of a drop in sales below a certain level. Strategies such as reducing employment, or reducing advertising expenses, or delaying construction are simply too broad. They must be more specific.

As an example, "If sales drop by 10% below expectations, our net income will decline by X million dollars. In order to reduce this loss by a specific goal—say, 75%—it will be necessary to defer the program or expansion of plant X, to reduce the number of employees by 500, and to cut variable production costs in proportion to the sales decline." Comparable strategies can be prepared to offset income loss by 50%, 100%, or whatever is considered appropriate.

Wherever possible, the expected results of the actions taken should be calculated in financial terms, or in other meaningful measures, such as market share, capacity, and available manpower.

Trigger Points

Contingency plans must specify trigger points, or warning signals, of the imminence of the event for which the plan was developed. For example, if a plan is prepared to deal with a 10% decline in sales below estimate, there will be warning signals. Sales may fluctuate daily, weekly, or monthly. What will trigger the implementation of the plan? In some cases, such as fire, the trigger point is obviously the event itself. In others, however, the point is not so clear. But a trigger point should be specified. Thus, the contingency plan should indicate the type of information to be collected and the action to be taken at the trigger point. Also, it is advisable to assign specific responsibilities to someone to collect relevant information and pass it on to the person who makes the decision to implement the plan.

Distribution managers should consider developing contingency plans for such potentially serious disruptions as the impact of strikes, energy shortages, labor shortages, natural disasters, and product recalls. We will use as an example contingency planning for the impact of strikes.

Planning for the Impact of Strikes—Your Own Firm's as Well as Others'

As the first step in preparing to face a strike, be fully briefed on contract deadlines. It should not be hard to keep track of such deadlines in your own firm, although it is easy to miss them in the case of your suppliers. Many purchasing agents require each supplier to list contract expiration dates, with this kept as part of the master purchasing file.

Every distribution operation has traditional alternatives for dealing with a work stoppage. These include the use of public distribution facilities, non-union transportation companies, or even non-union private distribution facilities.

If your firm is preparing for a strike, the company's main objectives in bargaining should be outlined, since they are likely to affect decisions made during the work stoppage.

If public warehouses are used as a strike hedge, the manufacturer should begin the public warehousing operation before the strike starts—dispersing the product so that it represents only a small fraction of the total capacity of each warehousing supplier. If such precautions are taken, the warehouse operator can file a secondary-boycott charge with the National Labor Relations Board should illegal picketing of a public warehouse occur. In selecting a public warehouse as a strike hedge it is important to appraise the labor stability and operational flexibility of the warehouse supplier. Properly applied, the "strike hedge" inventory can be a powerful bargaining tool in contract negotiation.

Strike Against Suppliers

The effect of other company strikes can be as disruptive as disturbances within your own organization. A first step in planning for this contingency is to determine which suppliers are critical to your operations. This may depend on whether they are sole sources or whether they supply a critical percentage of the purchasing volume.

Having identified the critical suppliers, the next step is to learn if those companies are unionized. If unionized, what is their labor relations record over the past decade? If there is a contract, when does it expire? Some information on contract expiration is publicly available, including the Negotiations Calendar issued by the U.S. Bureau of National Affairs.

The risk of a supplier's strike can be mitigated by developing an inventory hedge in the form of advance stockpiling of the critical items. At times, interim purchasing arrangements, such as the import of similar items, may provide a source of relief.

Strikes Against Customers

A strike against your customer also can have a serious effect on distribution operations. It is advisable to follow closely the contract expiration dates and negotiations affecting companies with which you have important business relationships. If a customer's strike seems imminent, consider accelerated deliveries, or such special accommodations as direct store delivery to bypass a strike in a retailer's distribution center.

Trucking Strikes

Transportation strikes require particular resourcefulness in planning. The traffic manager should be charged with finding alternative sources well in advance of the strike. The problem should be discussed with customers and suppliers, who may be able to use their own private trucking to move products in or out of your warehouse.

In times of trucking strikes, some small carriers will make a side agreement to settle on the master contract and agree to keep working during the general strike. The traffic manager should locate such firms and arrange to use them. Owner–operators of trucks also frequently ignore strikes. In addition to using customer and supplier private trucks, your own firm may have private transportation that can be used. Non-union transportation companies also can be used, though such firms naturally will serve their own regular customers first.

Similar steps should be taken in the case of strikes affecting rail unions or longshoremen. In such case consider diversion to other modes of transportation or to other routes that avoid the labor difficulty.

The approach outlined here for development of a contingency plan in the event of a strike can be applied to many similar emergencies.

Charting the Planning Process

Keep a record of planning actions taken, particularly on a complex plan in which several persons have been assigned tasks. Figure 17-1 shows a contingency planning action log that provides an example of how the plan can be charted.

Planning Steps

There are four steps in preparing a contingency plan:

1. Identify the contingency and describe your objectives. For example, if the computer goes down, will your major objective be to get the equipment back in operation, or will you switch from computer to manual systems?

2. Identify departments that will be affected by the contingency. For example, will the computer breakdown affect both office operations and warehouse operations? How will each of these departments cope with the situation?

3. Determine what action should be taken to lessen the impact

169

Contingency Planning Action Log

No.	Action to be taken	Assigned	Date	Due
1.	Determine expected length of labor stoppage.	AK	3/10/94	3/10/94
2.	Re-supply all using departments to maximum levels.	MM	3/10/94	3/17/94
3.	Notify key vendors of potential labor stoppage and advise shipping instructions if necessary.	PM	3/10/94	3/15/94
4.	Identify all available storage space at other locations	RS	3/11/94	3/14/94
5.	Move substantial reserve stock of critical high-usage items to available storage space and establish product locator system for same.	WS	3/14/94	3/18/94
6.	Determine probability of delivery vehicles crossing a picket line.	MJ	3/15/94	3/16/94
7.	Make up a work schedule for supervisors during the labor stoppage.	WS	3/21/94	3/21/94
8.	Continue to maintain maximum levels of inventory at using departments.	MM	3/21/94	3/31/94

Figure 17-1

of the event. For instance, would the maintenance and testing of standby manual systems ease the transition in the event of a computer breakdown?

4. Determine what connection one event will have on another. For example, how will the computer breakdown affect shipping and receiving operations?

No.	Action to be taken	Assigned	Date	Due
9.	Notify key vendors to implement alternative shipping instructions as of 3/29/94.	PM	3/22/94	3/24/94
10.	Notify neighboring warehouses of potential labor stoppage and establish a "lending" program.	HA	3/23/94	3/24/94
11.	Notify all using departments what actions have been taken and what alternatives have been established.	WS	3/24/94	3/24/94
12.	Set up a security plan for the facility.	WS	3/24/94	3/31/94
13.	Specify locations for crossing picket line for personnel and vehicles.	JW	3/24/94	3/24/94
14.	Review probability of labor stoppage as of 3/31/94.	WS	3/25/94	3/25/94
15.	Meet with all using departments to review alternative plans.	WS	3/28/94	3/28/94
16.	Implement alternative work schedule.	WS	3/31/94	3/31/94
17.	Keep department supervisors informed.	WS	3/31/94	

The increased attention to contingency planning is the result of a comparable increase in business uncertainty in recent years. Supply situations taken for granted in earlier years must be the subject of contingency planning today. In a time of increased government involvement in quality control, the contingency of product recall is more common. If the contingency event occurs—however unlikely it may be—the prudent manager will have available a plan of actions to be taken.

18

POSTPONEMENT

"Postponement" in distribution is the art of putting off to the last possible moment the final formulation of a product, or committing it to the market place.The theory of postponement is nothing new. It has been discussed in marketing and logistics textbooks for several decades.

But although the theory is not new, and the concept is simple, its full potential for saving money in the warehouse has not been reached. For example, you could make an inventory of 20,000 units replace an inventory of 100,000 simply by converting branded merchandise into products with no brand name. That's not only possible—it's being done right now.

Background of Postponement

Postponement was practiced long before marketing theorists started writing about it. One early example in industry—Coca-Cola—goes back to the early years of the 20th century. Postponement was described by a marketing theorist as the opposite of speculation. A speculative inventory costs more to maintain than the profit derived from using it to stimulate purchases.

Alderson was the first marketing expert to write about postponement. In 1950, he observed that "the most general method which can be applied in promoting the efficiency of a marketing system is the postponement of differentiation . . . postpone changes in form and identity to the latest possible point in the marketing flow; postpone changes in inventory location to the latest possible time." Alderson, Wroe—"Marketing Efficiency and the Principle of Postponement," Cost and Profit Outlook III,4 (September 1950) The risk of speculation is reduced by delaying differentiation of the product to the time of purchase. Merchandise is moved in larger quantities, in bulk, or in relatively undifferentiated states, thus saving on the cost of transportation.

In Marketing, James L. Heskett of Harvard Business School defines the principle of postponement as follows:

A philosophy of management that holds that an organization should postpone changes in the form and identity of its products to the latest point in the marketing flow and postpone changes of the location of its inventories to the latest possible points in time prior to their sale, primarily to reduce the risks associated with having the wrong product at the wrong place at the wrong time and in the wrong form. Heskett, James L., Marketing.1976, Macmillan Publishing Co., Inc., N.Y.

Types of Postponement

Postponement can take any of several forms:

- Postponement of commitment.
- Postponement of passage of title.
- Postponement of branding.
- Postponement of consumer packaging.
- Postponement of final assembly.
- Postponement of mixture or blending.

A typical product moves through four channels: manufacturer to distributor to retailer to consumer. Movement through each involves costs of transactions, as well as the costs of storage, transportation, and inventory carrying-costs. The principle of postponement prescribes reducing these costs by delaying or bypassing some of the stages between producer and consumer.

Postponement of Commitment

To delay shipping a product avoids risking its miscommitment. For example, using air freight enables some manufacturers to eliminate field inventories by keeping the entire stockpile at a factory distribution center located at the end of the assembly line. The product is moved into the market at speeds competitive with that of shipping from field distribution centers. This enables the manufacturer to avoid the potential waste of committing the wrong product to the wrong distribution center and later having to cross-ship products to correct an imbalance of inventory.

The most efficient distribution point is always at the end of the assembly line, since inventory at the point of manufacture will never have to be backhauled. If air freight is fast enough to allow all inventory to be kept at the factory, a manufacturer may trade off the premium costs of air freight by eliminating inventory losses from miscommitment or obsolescence.

With some merchandise, postponement includes dispensing with certain distribution channels. For example, an appliance retailer may have floor samples of each major appliance on the sales floor to show prospective customers. However, when this results in a sale, the actual appliance delivered to that customer may not even be in the same city. By using premium transportation from strategically located distribution centers, the appliance distributor will avoid committing inventory to each retail store. In this case, postponement is particularly useful because of the cost of handling, storing, and carrying an inventory of bulky and costly appliances. If the customer cannot take the merchandise home in a shopping bag, why move the item through the store?

Postponement of Title Passage

Delaying passage of title sometimes takes the form of "consignment." For example, a chain grocery retailer makes annual purchase commitments for canned vegetable items. The vendor's agreement calls for these items to be shipped to a public warehouse located at a point convenient to the retailer. However, the contract also specifies that warehousing charges will be paid by the manufacturer, and passage of title is postponed until the product is released from the warehouse.

Since the public warehouseman is an independent bailee, the manufacturer still has a measure of control over the inventory. However, the retailer has the power to release it upon call. By delaying passage of title, the buyer reduces the cost of carrying inventory. Yet, by positioning stocks at a convenient warehouse, passage of title is postponed without any sacrifice in customer service.

Postponement of Branding

Last-minute branding is another type of postponement. A vegetable canner may produce canned peas for his own house brand, as well as for the private brands of several different chain retailers. To meet these several brand requirements, the canner puts up all the canned peas "in bright"—blank cans that have no labels. Typically, these bright cans are held at the factory warehouse until orders are received for one of the private labels; then the order is run through a casing and labeling line. Thus, the manufacturer avoids the risk of putting the wrong label on a product and subsequently having to relabel it.

This type of postponement can be further refined if the manufacturer ships goods in bright to market-area distribution centers. By positioning bright cans (and casing and labeling lines) in distribution facilities close to the market, the canner may substantially improve customer service and still avoid the speculation of producing private label merchandise that may be slow to sell. Also, by postponing a commitment to perhaps six different private labels, one field inventory can do the work of six.

Postponement of Consumer Packaging

A fourth type of postponement is the delay of packaging. A pioneer in the manufacture of soft drinks, Coca-Cola, determined many years ago that its manufacturing strategy would be to produce a concentrated syrup that would be delivered to soda fountains or regional bottlers to be diluted with carbonated water and delivered in a package convenient and attractive for the consumer. Similarly, manufacturers of whiskey move the product across the ocean in bulk, then bottle it in the country in which it is to be sold. The risk of breakage and pilferage of glass bottles is eliminated, and a more favorable customs duty applies in some instances.

Postponement of Final Assembly

Postponement of final assembly is an accepted strategy in many manufacturing firms. For example, a manufacturer of office furniture may delay final assembly of desks or chairs until customer orders are received. Common components such as desk pedestals, seat cushions, frames, and hardware can be produced and held in stock. When a customer orders a desk or chair of a certain style or color, the order is filled by assembling and finishing common components on an assembly line. By doing this, the manufacturer avoids the risk of over-producing certain styles that cannot be easily sold.

Postponement of Blending

A sixth type of postponement is to delay mixing or blending. One widely recognized example is the multigrade gasoline pump developed by Sun Oil Co. By altering the mixture or blend of high- and low-octane gasoline, Sunoco offers its customers five different combinations of motor fuel at the time of purchase.

Some paint manufacturers offer a customer paint colors mixed at the store. This is a variation of postponement of final assembly, the final mixture or blending being made after the product is sold.

Increased Warehouse Efficiency

What does postponement do for warehousing efficiency? Basically, it enables one inventory to do the work of many. Where blending is postponed, the warehousemen can have a single stock-keeping unit that is convertible to many different products at the final brand is applied. To illustrate, in one line of automotive batteries, the only difference between private labels is a decal applied to the top of the battery. By delaying the application of the decal and the final consumer package until the last moment, the number of line items in the warehouse is dramatically reduced. (See Figure 18-1.)

Postponement can also reduce freight costs, which are usually the largest component of physical distribution costs. In the auto industry, the incentive for regional assembly plants has been to reduce freight costs. It is often cheaper to ship components than finished cars.

Another instance of postponement involves an Italian toy manufacturer that produces a light-density item. The item is shipped in bulk boxes to a warehouse at the port of entry where consumer packages purchased in the United States for the American market are also received, and the product is moved from bulk-shipping container to the final consumer package. Not only are freight costs reduced, but the consumer package is more attractive because it has not been subjected to the potential wear and tear of long-distance transportation.

Delaying final branding or packaging also reduces the ware-

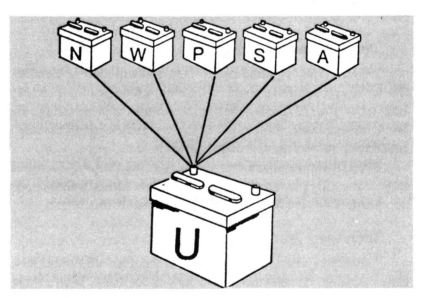

SOURCE: Warehousing and Physical Distribution Productivity Report, Vol. 14, #12, © Alexander Communications, New York.

Figure 18-1

houseman's risk of obsolescence. Certain brands may decline in popularity, reducing warehouse turnover and increasing the risk that the product will become obsolete from a marketing standpoint.

Potential Challenges

While productivity advantages in the use of postponement are substantial, the challenges are equally so.

Postponement at a warehouse level usually requires the warehouse operator to enter a whole new business—that of assembly and packaging. This creates additional risks in quality control. In the course of packaging, the product may be damaged or contaminated. It could even be mislabeled or put in the wrong package. Quality control in final assembly may be compromised.

A Manager's Questionnaire
For Warehouse Postponement

1. Do we produce identical products under separate brands?
2. How is the final brand name applied?
3. Can the product be remotely assembled?
4. What is the cost-saving in shipping the product in a knocked-down (KD) condition?
5. What is the cost of remote assembly?
6. What are the quality control risks of remote assembly?
7. What are quality control risks of remote branding?
8. What are customer service levels for the product?
9. Does our competition have a postponement strategy?
10. What is the cost-advantage of bulk shipment of our product?
11. What is the cost of remote packaging of the product?
12. What is the quality control risk of remote packaging?
13. What is the cost of moving the product through each channel of distribution?
14. Can the product skip one or more of these channels?

Figure 18-2

Ultimately, the widespread use of postponement could turn today's distribution centers into tomorrow's packaging and light-assembly centers. However, the management and labor skills involved at the warehouse level will need to be higher than those needed for receiving, storing, and reshipping.

Postponement in the Warehouse

Uncertain money markets and high interest rates have greatly complicated the problem of carrying inventory. An escalation of fuel costs and deregulation have affected transportation. With both capital and transportation costs increasing abnormally, it is likely that the use of postponement as a distribution strategy will grow in popularity in the near future.

The future possibilities of postponement are limited only by the imagination of the warehouse operator. Any product that has multiple private brands, a fragile consumer package, a bulky configuration in final assembly, or a very slight modification from one form to another is a candidate for postponement at the warehousing level. Products now packed only at the factory and frequently subject to breakage, obsolescence, or high freight costs may be shipped in an unbranded, unpackaged bulk state to a regional distribution and assembly center. Thus, there is no reason why the practice of postponement may not be far more widespread tomorrow than it is today.

19

SELECTING A THIRD-PARTY OPERATOR

Whether to outsource or not is no different than other make or buy decisions in business.

On the one hand, the option exists for you to operate your own proprietary warehouse, or warehouses, providing the level of service your organization requires to distribute your products. This may entail the purchase or lease of real estate, the assembly of material handling equipment, and the management expertise to provide the level of logistics service your business requires.

On the other hand, to outsource this service does allow the capital not invested in real estate or material handling equipment, or the credit not utilized in leasing real estate and material handling equipment, to be used elsewhere within the business. Similarly, the exposure of having employees both at the management and labor ranks would be transferred to a supplier as would some of the business risk and some of the management responsibility for these functions.

Use of a contractor to outsource the warehousing function therefore permits your organization to focus more on your core competency and allows a specialist with specific expertise in this area, namely warehousing, to provide such services for you. The selection of a warehouse operator should be conducted through a research process with the accumulated information base resulting in a clear choice.

When organizations consider outsourcing, loss of control is often mistakenly felt to be inevitable. There is no reason why your organization should lose control. The contractor's management along with your own organization's ongoing management attention should actually improve your control. A second misconception is that once your organization has outsourced its warehousing, it is no longer responsible for its management, and all of your organization's time and cost of such functions can be eliminated.

A Four Step Process

The gathering of information is a four-step process. First, define the scope of work that the contractor will perform for you. Second, gather order characteristics. Third, describe the shipping function. Fourth, assemble details on any special services to be performed.

Here are questions that should be asked:

- What are weights, dimensions, and other warehousing characteristics of each item?
- What are risks of spoilage, breakage, or theft?
- How is product packaged?
- How is product received?
- What are planned shipping and receiving volumes?
- Describe any seasonal peaks and valleys.
- How will orders be transmitted to the warehouse?
- How many lines per average order?
- How many units per line?
- What is average order size?
- What percentage is full truckload?
- Who will manage the transport function?
- Who pays the freight bills?
- Are any specialized receiving or shipping services needed?
- Describe other needed services such as: repack, labeling, price marking, wrapping, inventory checks, cycle counts and import/export documentation.

Once the data have been developed, an understanding must be reached in your organization regarding objectives that you expect the supplier to provide. These objectives should include your desired tradeoff between service and price.

After you have compiled this information, you are now prepared to assemble a list of potential warehouse operators who could perform the needed services. There are a number of directories and associations available that can be utilized as a starting point. Often these sources are incomplete. Enlist the assistance of people familiar with the third-party marketplace. These could be those within your own organization or other professionals with whom you are involved through associations. Some may engage a logistics consultant who has expertise in this area.

Once you have a bidder's list of third-party warehouses, the next step is the development of a Request for Proposal (RFP). The RFP should be based on the previously developed scope of work and submitted to the list of potential third-party operators, allowing a reasonable period of time for them to respond to the request. Responses should be in a predetermined format. This format should be easy for you to use in your evaluation process, keeping in mind that this is only the first screening of the warehouse operators. The evaluation criteria in Figure 19-1 could be used for an initial screening.

Evaluation Criteria

Prior to the submission of your RFP to prospective bidders, it is best to establish the specific items that will be evaluated and how each item will be weighted. Figure 19-1 shows a sample of eighteen evaluation items and how each will be evaluated.

Once you have received the responses from the RFP, an initial evaluation can be made using the criteria listed in Figure 19-2. You should reduce the candidates down to a workable number. After evaluating their credentials and experience you should be able to select a reasonable number to proceed with on-site evaluations.

In most on-site evaluations there are three areas that are reviewed: the corporate organization, the order entry activities, and the physical operations.

Figure 19-1

Evaluation Criteria	Third Party A	Third Party B	Third Party C
Service and Customer Satisfaction	1	1	2
Experience with Similar Business	1	2	1
Flexibility & Future Opportunities	2	3	3
Total Operating Costs	1	3	2
Ability to Improve Service to Distributors	3	2	4
Location, Physical Facilities & Operating Procedures	2	2	3

SOURCE: Thomas L. Freese, Vol. 11, No. 6, Warehouse Forum, © Ackerman Company, Columbus, Ohio.

Specific decision criteria must be established before the evaluation process commences. These criteria are then weighted in the same manner shown in Figure 19-1.

After having conducted on-site evaluations of each of the potential third-party candidates, an evaluation can be made from the criteria using rating numbers then displayed in a matrix for comparison as shown in Figure 19-2.

You may attach weighted values to these evaluation points. The process allows for an orderly approach that considers the scope of work and the operator's capability to perform it. After using this process there should be a clear indication of who the single best third-party operator is, or who the top two operators are. At this point, and only at this point, should you begin the process of evaluating the price.

The Contract

As with any significant long-term business relationship, a formal contract is customary. Eventually, the individuals involved in

1. MULTIPLE WAREHOUSE FACILITIES NATIONWIDE

If national coverage is of importance for your organization, ideally a third-party warehouse operator would have strong national coverage with regional locations available corresponding with your organization's markets. Such a third-party would have a strong network of capabilities to service all points within 24, 48 and 72 hour responses.

Rating and Description

1. *Strong national coverage with regional locations*
2. *Good national coverage with some gaps*
3. *Limited national coverage*
4. *Does not provide national coverage*
5. *Does not currently have multiple facilities*

2. INVENTORY MANAGEMENT AND CONTROL

In providing a complete offering it is often necessary to manage inventories. In order to do so the third-party should possess inventory management and control capabilities to include stock rotation, re-order management, procurement, and forecasting capabilities.

Rating and Description

1. *State-of-the-art inventory management systems in place*
2. *Fundamental inventory mgmt. applications in place*
3. *Basic structure exists*
4. *Minimal capabilities*
5. *Inadequate inventory management*

3. ORDER ACCEPTANCE AND PROCESSING

The ideal third-party operator has a state-of-the-art order acceptance and processing or "front end" system in place.

Rating and Description

1. *A state-of-the-art order acceptance system with satisfied reference users*
2. *An order acceptance system operating with some improvement opportunities*
3. *A partial order acceptance and processing system*
4. *Currently examining purchase and/or start-up of an order acceptance system*
5. *No order acceptance system in place*

4. PICK AND PACK OPERATIONS

Your organization may require a pick and pack fulfillment service.

Rating and Description

1. *State-of-the-art*
2. *Individual pick-pack operations currently in operation*
3. *Partial pick, pick-pack operations*
4. *Experience with pick-pack/currently none in operation*
5. *No pick-pack operations exist*

5. ORDER FULFILLMENT

In addition to each pick or pick-pack operations the existence of current capabilities in individual order fulfillment are crucial to the success of such a third-party. Such individual order fulfillment would include the operations of order capture, processing, picking, packing, shipping, tracking, and order acknowledgment activities.

5. ORDER FULFILLMENT (Continued)

Rating and Description

1. *State-of-the-art capabilities are in place and operating to handle individual order fulfillment to include systems, order capture, acknowledgment, manifest tracking, etc.*
2. *Existing order fulfillment activities are in place and operating for a number of potential reference accounts.*
3. *Management is experienced in the order fulfillment activities and, although not currently operating order fulfillment activities in all its regional locations, has the management depth to do so.*
4. *Limited order fulfillment capabilities and/or management experience.*
5. *No existing order fulfillment capabilities and/or management experience.*

6. ASSEMBLY/PACKAGING/ VALUE ADDED ACTIVITIES

In addition to the normal four steps of order capture, inventory management, fulfillment, and transportation generally associated with the outsourcing of logistics, there is a growing need for value added services to be provided at the point of distribution. These activities would include such things as packaging, kitting, sortation, burn-in, testing, repair, modifications and bench work activities.

Rating and Description

1. *Existing operations are underway to provide numerous value added activities for a number of different customers to include many of the above noted value added services.*
2. *Value added activities are being performed in a number of locations for a number of different customers.*
3. *Limited value added activities currently are in place.*
4. *Management understands and is capable of providing value added activities although none are currently being conducted.*
5. *No evidence exists of value added activities and/or management capabilities to set up such activities.*

7. CREDIT CARD VERIFICATION

As a subset of the order capture activities your organization may require credit card verification. If your organization requires full fulfillment activities such credit card verification may be essential.

Rating and Description

1. *Current state-of-the-art computer tie-in capabilities exist as a part of the order acceptance and processing systems to handle all credit card verifications.*
2. *A credit card verification system exists and is presently operating.*
3. *Limited credit card verification is available through a manual or semi-automated process.*
4. *Credit card verifications could be accommodated through individual contact with credit card companies to verify credit.*
5. *No credit card verification exists, nor is the third-party capable of accommodating same.*

8. INVOICING, CREDIT AND COLLECTION

In addition to credit card verification, it is often necessary with sales on a business to business basis to have the capability of a credit and collection activity to include invoicing of individual shipments.

SOURCE: Thomas L. Freese, Vol. 11, No. 6, Warehouse Forum, © Ackerman Company, Columbus, Ohio.

Figure 19-2

putting together such a relationship move on (they are promoted or transferred, or retire). Without a well detailed agreement, those who follow may have a totally different set of expectations and understanding of the arrangement than was intended. Any such agreement should address:

- Scope of work
- Responsibilities
- Extra services
- Damages
- Risk management
- Termination
- Remedies
- Agreement modification
- Rate adjustments
- Compensation
- Service expectations

Of all these points, the most important is termination. Every partnership must end sometime, and an orderly process must be anticipated. The agreement must be sufficiently flexible to accommodate normal changes in prices and procedures.[8]

Managing a Continuing Relationship

The most important step in maintaining a good continuing relationship with the new supplier is to constantly communicate your expectations. One RFP contained a first page that was titled "What We Expect of You and What You Can Expect of Us." Many third-party relationships that fail do so because of inadequate communication of expectations.

Since the consequences of misunderstanding are so critical, it is best to communicate your expectations in more than one format. You should do so through written correspondence, personal visits, and periodic phone calls. An operations manual is a formal way to outline your expectations, but the right people won't always read it carefully. You may wish to use checklists, which are shorter and easier to control. A personal visit is the most effective way to communicate, but this is expensive and time consuming. Visits should not just be in one direction. Ask representatives of your supplier to visit your operations so they can meet your people and see how you run your own warehouses.

After the decision is made and the new warehouse is in operation, you owe it to yourself and your supplier to constantly measure the performance of that warehouse. Performance measurement is a big topic, and we cannot begin to cover all of the options. As you look at those measures that you plan to use to gauge the performance of a warehouse, consider five rules:

1. Keep it simple. A complex method is hard to understand and expensive to administer.

2. Be sure the thing you are measuring is critical to the operation. Measuring trivia wastes time and causes people to get lost in detail.

3. Be sure that the measurement is timely. Stale information is a waste of time.

4. Be sure that you measure quality as well as productivity. Sometimes one is achieved at the expense of the other.

5. Consider multiple variables when you measure either quality or productivity. For example, consider both the number of pieces shipped accurately and the dollar value of merchandise shipped correctly. A single error involving very expensive merchandise could be more serious than normal statistics would reveal.

Once you have done the measuring, be sure that regular feedback is provided to the warehouse supplier. The manner in which it is supplied can be a motivator or a demotivator. If you emphasize the positive and encourage competition with other warehouses, your warehousing people will be favorably motivated.

Managing for maximum efficiency is an ongoing process. It is enhanced by periodically meeting and setting new objectives. Any operation that is not continuously improving is actually losing ground since the competition is presumably moving forward. It is valuable to meet at least once a year with management at each public warehouse to agree on attainable goals for the year ahead.

By making the right start, building and maintaining excellent communication, and periodically reviewing objectives, you can make the experience of picking public warehouses an orderly and productive process.

Part V

PROTECTING THE WAREHOUSE OPERATION

20

PREVENTING
CASUALTY LOSSES

There are three kinds of major losses that could affect your property:

1. A business interruption interferes with the ability of the warehouse to function normally. This includes any occurrence that makes it impossible to ship, receive, or move materials.

2. Some casualty losses affect both the building and its contents.

3. Other losses may affect just the warehouse structure.

Types of Casualty Losses

The most catastrophic loss is fire. A warehouse fire, once out of control, is likely to destroy both the building and its contents.

Windstorm is a common threat to warehouse buildings. The modern warehouse is particularly vulnerable to windstorm damage because its wide expanses of flat roof are readily damaged by high winds. When such roofs are constructed with asphalt, insurance carriers limit the amount that can be used and thereby reduce risk of fire. Because asphalt is an adhesive, reducing its use to cut fire risk increases the risk of damage by windstorm.

The risk of water damage includes both floods and sprinkler system malfunction. Sometimes ill-trained emergency fire brigades may set off the sprinkler systems, and the leakage can cause as much damage as fire.

Overloading the building, structural failure, or earthquake may all cause a building to collapse. The risk of explosion is influenced by the products stored in or near the building.

Vandalism and malicious mischief clauses cover losses caused by sabotage or other acts of malicious destruction. A more general risk is that of "consequential loss"—the losses that may come as the result of an earlier event. One example is a power failure that disables the refrigeration system in a freezer, with spoilage of stored products as a consequential damage.

Controlling Fire Risks

One of the best ways to control fire risk is to purchase both insurance and *loss-prevention advice* from a carrier who emphasizes loss prevention. The more progressive insurance companies such as Factory Mutual (FM) and Industrial Risk Insurers (IRI) have continuing research and development on loss prevention. These companies control losses by making every employee aware of how to prevent fires.

Sprinkler Systems

Means of controlling the risk of fire vary in different countries, but in the United States today the most widely accepted method of fire risk reduction is the automatic sprinkler system. The automatic sprinkler is a series of pipes installed just under the ceiling of a warehouse building, carrying sufficient water to extinguish or prevent the spread of fire. There are two widely used types of sprinkler systems—dry-pipe and wet-pipe systems.

Dry-Pipe Systems

The dry-pipe system uses heated valve houses, each with a valve containing compressed air that keeps water out of the sprinkler pipes. When the air is released, water is permitted to flow. Dry-pipe systems function in unheated warehouses or in outdoor canopies where temperatures fall below freezing.

Sprinkler heads are installed on the pipes at regular intervals.

Each of these heads has a deflector to aim the flow of water and a trigger made of wax or a metallic compound that will melt at a given temperature. When heat reaches that temperature, the trigger snaps and allows water to flow through the system and onto the fire.

Most systems include a water-flow alarm that will ring a bell and send an electronic signal to summon the fire department. The electronic signal is particularly important for controlling a false alarm or an accidental discharge.

Wet-Pipe Systems

In the wet-pipe system, water fills all pipes up to the sprinkler heads, which means the entire warehouse must be kept above freezing. The wet-pipe system works faster since compressed air does not have to be released first and this speed makes it a better system for fire control. Furthermore, the wet-pipe system is less susceptible to false alarms than the more complex dry-pipe system.

All sprinkler systems are equipped with master control valves, which present two risks: accidental closure and deliberate shutoff by a vandal. Nearly a third of dollar losses from fires in buildings with sprinklers occur because the control valves were closed.

A New Type of Sprinkler System

Approval of the Early Suppression Fast Response (ESFR) sprinkler system is one of the most significant advances in a long history of the use of sprinkler systems for fire protection in the United States. In most of the world, sprinkler systems are regarded as an American invention and are often considered unacceptable because of the danger of water damage. In the United States, such systems are required by building codes in all warehouses but the smallest.

Two developments of the past few decades have made sprinkler system protection more complicated: the increased use of storage racks, and storage of more products that contain volatile or poisonous chemicals.

Protection experts consider rack storage to be hazardous because a typical rack arrangement leaves vertical spaces that act as a flue in the event of a fire. The insurance underwriters' solution to the problem was to specify installation of intermediate sprinkler heads within the rack.

This brought cries of dismay from warehouse owners because intermediate sprinklers are expensive and they destroy layout flexibility. Racks equipped with intermediate sprinklers can never be adjusted or moved. Furthermore, it is relatively easy to strike the in-rack sprinkler heads by accident while handling freight. For these reasons, many warehouse operators have stubbornly resisted underwriters requests for in-rack sprinkler systems.

Eliminating the need for in-rack sprinkler systems is the primary virtue of the ESFR sprinkler system. ESFR has two unique features: a quicker response time and a heavier sprinkler discharge. When a sufficient quantity of water is dropped onto a fire at an early stage, the fire will be suppressed before it presents a severe challenge to the building itself. Furthermore, the ESFR system typically opens fewer sprinkler heads than a conventional system and therefore reduces water damage. Reducing the size and duration of the fire will also reduce smoke damage.

There are two ways to deal with fire—suppression and control. Standard sprinkler systems operate on the control principle. The sprinkler system allows a fire to develop, but retards its spread by opening 20 to 30 sprinkler heads. This widespread discharge soaks surrounding storage piles so they will not ignite. The water discharge protects the building by cooling the steelwork and preventing structural failure and collapse. Suppression is left to firefighters.

The ESFR system is designed for suppression, rather than control. The goal is to apply enough cooling water to a heat source before a fire plume has any chance to develop.

Other Protection Against Fire

Fire extinguishers are an important first response in fighting fires. A portable extinguisher often prevents a small blaze from becoming a major fire. The best way to be sure that fire extinguishers will be used effectively is to have periodic training and drills.

Most fire protection systems also include fire hose stations to be used before the fire trucks arrive. When fire drills are held, instructions on how to use these hoses should be included. Many major fires are prevented by teams trained to act effectively the instant a fire is discovered, and well before the fire department can arrive.

The warehouse operator's proficiency in loss prevention is usually reflected in the rate insurance underwriters assign to the facility. Some public warehouse operators will advertise their fire insurance rate, knowing that it reflects their success in loss prevention. Fire underwriters measure both management interest and employee attitudes toward fire protection. They recognize that loss risks can be increased by careless smoking, poor housekeeping, or even poor labor relations.

Wind Storm Losses

Certain parts of the country are recognized as having a higher hazard from storms. Gulf Coast areas are notable for hurricanes and parts of the midwest have repetitive problems with tornadoes. Probably no economical warehouse design could withstand a direct hit from a tornado or the full force of a hurricane. However, certain kinds of construction have proven more resistant to such perils.

A preengineered metal roof, for example, has greater wind resistance than a built-up composition roof, simply because the metal roof does not rely on asphalt or any other chemical adhesive. Overhanging truck canopies should be avoided in high-wind areas, since they are particularly susceptible to wind damage.

Flood and Leakage

One way to control the risk of loss from either flood or leakage is to be sure that no merchandise is stored directly on the warehouse floor. The use of storage pallets or lumber dunnage to keep all merchandise a few inches above the floor can provide a measure of protection in the event of leakage.

Mass Theft

Because any burglary protection system designed by one human being can be thwarted by another, warehouse managers should never assume that a protection system will always prevent mass theft.

Periodically, warehouse managers should deliberately breach the security system just to learn how fast the police will respond. Holding this kind of burglary drill is potentially dangerous, and managers who do it must be sure that they do not become victims of a police accident. One way to reduce this risk is to inform the police chief, but not the alarm company, that a drill is taking place.

Since the technology for electronic protection is constantly being improved, the alert warehouse operator should always look for new equipment better than that currently installed. But all such equipment should be "fail-safe"—meaning that it will ring an alarm if wires are cut or power is interrupted.

Outside lighting is an important deterrent to mass theft. Intense outside light will make a thief feel conspicuous. Rather than run the risk of being seen, the thief probably will select a less well-lit building.

Vandalism

Vandalism, sabotage, or other deliberate destruction is difficult to control. The most serious vandalism risk is the intentional closure of sprinkler valves. There are electronic devices to counter this. Also, it is possible to apply a seal that shows at a glance if a valve has been turned by an unauthorized person. The risk of vandalism increases during a period of labor strife, and management should be especially alert at those times.

Surviving an Insurance Inspection

The insurance inspector is far more concerned about the materials in your warehouse than about the building itself. With few exceptions, warehouse buildings are fire resistant. If you are fortunate enough to be warehousing noncombustibles, and if those materials are in bulk with no combustible packaging, then you have a very safe warehouse. However, almost every warehouse is filled with products that contain some paper or corrugated packaging, plastics, wood, and combustible chemicals.

The insurance industry rates combustibles in five classes:

1. A special-hazard product is one in which the plastics content of packaging and product is more than 15% by weight or more than 25% by volume. Expanded plastic packaging, such as that used to protect computers and television sets, is particularly hazardous.

2. A Class IV commodity can be any product, even a metal product, that is packaged in a cardboard box if nonexpanded plastic is more than 15% of the product weight, or if expanded plastic is more than 25% of the product's volume. An example of such an item is a metal typewriter with plastic parts that is protected by a foam plastic container.

3. A Class III item is one in which both product and packaging are combustible. An example is facial tissue in cartons.

4. A Class II product is a noncombustible item that is stored in wood, corrugated cardboard, or other combustible packaging. While the contents will not burn, the package will. Examples include refrigerators, washers, and dryers.

5. A Class I item is the least hazardous commodity—noncombustible products stored in noncombustible packaging on pallets. The pallets themselves will burn, but if everything on the pallet is noncombustible, it is rated as Class I.

The fire inspector will consider the quantity of material in your warehouse that falls into each of the five classes and the manner in which you store materials. From a protection standpoint, solid-pile storage—freestanding stacks that are snugly against each other—provide the least hazardous storage arrangement. The exception would be a commodity that is in itself quite dangerous, such as a flammable chemical.

You can make even racks safer from the underwriter's point of view. Allow a vertical space between rack structures, to permit sprinkler water to run down into the racks. Pallet racks or installations that use metal grates are considered safer than solid steel shelving, simply because the openings allow water to run down through the levels.

The fire inspector will also look at the number and width of your aisles, and the height at which goods are stored. Nearly all underwriters request that goods be at least 18 inches below sprinkler heads; but with more hazardous materials, an even greater space between product and sprinklers will increase the possibility that the sprinkler system can control a fire.

The problem with high-piled storage is a chimney or flue effect. If you have a 30-foot pile of combustible boxes with a 6-inch space between rows, a fire will move up the 30-foot flue like a blowtorch. The flame at the top will be extremely hot and violent, and this heat can expand the steel framework of the building and cause a collapse.

Fire safety for the most dangerous materials is improved with ample aisle separation. If the hazardous products are separated from the rest of the merchandise by wide aisles, the chance of controlling a fire is increased. In some cases, extremely hazardous materials are put in a special-hazard building far from the rest of the facility. Thus a fire in that building will not threaten the larger warehouse.

Plant Emergency Organizations

A plant emergency plan can greatly reduce the risk of major loss. For this reason, casualty underwriters strongly encourage well-trained plant emergency teams. The emergency team fills the gap between when an emergency is discovered and the fire department or other professionals arrive.

The emergency organization should always include representatives of plant *maintenance* or *engineering* departments, since these people are the most familiar with the protection systems, control panels, and circuit breakers. The plant emergency team should know how to operate all protective equipment and be trained to react with confidence and accuracy.

The most important aspect of training is practice drills. Such drills should include sounding the alarm; moving to a prescribed emergency station; and handling fire hoses, extinguishers, and sprinkler system controls. The last of these is the most complex, since a large sprinkler system usually has many valves as well as a booster fire pump. A good emergency organization also is schooled in First Aid, since the first priority is always saving lives.

Emergency teams need regular drills and preparation. When emergency plans are made and then forgotten, the team rapidly becomes disorganized and useless.

While the prime goal of a warehouse manager must be to produce a profit, insurance underwriters tend to evaluate management interest in reducing fire risk as much as any other factor in rating a facility.

The best insurance policies will not adequately cover all the losses in a major casualty disaster. The ill will from customers and suppliers is difficult to measure and nearly impossible to recover.

When the chief distribution executive is not overly interested in protecting property, it is unlikely that his subordinates will be either. On the other hand, when senior management interest in property conservation is strong, this attitude will pervade the organization. The cornerstone of a program to control casualty losses is senior management's commitment to it.

21

"MYSTERIOUS DISAPPEARANCE"

Protecting merchandise from "mysterious disappearance" requires a careful blend of the physical techniques of security with personnel and procedural precautions.

The physical techniques—the easiest to put into practice—too frequently are the only ones used in a warehouse. However, the best security procedures in the world cannot defeat collusion between two dishonest individuals—one employed inside the warehouse, one outside. And while your goal is to hire only honest, trustworthy employees, even a warehouse with 100% honest employees is vulnerable to dishonesty from outsiders. So procedural checks become important to ensure that honest employees remember to use security systems. Thus, a combination of all three—physical, personnel, and procedural precautions—must be considered when you look at warehouse security.

Electronic Detectors and Alarms

A first line of defense in theft security is electronic alarm systems, which generally are more reliable than a watchman. But it is important to remember that an electronic system can always be defeated. Any system designed by one human can be overcome by another, especially since a sophisticated theft ring may include a former employee of an alarm company. Furthermore, some alarm companies use outside electricians as subcontractors to install equipment. And one dishonest electrician can defeat most systems.

However, it's harder to defeat the electronic systems being

installed today, thanks to the increasing sophistication of the new equipment.

Door Protection Alarms

The least costly electronic alarms are those that provide protection for doors only, but they are not adequate for most warehouses. With current construction techniques, the outside walls of a warehouse are seldom load bearing, and therefore are relatively thin. Whether the walls are constructed of masonry or metal, a determined thief will have little trouble cutting a large enough hole in the wall to allow a mass theft.

In one such case, thieves backed a truck against a wall of the warehouse. Working from inside the van, they removed concrete blocks from the exterior wall to gain entry, and loaded the van with the highest value product in the building. Fortunately, one of the thieves tripped an electronic beam in the rail dock area and sent in an alarm.

Because of the vulnerability of nearly all warehouse walls, the best electronic alarms include a "wall of light" surrounding most of the storage areas. With this, the thief who manages to penetrate an exterior wall will still sound an alarm when entering the storage areas.

Skylights and roofs are also easily entered, primarily because of lightweight materials used in modern construction. Protecting an entire roof area with electronic beams is not practical. Fortunately, however, it is equally impractical to remove any significant volume of stored product over a roof.

One answer to the compromises necessary with alarm systems is to install sectional alarms, providing one standard of security for most inventory of normal value, and a high-security storage area for high-value items. Even the ultrasonic system can be practical in an isolated room designed for a small portion of the inventory.

The typical alarm system used in warehouses usually must be switched on or off as an entire system, so if there is a breach it's difficult to pin down where it occurred. Warehouse supervisors may have difficulty "setting up" the system at night when one of many doors is partially ajar, or one electronic beam is blocked. However, more sophisticated systems have an "annunciator panel" to show exactly which doors or beams have been breached.

Such a system may also permit you to keep certain doors on alarm, but take others off. This enables you to activate the system for portions of the building not in daily use. In a very large or lightly staffed building, this system may be worth its additional cost.

Fire codes in most communities require many safety pedestrian doors along the walls of the warehouse to allow workers to escape from the building in case of fire. Since such doors can also make it easier to pilfer merchandise, you should stop any casual use by having an alarm on each personnel door which rings when it is opened.

Grounds Security

The physical arrangement of grounds and access to your warehouse will either invite or discourage theft. While it would be ideal if you could restrict access to the property surrounding the warehouse, this often is not practical. Because most warehouses are involved in high-volume movements of freight, common carrier truck drivers and railroad employees must come on the property to spot or remove vehicles. Such access usually is required at night or after working hours.

High fences and gates help control access to the property. After-hours patrol of the grounds by a guard dog and trained handler is an excellent psychological deterrent. Some guard dog services operate on a roving basis, with protection provided for a number of facilities by the same guard team.

Where you locate employee and visitor parking can also either deter or encourage theft. All persons entering or leaving a warehouse should use a single entrance, with other personnel doors available only for emergency. Parking should never be permitted adjacent to warehouse walls or doors.

Seals

A seal is a thin metal strip (with a serial number) that is fastened so that it cannot be restored once it is broken. Properly used, the seal is a good physical deterrent to cargo theft. Because theft in interstate commerce may be investigated by the FBI, for years the mere presence of a seal on a box car was enough to discourage unauthorized entry.

If a box car is broken into while spotted at the shipper's or receiver's plant, the railroad has a legal right to disclaim responsibility, as the car was not under its control. For this reason, the use of an inside rail dock may be the only means of protecting against theft from a railroad car. A parked truck trailer presents similar problems. You can buy electronic alarm systems that extend coverage to spotted trailers.

To use seals effectively, you must have procedures that are carefully followed, as well as good communication between shipper and receiver. You can't allow seals to be either applied or removed by unauthorized personnel if you expect them to serve their purpose. More recently, however, bolder thieves have attacked the box cars themselves.

Storage Rack

One way to reduce losses of highly vulnerable products is to place them on the upper levels of storage racks where they can be readily removed only by a trained lift-truck driver. You should use storage racks to improve security as well as storage productivity.

Controlling Collusion Theft

Part of the overall national breakdown in morality is a significant increase in collusion theft. Warehouse collusion theft is stealing which is usually accomplished by a partnership consisting of a truck driver and a warehouse employee. Collusion theft is the most difficult warehouse crime to control, since there is no electronic system and no paperwork system that is certain to expose collusion when it exists.

Furthermore, collusion thefts are frequently quite large, and one report indicates that 40% of warehouse thefts are by employees. It is self evident that the best way to avoid collusion theft is to hire honest people, but the task of accomplishing this has become much more difficult in recent years. There was a time when the use of a polygraph (lie detector) test was a permissible practice in most states, but Federal law now prohibits its use.

There are two ways to defend against theft and pilferage. One is a combination of physical deterrents and systems that make it difficult to break security. The second defense is by confirming the honesty of all employees.

Physical Deterrent

Many warehouses use physical means of discouraging pilferage. These include careful supervision of loading and unloading of trash containers to be sure that they are not a vehicle for warehouse theft. Some require that employees park a significant distance away from the warehouse to discourage the carrying of product from warehouses to automobiles. Television cameras are frequently mounted on loading docks to demonstrate that security people have the ability to closely observe every action taken on the dock. Many companies have uniformed guards at docks and at truck entrances to demonstrate that there is close surveillance on the property.

Restricted Access

Only warehouse employees should have access to storage areas. Bank customers aren't offended by barriers that prohibit them from walking behind the counters where the money is stored. Similarly, no user of a warehouse would be offended if his or her movement is restricted in the same manner. Even warehouse employees should understand that they are authorized to enter only those areas involved in their own work.

Yet many warehouses have no physical restriction that prevents visitors from wandering at will. Furthermore, many warehouses don't prevent visiting truck drivers from walking or loitering in storage areas. Posting signs or painting stripes may not be good enough. Look into how fencing, cages or counters can be used to prevent unauthorized persons from entering storage areas.

Confirming Employee Honesty

A major factor in collusion theft is management's perceived lack of concern for security. The number one failure is preemployment screening. It is difficult to screen applicants to include only those people who are absolutely honest, have the required skills, and are drug-free. The cost of hiring the wrong person can easily be in the thousands of dollars.

There are many ways to do preemployment screening. Personal interviews, drug screening, reference checks, and background reviews are all necessary steps in the process.

Preemployment tests can help you screen candidates for the job. All prospective warehouse and delivery employees should be tested for both skills and integrity. The cost of accomplishing this is low, ranging from $15.00 to $40.00 per candidate. Researchers at the University of Iowa conducted a study of more than forty tests and a half-million testees. They reported that the tests can identify irresponsible and counterproductive behavior that drives bosses crazy: disciplinary problems, disruptiveness on the job, frequent tardiness, and absenteeism. Dr. Frank L. Schmidt of Iowa stated, "It is very difficult for dishonest people to fake honesty." Why would a thief expose bad habits in a test? Partially because thieves have a propensity to believe that "everybody does it" and it would be implausible to claim that they never do it themselves.

Though a growing number of experts have demonstrated a significant correlation between people who fail the test and steal from employers, other professionals are concerned about the large number of "false-positive" people who are honest but are incorrectly identified as dishonest because of test scores.

Florida Atlantic University conducted an independent study of employees at convenience stores, particularly stores with substantial pilferage. Prospective employees had taken one of two honesty tests, but they were not used in the hiring process. A group of fifty-four employees who were terminated for theft was carefully reviewed. These fired employees had low test scores while the still employed group had passing scores.

Two leading suppliers of honesty tests are London House and Reed Psychological Systems.

Figures 21-1 and 21-2 show sample assessment reports used by London House, a division of the McGraw-Hill Companies. Both London House and Reed consider their questionnaires to be sensitive and will not allow them to be published. The test designers claim that their instruments are in compliance with all Federal and state laws prohibiting discrimination on the basis of race, color, religion, age, sex, national origin, disability, or record of offenses. These surveys are designed to be easily given and administered. The tests can be completed in less than an hour and can be scored with the aid of a personal computer. Raw scores are faxed to the testing company or processed by touch-tone telephone.

Significant Behavioral Indicators
—Interview Questions—

SCALE: Work Values

> 514: You've indicated that you're not always very prompt. What types of situations typically cause you to be late for work?

SCALE: Supervision Attitudes

> 614: Your response indicates you frequently break rules to do the job right. Please describe some situations when you did this.

> 621: Based on your response, you've occasionally gotten into trouble for fooling around at work or school. Please describe some examples.

SOURCE: From VII, No. 5, Warehousing Forum, courtesy of London House Division, McGraw-Hill.

Figure 21-1

Significant Behavioral Indicators
—Positive Indicators—

SCALE: Honesty

- In recent years, has never though about
  stealing money.. (89)
- Thinks he/she will never take something of
  value from future jobs without permission..................... (88)
- Believes he/she is definitely too honest to steal .... (39)
- Has never taken company merchandise or
  property from jobs without permission .... (96)
- Has not taken any money, without authorization,
  from jobs in the last three years.....................................(110)

SCALE: Drug Avoidance

- Believes that someone who drinks alcohol
  is unacceptable..(115)
- Believes that someone who uses uppers
  (amphetamines) is very unacceptable ....(116)
- Believes that someone who uses marijuana
  is very unacceptable...(118)
- Believes that someone who shares marijuana
  with others is very unacceptable ....(119)

SOURCE: From VII, No. 5, Warehousing Forum, courtesy of London House Division, McGraw-Hill.

Figure 21-2

Giving and Checking References

A growing number of people are fearful of giving negative feedback about a former employee. Facts offered in good faith or opinions honestly held are a solid defense against defamation claims. If you check references, your most important question is "Would you rehire?" The Society for Human Resource Management (SHRM) estimates that nearly one-fourth of all resumes and applications contain at least one major fabrication. Reference checking will reveal such dishonesty even if the person giving feedback is attempting to avoid anything negative. You are more likely to get a useful response if questions are in the form of a choice between two options. For example, "Would you consider this individual to be more technically inclined or a more people-oriented person?" It is best to ask the most interesting questions first, finishing the interview with basic verification of employment dates, salary, and title. Reference checking is time consuming and therefore expensive. Major companies indicate that they spend about 10% of their hiring time in the process of reference checking, and the cost of hiring each employee is approximately $9,200.00 While this seems expensive, consider the potential costs incurred by a dishonest employee.*

Other Collusion Theft Controls

It may be necessary to use undercover contract employees to detect collusion theft. Undercover contractors are available from a variety of sources, and they may disclose other problems as well as collusion theft. One example is drug dealing in the warehouse.

Another proven deterrent for collusion theft is to conduct random test unloadings for outbound vehicles. Choose at random one or two outbound loads each month, and call the trucker back in for the load to be completely rechecked. When potential thieves learn that this kind of surveillance is underway, they may conclude that it is too risky to steal at your warehouse.

Outbound loads are not the only place collusion theft can occur. It is equally likely that collusion will take place on in-bound shipments, with a warehouse employee signing for the full load but actually leaving a portion of it on the truck to be hauled away and sold.

Another key to controlling collusion theft is excellence in inventory management. If employees gain the idea that management has no idea how much material should be in inventory, some people will get the idea that if something disappears it will never be missed. On the other hand, when inventory discrepancies are regularly and carefully checked, the message is out that management can and will maintain control over the stocks in the warehouse.

---

* From VII, No. 5, Warehousing Forum, by Daniel Bolger, P.E., President, The Bolger Group.

Collusion is extremely difficult to control, but it will be minimized if you take care to hire honest people and to demonstrate that management has a keen interest in protecting the property that is in the warehouse.

Your best defense is the fact that the great majority of workers are honest and do not want to work with a thief. When you project your interest in loss prevention and your willingness to go to any length to prevent dishonesty, you can enlist nearly every employee to be part of the security team. Stolen merchandise means lost profits and a danger to the future of the organization. When senior management demonstrates its concern about maintaining the security of all of the property in the warehouse, it is usually relatively easy to gain similar cooperation from the great majority of warehouse workers.[9]

Marketability of Product

In analyzing your theft security, study the marketability of each product in the warehouse: Some merchandise may be valuable, but not easily sold; other products may be of lower value, but easily sold. Thieves generally will consider marketability of product more closely than value. For example, basic food products, paper products, and other products that do not have a high value but do have a ready market have been the object of mass thefts. If you're not sure of the marketability of certain products, law enforcement officials can give you a good idea of how frequently various products are stolen for resale.

After you determine marketability, apply selective security standards, with the highest security measures used for the most marketable merchandise.

Hiring Honest People

A newer method of testing a job applicant's honesty is preemployment screening through voice stress analysis. Some claim that the results are superior to those obtained with the polygraph. When the person being questioned intends deceit in his or her answer, the voice will contain stress which may not be audible to the human ear, but which can be measured by a trained analyst using a special recorder. While most interviews are conducted in the presence of the speaker, it is also possible to analyze a tape recording or a telephone conversation.

The 1988 ban on the polygraph for preemployment screening caused renewed interest in handwriting analysis, a personnel technique that is more widely used in Europe than in the United States. The effectiveness of this technique has been debated, yet many claim that handwriting analysis will reveal a great deal about the personality of the writer.

Using Supervision for Security

A good way to defeat collusion is to have a "second-count" check. Ideally, the person who does this should be a foreman or supervisor who is a member of management.

There are several ways you can use a foreman as a first line of defense against collusion theft. First, the placement and breaking of all seals should be the responsibility of a foreman, accompanied by the truck driver whose vehicle was sealed. Second, upon receipt of any load, the foreman as well as the receiver can verify merchandise in the staging area. In a busy or crowded warehouse, this verification may take place in a storage bay, since the load is taken directly from the vehicle to storage. Third, as only a random check of certain loads can be made, that check is best done by a member of management. This does not mean that a collusion ring could not extend to management, or could not include more than two persons. However, experience has shown that few collusion thefts involve more than two individuals.

Procedures to Promote Security

Personnel policies also can be structured to encourage honesty. Strict policy on the acceptance of gifts is one example. Truckers report that getting a convenient unloading appointment at many busy grocery warehouses depends on a "gift" to the receiving clerk.

When this kind of corruption exists, the moral atmosphere that encourages pilferage of merchandise is also likely to be present. Even honest persons and good security systems can be overcome by creative outside thieves. Therefore, strict adherence to security procedures must be carefully enforced.

For example, if only foremen are to break and apply seals, you must create—and use—paperwork that identifies the foreman who performed the task.

Every empty container is a potential repository for stolen merchandise. An empty trailer left at a dock could be the staging area for a theft. The same is true of empty box cars, and even trash hoppers. One approach is to restrict access to such containers. Empty freight vehicles should either be kept under seal, or inspected as they leave the warehouse. Furthermore, this inspection must be covered by a written record that it took place.

Customer Pick-Ups and Returns

Customer pick-ups create an additional risk of theft. One way to control this risk is to have merchandise for pick-up pulled from stock by one individual, with delivery to the customer made by a different worker. Again, this procedure is based on the fact that there usually are only two people involved in collusion thefts.

Customer returns present an unusual security problem—particularly if the merchandise is returned in high volume or in nonstandard cartons. If this is the case in your warehouse, be sure customer returns are checked thoroughly as soon as the merchandise arrives at the dock. If you don't keep your returned-goods processing current, you're inviting pilferage.

Control of documents can be as important as control of freight. Since fraudulent parcel labels can be used to divert small shipments, the best way to prevent this is to supervise the issuing of labels.

Many warehouse crews break for lunch or coffee at the same time, with no one available to inspect exposed docks or entrance doors. It's a good idea to stagger rest breaks, with a few individuals always on duty to protect against unauthorized entry.

Security Audits

Good management must actively combat theft, focusing on the three aspects of defense: the physical, personnel, and procedural.

Security Checklist

1. When was your alarm system last reviewed and checked?
2. Have you considered sectional alarms?
3. Are restricted areas defined in writing?
4. Are the boundaries of the restricted areas obvious to employees and visitors?
5. Are the means of restriction adequate (do you use fences, counters, cages as necessary)?
6. Does the warehouse have a reputation for tight security?
7. Are rules enforced strictly and impartially?
8. Does management set the proper example for loss prevention?
9. Do you have the necessary checks to ensure that procedures are followed correctly?
10. Is employee/visitor parking located away from the building?
11. Do you make spot checks to ensure that watchmen and security guards do their jobs properly?
12. Are walls and fences checked periodically for evidence of entry?
13. Are employees and visitors required to use one door only, with all other doors equipped with alarms?
14. Are docks and doors never left unattended?
15. Are rest breaks staggered?
16. Is your returned goods processing up-to-date?
17. Do you deliberately overship goods to see if the overshipment is reported by receiving personnel?
18. Have you determined which products are most attractive to thieves?
19. Do you have different levels of security for different value products?
20. Do you emphasize randomness in your security checks?
21. Do your supervisors look for patterns that might indicate pilferage or collusion?
22. Is access to empty freight cars and trailers restricted?
23. Do you check trash containers randomly?
24. Do you have a systematic, rigorous audit of your security procedures and policies?

Figure 21-3

No one of the three will be effective without the other two, and a "Maginot Line" complex can be as fatal in warehouse security as it was in military history. The greatest enemy of security is complacency. An outside security audit is a good way to guard against management complacency.

22

SAFETY, SANITATION, AND HOUSEKEEPING

Because so much lifting, pushing, pulling, and human interaction with power equipment and large mechanical devices is involved, the potential for accidents in warehouses is high. Any of the hundreds of different tasks performed daily in a warehouse can create a condition where serious accidents may happen.

Accidents in the Warehouse

Of all injured warehouse workers, most are not the newcomers. A Federal study shows that workers with five or more years' experience in warehousing are more likely to be injured than others. Those with one to five years are second, and those with less than one month have the least likelihood of injury. Apparently long experience causes carelessness which leads to injuries. The same study shows that accidents happen early in the day, so fatigue may not be a factor. Twenty-nine percent of accidents occur after only two to four hours on the job, and another twenty-three percent occur after less than two hours on the job. Only eight percent occur after eight or more hours in the workplace.[10]

The most dangerous work in a warehouse is loading and unloading of freight vehicles. Warehouses having the most accidents are those occupied by wholesalers and retailers. Sixty-eight percent of all warehouse injuries occur in these warehouses, as contrasted with only eight percent in warehouses devoted to transportation and public utilities. Presumably public warehousing is in this smaller category.

- Injuries are more frequent in larger warehouses. Those with eleven or more employees had 82% of injuries.

- Older workers are safer. Seventy-four percent of injured workers were thirty-four or younger.

- Though personal protective equipment will prevent injuries, the majority of warehouse workers do not wear such equipment. The most commonly used protectors are gloves and steel-toed safety shoes. Hard hats and goggles are also frequently used.

- While safety training is common, many workers never receive it. Training in proper manual lifting is the most popular, but only 28% of injured workers had been exposed to it. Only 23% had received a forklift operator training course.

Two types of accidents constitute nearly two-thirds of all injuries:

1. Overexertion is most common, accounting for 38% of injuries.
2. Being struck by a falling or flying object accounts for 26% of injuries.

Three types of injuries account for nearly all those that occur in warehouses:

1. Muscle sprain or strain is most common.
2. Bruise or contusion is second.
3. Cut, laceration, or puncture is third.

Although injuries requiring hospitalization were only nine percent of the total, more than three out of four accidents resulted in some time lost.

When workers were asked why accidents happen, fifty-four percent felt that it was not conditions at the work site that caused the accident. However, twenty-two percent felt that lack of space in the workplace was a factor. Some felt that working too fast or working in an awkward position was a contributing cause. Figure 22-1 shows what activity workers are engaged in when accidents take place.

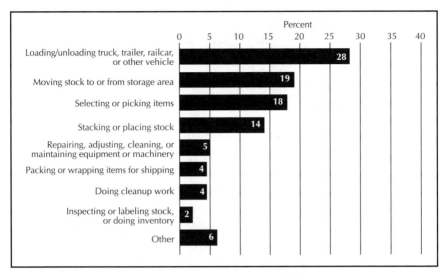

SOURCE: Injuries to warehouse workers by US Bureau of Labor Statistics

Figure 22-1

Common Sense for Common Warehouse Operations

A recurring problem in warehouse safety is the stacking of corrugated boxes. Humidity in a warehouse can break down corrugated boxes and, if not well stabilized, they can topple down at any time. Therefore, it is important to inspect high-piled merchandise for any sign of carton failure and to take down stacks before they fall.

Metal or plastic banding used to seal many types of containers is also a source of injuries. When this banding is cut away to open the containers, it can fly back, causing serious lacerations or eye injuries. Eye protection and gloves could prevent this. In addition, employees should be trained to stand to the side of the container, away from the banding, so they will not be hit if the banding flies. After the banding is cut, it should be disposed of so it doesn't trip someone or wrap around the wheel of a forklift.

Conveyor systems are another concern in the warehouse. Most conveyor systems have built-in safety features such as guards, pop-out rollers, and automatic shut-down devices. However, these features can be easily avoided by workers who are in a hurry or simply careless. Warning signs should be posted to alert all workers to the potential danger of the conveyor.

Redesigning Lifting Tasks

Overextension of muscles, particularly the lower back, is an injury that accounts for thirty to forty percent of health claims totaling up to nine billion dollars a year. With these figures in hand, many employers have begun to seriously address this problem and have devised methods to reduce the risk.

The substantial amount of research and training applied to the manual lifting problem has not been very effective. Employers train workers in the traditional "squat lift," yet injuries still occur. Many workers find this method very unnatural, since it requires more effort than a natural lift. The more effective approach to reduce injuries for these tasks is to apply mechanical aids such as lift trucks, conveyors, and dolly wheels.

The most effective way to eliminate back injuries is to redesign the workplace using ergonomics. Ergonomics is a big word to describe a simple process—arranging work for human comfort. In this situation, the goal is to eliminate the need for employees to lift. Although all lifting can never be eliminated, many tasks can be made less strenuous. Each task where employees are experiencing back injuries should be analyzed to determine how it can be redesigned to eliminate the cause of injury. Some solutions are simple. For instance, the height of off-loading can be changed so the worker doesn't have to bend over as far or reach up as high. Other solutions are more complex and may require mechanical devices such as conveyors or power hoists to make the job easier. These solutions are best implemented during the design phase of the facility so that costly redesign is not needed further down the road. Studies indicate that the ergonomic approach will reduce about one-third of back-related injuries. When you have done all you can to engineer the problem out, refocus your efforts on training.

Training

Safety training should be implemented in every warehouse for three reasons. First, and most important, every employer has a moral obligation to provide a safe environment for all employees and visitors. Second, Federal, state, and local laws mandate a vast array of specific responsibilities for every employer in regard to safety. Failure to uphold those responsibilities can result in substantial fines and penalties. Finally, from an economic standpoint, a safety program is cost effective. Reducing accidents should result in lower insurance premiums, reduced cost for worker's compensation coverage, and less absence due to lost time injuries. Secondary benefits include higher productivity because morale is improved, less damage to products moving through the warehouse, and reduction of fines or penalties for safety violations.

An effective safety training program should satisfy these seven key objectives:

1. Ensure that the building, equipment, and products are safe for all employees and visitors.

2. Instruct all employees on the advantages of working safely, injury and disaster response, and all applicable safety regulations.

3. Provide adequate equipment and clothing to protect employees from injury.

4. Inspect to confirm compliance with all rules, procedures, and training covering safety.

5. Comply with all laws relating to safety.

6. Provide a reporting system for accidents, injuries, and legal compliance.

7. Provide information on all hazardous chemicals and materials in the warehouse.

The above objectives relate to the employer's responsibilities. However, employees also must be made aware of their individual safety responsibilities. Warehouse employees should be responsible for the following safety actions:

1. Know all company safety rules.
2. Observe all safety procedures while in the warehouse.
3. Report injuries, unsafe conditions, and unsafe practices.
4. Wear all protective equipment and clothing required.
5. Conduct all work activities in a safe manner.

Safety Behavioral Factors

Motivation and attitude have a lot to do with safety. Good work habits are developed by warehouse people who are well coached and supervised by effective supervisors. Lack of alertness, boredom, or fatigue may reflect the worker's attitude toward himself and the people around him.

To reduce the impact of behavioral factors on the incidents of accidents, there are a number of principles that warehouse managers can follow. An important start is employee selection. Safety should be stressed during employee selection and orientation. This can be strengthened by emphasis from top management. A formal reward system should be established so that workers feel compensated for the amount of effort or skill required of them in following safety practices. The assignment of jobs should be based on the similarity between the requirements of the job and the characteristics of the worker. Increased worker satisfaction, improved quality, and improved safety records will result. Employees should not be assigned to tasks for which they have not been properly trained.

Some help is available from medical tests. Applicants can be prescreened through lower back X-ray films. An examination of medical records will often identify back-related problems in the medical history of an applicant.

Positive group norms and attitudes that increase worker acceptance of safety practices can be developed through planned group discussion or roundtables involving active participation by workers. A company safety committee can be an effective means of providing this environment.

Manual Handling and Safety

The warehouse manager has three approaches available to reduce the risk of manual handling tasks. These involve the job itself, training workers, and employee selection. The first approach is to redesign the job so that the physical and mental demands can be met safely by the average worker.

The second approach to reducing job risk is to train workers in techniques that minimize task hazards. Training can be used to make the employee aware of the dangers associated with various tasks performed in a warehouse. At the very least, the idea that injury risk "goes with the job" should be discarded. Show the employee how to take the time and the precautions to perform a function safely. The effectiveness of the training can be measured by comparing injury rates after the program with those that existed before training started.

The third approach involves employee selection. Screen potential employees to exclude those individuals who cannot perform safely under the demands of the job. Screening can involve physical fitness evaluations, strength and endurance tests, mobility tests, vision tests, and lower back X-ray films.

Reducing the injury risk of manual handling tasks should be viewed by the warehouse operator as an investment in worker safety, employee morale, reduced worker's compensation costs, and increased productivity. Now, instead of protecting workers from hazards, the emphasis is on removing the hazards from the work.

Management's Role

Management's attitude and commitment to the safety program is its most critical dimension. Employees attitudes will mirror management's, and if management does not show concern for accident prevention, then the rest of the workforce will be apathetic. Management should be proactive in identifying risks and reducing them through better equipment, new training, or new procedures and rules.

Management is ultimately responsible for protecting the on-the-job health and safety of workers. Leave no doubt about your personal concern for the safety and health of your people. When you show that you care about them, they will show more care for each other.

Sanitation

Although the Federal Food, Drug and Cosmetic Act was passed in the 1930s, strict warehouse inspection by representatives of the Food and Drug Administration (FDA) was not a source of serious concern until the 1970s and beyond. Everyone involved in warehousing took notice when the chief executive of a major east coast food chain was convicted of a felony in 1975 in connection with extreme contamination in one of his company's distribution centers. This case, which went to the U.S. Supreme Court, certainly provided general recognition of the power of FDA to enforce its citations, targeting its prosecution on the senior management of a company if necessary.

The result was greatly increased interest on the part of food warehousemen, in particular in ensuring their warehouses maintained conditions that would prevent similar litigation by the FDA. Some food companies hired professional inspection services, firms that could function as independent auditors of sanitary conditions in both private and public warehouses.

FDA inspectors may hold the warehouseman responsible for any evidence of contamination, even when there is some doubt as to whether it originated within the facility. Rodents are not the only sanitation problem, though they have generally caused the greatest concern. Insects and birds also are a source of contamination. And, because they fly, they are more difficult to control.

Any sanitation program must consider the following:

- Building design and construction
- Rodent exclusion and extermination
- Bird control
- Insect control

- Inbound inspection

- Cleanup methods and equipment

Other types of contamination can occur by odor transfer from an odoriferous product to one that has absorption qualities. Leakage, staining, and chemical reaction can be caused by incompatible items stored in close proximity.

Many companies have developed internal sanitation guidelines listing quality standards to be followed. Some companies retain an inspection service even more rigorous than the inspections typically performed by the FDA or local health inspectors. The American Institute of Baking is one of these inspecting organizations, and they have listed conditions that will automatically rate a warehouse "unsatisfactory" (Figure 22-2).

Excellence in Housekeeping

Excellence in housekeeping not only affects how the place looks, but it also affects safety, security, and overall performance. If the positive benefits so clearly outweigh the negatives, then why is this area so frequently neglected? The only conceivable reasons are a lack of appreciation and support from senior management or an inability to clearly quantify the benefits. Housekeeping, like quality, requires a belief that in the long run it will support your firm's goals to grow and be profitable.

Too often, management forces housekeeping on employees. This is not an effective way to achieve long-term excellence. As a basic premise, we have to believe that all employees want their companies to be successful and that if they are confident their housekeeping efforts will contribute to this objective, they will eagerly support it. The key here then is communication. As you have justified this effort to your management you must now do the same for the people performing the daily tasks within your warehouse. You must sing the praises of housekeeping excellence frequently, loudly, and with passion. Send the message that housekeeping is high on your priority list and that a failure in this area has serious consequences. When you demonstrate your interest, your

Conditions that Automatically Rate a Warehouse 'Unsatisfactory' to the F.D.A.

1. No sanitation program or pest control in effect.

2. Presence of live rodents or birds inside the distribution center.

3. Any significant insect infestation, internal or on the surface of the stored product.

4. Rodent gnawed bags or rodent evidence on pallets of stored stock, or numerous excreta pellets on the floor-wall junction of the food distribution center.

5. Cockroach infestation in the food distribution center that could cause regulatory action or could adulterate product.

6. Stored product contaminated with poisonous or hazardous chemicals (i.e., pesticides, motor oil, etc.)

7. Perimeter floor-wall junctions and storage areas completely restricted, preventing cleaning, inspection and pest control procedures.

8. No internal rodent control program. As an example, mechanical traps, snap traps or glue boards not present or evidence of dead and decomposed rodents in traps, or an insufficient number of traps present.

9. Rodent bait stations used inside the stock storage area of the food distribution center with evidence of dead rodents or spilled bait present.

10. Dock doors and pedestrian doors not rodent proofed and significant rodent evidence noted inside the food distribution center.

11. Severe roof leaks that may have contaminated product in storage.

12. Insect or rodent-infested transport vehicles.

13. Freezers and coolers, without thermometers, and temperatures above required minimums except during periods of heavy traffic.

14. Poor personnel practices, such as eating and smoking in undesignated areas, that could contaminate food in storage.

15. Illegal pesticides used in the distribution center or transport vehicles.

16. Non-certified persons using restricted-use pesticides.

17. Transporting food ingredients and product in the same trailer or rail car with toxic chemicals except when they are physically separated on the load. Separation can be by partition or location provided that leakage or spillage could not contaminate stock.

From a publication of the American Institute of Baking.

Figure 22-2

employees will know they cannot let you down. If you do this well, audits may in time become unnecessary.

Benefits

Though it may not be obvious that management is demanding dramatic improvements in both productivity and quality, excellence in housekeeping will support these goals. Following are some examples:

Productivity will improve as a result of:

- Clearer aisles (faster put-a-way, order picking, shipping)
- Improved inventory control (accurate stock location)
- Reduced damage (reduced scrap, accurate inventory records)
- Fewer accidents (less lost time)
- Less equipment downtime (more dependable work plan)
- Less pilferage (accurate inventory)
- Higher morale, greater pride (motivated workforce)
- Improved physical inventory taking (less time, more accurate)

Quality will improve as a result of:

- Shorter order cycle times (warehouse is more efficient)
- Accurate order shipment (correct quantity, right product)
- Fewer returns (less handling, paperwork and reshipments)

Many of these benefits can be linked to your housekeeping efforts. Others may simply help to build a stronger, more dedicated warehouse team.

Housekeeping excellence can and will help your company to become more profitable and provide greater levels of customer satisfaction, and as a result support growth. Housekeeping should not be compromised in the name of productivity for they are too closely related. If you feel you have already achieved excellence in this area, don't become complacent . . . take a fresh look at your operation with the mind set that "it can be improved." If you want to improve your housekeeping, develop and start using a checklist like the one shown in Figure 22-3.[11]

HOUSEKEEPING AUDIT

(1 = Unacceptable, 2 = Needs improvement, 3 = Good, 4 = Excellent)

Circle One   Circle One

- Is the dock area clear of stored materials or other obstructions?...................... 1 2 3 4
- Are the battery charging areas marked adequately and kept clear?.................................. 1 2 3 4
- Are exits secured, clearly marked, lighted, and accessible?...................................... 1 2 3 4
- Is the floor surface maintained in a clean, dry nonslip condition?............... 1 2 3 4
- Are all indoor and outdoor lighting fixtures adequate and functional?........................... 1 2 3 4
- Are all work areas and work benches clean and orderly? .. 1 2 3 4
- Are hose stations and fire extinguishers properly marked, inspected, hung, and kept clear of obstructions? 1 2 3 4
- Are employees properly attired and appearance neat?..................................... 1 2 3 4
- Are office areas, breakrooms, and restrooms clean?........... 1 2 3 4
- Are pallets stacked neatly in appropriate locations?.......... 1 2 3 4
- Are any storage rack or shelving components bent or damaged to the extent that replacement is necessary? 1 2 3 4
- Are aisleways clean and clear of obstructions and trash?.... 1 2 3 4

- Are materials stored in their properly designated areas only?.................................... 1 2 3 4
- Are there any leaning stacks or crushed cartons of products in storage racks, shelving, or on the floor? 1 2 3 4
- Are pallets or cartons protruding from the rack into the aisle?............................. 1 2 3 4
- Are hazardous materials waiting for disposal sealed properly and stored in one location?............................. 1 2 3 4
- Is the dock approach well drained, free of potholes, and clean?........................... 1 2 3 4
- Does the warehouse equipment appear to be clean and in good operating condition?........................... 1 2 3 4
- Is the returned goods processing area neat and organized?.......................... 1 2 3 4
- Are broken pallets and trash staged neatly in designated areas? 1 2 3 4
- Are the safety cabinets and first aid stations properly marked and adequately stocked with supplies?........................... 1 2 3 4
- Are trucks secured in place at the dock while loading and unloading the trailer?.......... 1 2 3 4

ADD CIRCLED POINTS TOGETHER FOR TOTAL SCORE = _____

Auditor's Name: _____ Date: _____ Location: _____

Figure 22-3

23

VERIFICATION OF INVENTORIES AND CYCLE COUNTING

If there is any warehousing job that is usually taken for granted, it is that of taking physical inventories. This job is considered a "necessary evil," yet it involves a considerable investment in terms of resources—people, equipment, and time. While there are no magic solutions, certain systems, most notably bar coding, can help improve the accuracy and shorten the time needed for physical inventories.

A newer alternative to taking physical inventory is cycle counting, a technique that many people believe to be significantly more efficient than the traditional approach.

Purpose of Inventories

Why even take physical inventories? An obvious answer lies in the impact that inventory has on the corporate financial statement. Finance people are certainly concerned about accountability, and public accountants are retained to verify the accountants' work. The very fact that we need to take physical inventories is a symptom of our failure to receive, store, ship, and record inventory correctly.

Annual physical inventories are similar to annual medical examinations—they provide a look at what is potentially wrong and may

This chapter is based on an article by W.G. Sheehan, The Griffin Group, in Warehousing and Physical Distribution Productivity Report, Vol. 16, No. 11, ©Alexander Research and Communications, New York.

provide the opportunity to correct past errors. Considering the impact that physical inventories have on operations, they deserve more attention and emphasis than we normally give them. Inaccurate inventories can impair a company's ability to plan materials requirements, and cause uncertainty in financial planning. Inaccurate inventories lead to lost sales, lower fill rates, and lost productivity in the warehouse.

How often should the warehouse take physical inventory? Only you can decide for your company. But bear in mind that loss of assets is not the only loss suffered by the company when the physical inventory and the book inventory do not balance.

Inventory Carrying Cost

Inventory carrying cost is one of the major costs of doing business—anywhere from 30% to 60% of the value of the inventory carried. Inventory cost as a percentage of sales is largely regulated by the rate of turnover. For example, if it costs 30% of the value of the goods to carry the inventory and you are turning the inventory over at a rate of 10 times each year, inventory carrying costs are about 3% of sales. Failing to turn the inventory can erode profits.

Benefits of Improvement

Have you measured the investment in resources you make each year in performing physical inventories? How much time is spent on your physical inventories in planning, counting, recounting, and reconciling?

What would be the potential savings for a 10% to 20% reduction in the hours required for physical inventories? In many cases the savings can be substantial.

Problem Areas

Neglect may be the biggest problem. Too many things are taken for granted. Too many potential problems are overlooked. Too much time is spent "fighting fires." Many counts are not accurate, and hence have to be redone. Too many counts are impossible to trace, including who did what, where, when, and how. There is a lack of communication between the warehouse, the office, and supervisors. This is often compounded by inaccurate counting on the part of those persons actually taking the physical inventory.

How Can a "Physical" Be Improved?

Remember that each department or function in your company needs the results for different reasons. Finance needs the numbers for accountability. Marketing needs to be sure what is available to sell and ship. Everybody is concerned about accuracy!

The primary purpose of the physical inventory is to determine the correct quantity and location of material in storage. Accuracy in the location system alone will provide tangible improvement in warehouse productivity.

Fiscal responsibility for stored goods is an inherent responsibility of the warehouse manager. In addition to confirming the accuracy of inbound and outbound shipments, completing a reliable physical inventory is a measure of management's success in protecting stored merchandise.

To achieve a reliable count, it is best to conduct two separate and independent counts. Reconciliation of the two counts is paramount in proving the reliability of the physical inventory without regard to differences between the physical and "book" balances. Reliability is enhanced through proper scheduling of the inventory. If the warehouse manager can be completely sure a thorough and reliable physical count has been taken, much more credence is provided to subsequent reconciliations with the book inventory.

To be sure that the counts are truly independent, one or more persons should be assigned to different areas for the first and second counts. If the inventory is to be effective, it must be accurate, fiscally responsible, and reliable.

Efficiency in inventories is achieved by minimizing resources required. Not only do we want an accurate count, but we also want the inventory conducted within a reasonable or minimum time.

Incentives

How can efficiency and effectiveness be combined? One suggestion is to provide for an incentive system. During the actual count it is vital to achieve a balance between effectiveness and efficiency. There is little to be gained by rushing through a physical count and then doing an inordinate number of recounts. Nor is it sensible for the counters to take so much time ensuring an infallible count that taking the inventory becomes extremely time consuming.

To achieve the proper balance it is not only appropriate but also necessary to establish a goal for the number of required recounts or third counts. While this goal will fluctuate, a goal of 98% accuracy between the first two counts is certainly realistic. Considering the total number of man-hours invested in a physical inventory, it is feasible to provide some incentives to workers who do the count for meeting performance goals of quantity and quality. One incentive program provided money prizes to each member of the team who counted the most items or turned in the greatest number of inventory tickets. Another prize went to team members who achieved the most accurate count. One criterion, however, was that the winners in each category had to rank in the upper third in the other category. The team that counted the most tickets also had to have a reasonable level of accuracy, with the most accurate team also having to achieve a given level of productivity.

Achieving efficiency and effectiveness through motivation— the incentive system—will ensure the reliability necessary for the reconciliation process. Reducing the number of third counts can materially reduce the hours devoted to the inventory-taking process.

Managing the Inventory-Taking Job

To improve the taking of physical inventories, we draw on the classic management principles of planning, staffing, organizing, coordinating, and controlling. Plan for physical inventories. An efficient physical inventory starts weeks or even months before the actual inventory date.

The first step is to assign an inventory coordinator for each of the physical inventories to be conducted. This appointment pinpoints responsibility and establishes a sole source for information flow. An annual wall calendar can be posted and marked with dates (whether tentative or firm) for inventories. The calendar should include specific dates for preparation for the inventory.

Involve the Whole Team

After suffering several excruciatingly painful physical inventories, a group of warehouse and office employees met to share ideas on the experience. It was interesting to note the enthusiasm generated simply by including employees in the planning process. And good standard inventory procedures can be developed as a result of this type of meeting.

Getting Organized

The next step to take prior to the actual physical inventory date is to make every effort to get rid of obsolete items, waste, or excess material that can only lengthen the time required to take the count. Then, two weeks before the inventory date, get organized by bringing key personnel together to touch base on responsibilities, availability of workers, equipment requirements, supplies required, etc.

Advance Preparation

Decide whether the inventory should be pretagged, or tagged during the first count. Some computer systems can easily produce inventory tags preprinted by location. You may want to consider filling out inventory tickets in advance—but not the quantity—in order to facilitate the actual taking of the physical inventory. Damaged goods and items in the repack area should be counted in advance. Employees must be trained how to fill out tags and who to contact regarding problems with the tags.

The Importance of Cutoff

A cutoff date for receipts and shipments must be established for an accurate reconciliation after the physical count has been completed. Certain areas should be zoned off (and not included in the actual physical inventory) for material received after the cutoff date, or material shown on the inventory records as shipped, even though it has not left the shipping dock. Documents for merchandise received or shipped after the cutoff date should be stamped so they are identified as not part of the reconciliation process.

Zones

A useful technique is to zone the inventory area into sections that require similar amounts of time to count. Teams should then be assigned to those sections. After the physical inventory coordinator has developed the plan, zoned the area to be counted, and assigned inventory teams by zones for first and second counts, conduct a physical inventory briefing for all persons concerned. This briefing should be for the counters in the warehouse, as well as clerical personnel who will work on the physical inventory and subsequent reconciliation. Highlight any difficult-to-count items, potentially difficult product codes, or other unique identification problems. Have examples of such items available for everyone to see before starting a count. The meeting should cover any unique requirements; team assignments; designated zones; an expected time schedule; how the tickets are to be filled out; and requirements to record serial numbers, lot numbers, or production date codes.

This meeting also is the time to warn personnel conducting the physical inventory about possible differences in pallet patterns and the difference it can make on the actual quantity on hand. This is particularly important since so many companies have standard unit-load programs.

Arithmetic Aids

If bar code readers cannot be used, encourage your warehouse personnel to use hand-held calculators. Issue guidelines on how or where counters are to make calculations. Don't allow them to write such calculations on the carton or, worse yet, on the inventory ticket, because if they do, it becomes impossible to ensure the independence of separate and distinct physical counts. Provide inventory teams with scratch paper and have them make their calculations on the back of the ticket that is turned in. Keeping these calculations assists in providing an audit trail when reconciling discrepancies.

Controlling the Count

Having a supervisor present during the count is valuable in ensuring an effective and efficient inventory. This individual should constantly spot-check procedures, read inventory tickets to ensure they have been properly filled out, and then sample some counts for accuracy.

We can't overemphasize the necessity for proper control of inventory tickets. This control begins with each inventory ticket having a unique serial number. The tickets should be handed out in sequence, according to zones, and should be placed on the product, also in sequence, so that an audit trail is maintained. This also expedites any third counts needed, since the counter will know precisely where to find the inventory ticket in question. The coordinator must ensure that inventory teams turn in all tickets, then verify that every ticket is accounted for at the conclusion of the inventory. One simple way to provide for this is to void the first and last tickets for each inventory.

Reconciliation

Remember that the end of the count is not the end of the inventory. Office procedures are just as important as warehouse procedures. Exercising proper control over the inventory in the warehouse is no more important than exercising proper control over the inventory within the office. The inventory is not complete until all counts have been reconciled and inventory records updated.

An important last step of inventory-taking is a review among team leaders, including representatives from the warehouse and office, which allows participants to cover problem areas and to solicit suggestions for improving future inventories. Notes should be kept concerning the number of hours required for the inventory, the number of persons, the equipment, the items counted, and the quantities of merchandise involved. This information is useful in improving planning for the next inventory. You can also use this session to recognize effective performance.

Eliminating Inventory Errors by Cycle Counting*

The goal of every warehouse and distribution center manager is inventory accuracy. The best way to achieve this is to count a small percentage of the inventory on a regular cycle basis. This sample of the total inventory can be easily compared to the inventory records, and since you are continually counting, when errors are identified, you have an excellent opportunity to define the cause of the error and take corrective action. This process of regular cycle counts, reconciling with the inventory records, and identifying and correcting these errors is called cycle counting.

It is important to realize that the immediate objective of cycle counting is not accurate inventory; it is the identification and elimination of errors. A byproduct of identifying and eliminating errors will be an accurate inventory.

To respond to the question, "What to count?" one must realize that this question is really two questions in one. The first question is, "How often should an item be counted?" and the second is, "When should each specific item be counted?" The answer to the first question lies in Pareto's Law. Pareto's Law, or the A–B–C Concept, has many different applications but the A–B–Cs of cycle counting are:

- Approximately 80% of a warehouse's dollar throughput is typically attributed to 20% of the stockkeeping units (A items).

- Approximately 15% of a warehouse's dollar throughput is typically attributed to 40% of the stockkeeping units (B items).

- Approximately 5% of a warehouse's dollar throughput is typically attributed to 40% of the stockkeeping units (C items).

---

* This section is by James A. Tompkins, Ph.D., President of Tompkins Associates, Inc., Raleigh, NC.

The obvious cycle counting application is that the A items should be counted much more frequently than the B items, and the B items much more frequently than the C items. A typical scenario would involve counting 6% of your A items each week, 4% of your B items each week, and 2% of your C items each week.

The second question, "When should each specific item be counted?" is relatively straightforward. An item should be counted when it is easiest and cheapest to get the most accurate count. When is this?

1. When an item is reordered.
2. When an inventory balance is zero or a negative quantity.
3. When an order is received.
4. When the inventory balance is low.

Thus, the answer to the question, "What to count?" is that you should count on an A–B–C basis and count when it is easy to count.

The number of people who should do cycle counts depends upon the number of items in inventory, the desired count frequency, the number of storage locations for each item (and the number of count irregularities such as number of recounts), accessibility of items, and physical characteristics of the items. A realistic standard is that a cycle counter can count forty items per day. The cycle counters should be very familiar with the stock location system, the warehouse layout, and the items being counted. Cycle counters should be assigned to the job on a permanent basis. This does not mean that cycle counting is necessarily a full-time job, only that the persons assigned to cycle counting should be part of a permanent team.

Cycle counters must recognize the probability of crossovers. A well-established cycle counting procedure will include checks of items that are likely to be misshipped. When a cycle count reveals an overage in one item, there is an immediate check of those items that would normally be confused with the one that is not in balance.

Some auditors are concerned with the "proper checks and balances" within a warehouse, and believe that only people from outside the warehouse should be assigned full-time to cycle counting. In fact, to the contrary, it has been shown that true inventory management will begin when management realizes that the job of maintaining accurate inventory is not that of accountants and auditors, but that of warehouse and distribution managers. Thus, the whole "checks and balances" discussion is irrelevant and it doesn't matter to whom the cycle counters report.

Having determined what to count and who should do the counting, the remaining question is "What are the procedures for cycle counting?" Valid cutoff controls are critical. As cycle counting should take place without affecting normal operations, a very careful coordination of counting and transaction processing must be planned. Any transaction that takes place after the inventory balance is reported and the actual count is made must be isolated so that an accurate inventory reconciliation may take place. This can be done in many ways:

- Record all inventory transactions and cycle counts in "real time."

- Record the time when each location is counted and report transaction times.

- Coordinate transaction with the count process so that counts are taken when transactions do not occur.

- Don't process any transaction for items scheduled to be counted until after the items are counted.

- Once the cutoff controls have been set up, the next procedure is count documentation. The information that should initially be recorded on the cycle count document is:
    part number
    part location
  The cycle counter should then record:
    count quantity
    counter's name
    date and time of count

The cycle counting document should then be forwarded to a reconciler. Counters should not reconcile their own counts. Reconcilers should compare the cycle count to the inventory record and determine if it was a good count. For this, one needs to have established a cycle count tolerance limit. Cycle count tolerance limits should be based upon the value of the items being counted. For example, if a $1,500 item is counted and a cycle count of 490 is made, and this is reconciled against an inventory record of 500, it is clear that this is a bad count. But if a $.02 item is counted and a cycle count of 490 is made, and this is reconciled against an inventory record of 500, it is clear that this is a good count. A typical count tolerance limit would be $50. That is, if the cycle count is within $50 of the inventory record, the count is considered to be a good count. If the cycle count indicates a discrepancy of $50 or more, it is considered a bad count. When a bad count occurs, the item should be recounted. The recount will verify if the original count was truly a bad count or if an error was made and the item count is within the tolerance limit. Whether the recount variance is within the count tolerance limit or not, the reconciler should adjust the inventory record to agree with the verified recount. If the variance of the recount is outside the tolerance limits, further investigation is necessary.

The further investigation of a bad count is required to determine why the inventory record is incorrect. If the counted quantity is verified as less than the inventory record, there are two easily checked possibilities:

1. An outstanding allocation of this item may have been filled without recording the transaction, or

2. A recently completed sales or production order may have been incorrectly subtracted from inventory.

If the variance shows that the count quantity is higher than the inventory record, replenishment orders should be checked to see if any product was received and not properly recorded, and recently completed sales or production orders should be checked for erroneous subtractions from the inventory records. It is important to recall that the objective of cycle counting is the identification and elimination of errors. Thus, the error investigation process is critical in cycle counting.

At the end of each month, a cycle counting accuracy plot should be given to management. Cycle counting accuracy is defined as:

$$Cycle\ Counting\ Accuracy = \frac{Good\ Counts}{Total\ Counts}$$

Figure 23-1 presents a graph illustrating the typical progress made once cycle counting is implemented. If a problem occurs, such as that shown in Figure 23-2, it is crucial that either the error investigation identify the problem and resolve it, or that the cycle counting quantities be increased until the errors can be identified and eliminated.

Cycle counting should be implemented in two phases. The first phase:

1. Identify approximately 100 items. These items should be A, B, and C items.

2. Repeat counts on these items. Count frequency should be whatever is necessary to identify errors.

3. Eliminate errors.

Once the first phase results in 100% accuracy for several weeks, the full-cycle counting system should be implemented. But the origi-

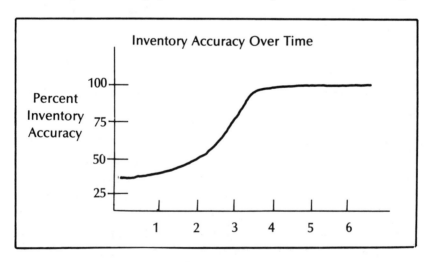

Figure 23-1. Typical Inventory Accuracy Progress

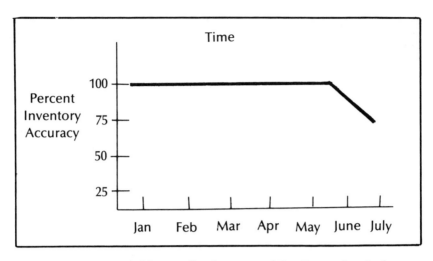

Figure 23-2. A Problem Indicating a Need for Corrective Action

nal 100 items should continue to be counted for at least two months. The special counts on these 100 items should be continued until a 98% + inventory accuracy is achieved.

Some have questioned whether or not public accountants who are responsible for audits will accept a cycle count. The fact is that cycle counting was developed because of frequent and severe problems in the accuracy of traditional physical inventories. As cycle counting has matured, auditors have recognized that the procedure is usually far superior to the traditional physical inventory. As long as the cycle counting is cleared with the audit team, a cycle count is recognized as a superior means of checking inventory.

Cycle Counting, Another Approach

If you are uncertain about cycle counting consider this approach: Whenever an item appears to be at zero, either in the warehouse or in the books, check *only that item* to see if the physical and the book agree. Alternatively, whenever an item falls to fewer than ten units, cycle count that item to be sure that the book and the physical are in agreement. By highlighting those items that are exhausted or nearly exhausted, you can make cyclical counts relatively quickly and easily. Furthermore, you will solve the critical problems of items that are nearly out of stock.

What Does the Future Hold?

Considering where we stand today with computer technology and automated systems, it is not difficult to project where we might be a few years hence with respect to taking physical inventories. By using universal product codes, as well as optical character readers, the job of taking physical inventories will become less difficult and less time-consuming, as well as more accurate.

If inventory changes can be recorded and records updated on receipt of material and shipment (or sale of material), then it is a natural extension of that concept to incorporate existing technology, equipment, and systems into a truly automated physical inventory system.

Product codes or item identification can be input to computer through the "wand" or readers now being used with several distribution systems. With portable keyboard/wand equipment, it is possible to input the count accurately either manually or on the basis of a standard unit-load program.

If a warehouse handles items in a standard pallet pattern, and this information is preloaded in the computer, the reading of a code with the standard pallet quantity will eliminate many calculations now done by the inventory team.

Part VI

THE HUMAN ELEMENT

24

ORIENTATION AND TRAINING

Some companies do little more for the new employee than to explain who the boss is and where the restroom is located. Since every new employee will be "filled-in" on the company by somebody, management should see that this orientation is directed by somebody knowledgeable.

Orientation should include:

- Description of the industry
- History of the company
- Unique advantages of the company and its industry
- Information about company management and ownership
- Identification of company customers and products handled
- Information about unions and labor agreements (if any)
- Procedure for resolving problems or misunderstandings
- Location of restroom, lunch facilities, parking, time clock
- Details about pay and benefits
- Recommended work clothing
- Work and safety rules

This information is often best provided by the employee's supervisor. This allows the new employee the chance to become acquainted with the supervisor, and it establishes him or her as the best source of information about the job and the company. An employee handbook is another source, and a sample is shown in Figure 24-1.

Figure 26-1
Excerpts from Personnel Handbook

Personnel Relations Policies

Personal honesty, responsibility, enthusiastic cooperation, loyalty and productivity are expected of everyone.

It is the policy of the company to treat everyone with respect, provide fair treatment for everyone, good, safe working conditions, superior benefits, and wages comparable with other companies in the public warehousing industry.

Anyone should feel free to discuss his work or work relationships with his supervisor or, through his supervisor, with any other member of management.

Whenever possible, promotions are made from within the company thus providing the opportunity for all to advance to the limits of their capabilities.

Work Schedule—The normal work week begins at 12:01 a.m. Sunday and ends at 12:00 p.m. the following Saturday. Work schedules vary at each location due to local business hours or shift schedules. Our normal warehouse hours are 8:00 a.m. to 4:30 p.m. with a one-half hour lunch period and a morning coffee break.

Overtime—Overtime, in excess of forty hours per week (or in accordance with local labor agreement) for hourly employees, will be compensated at the rate of time and one-half. Supervisors and management trainees will work some overtime.

Pay Categories—Hourly warehouse and clerical employees are paid at a specific rate per hour and are paid time and one-half for all hours over 40 per week. Hourly employees use the time clock and time cards.

Weekly salaried employees are paid by the week. The hours worked, lost time, vacations, etc. are recorded on a weekly time card.

All semi-monthly salaried employees are paid twice per month. They have no time cards and overtime is not paid.

Time Card Regulations—Each employee is required to punch his or her own time card not more than 12 minutes earlier than the scheduled start of the shift. At the end of the shift the time card must be punched immediately. When an employee leaves the premises for lunch or other personal business, the employee must clock out upon leaving and in upon returning.

Adjustments—All pay adjustments are made on the basis of 1/10 of an hour. You must work a full 1/10 hour (6 minutes) to earn 1/10 hour of overtime pay. If you are late from 1 through 6 minutes, you will lose 6 minutes pay. If you are late 7 minutes, you will lose 12 minutes pay, etc.

Holidays—The entire company observes holidays as established between management and the collective bargaining unit in local collective bargaining agreements. Each full time employee will receive eight hours pay for established holidays which generally include:

Figure 24-1

New Years Day
Memorial Day
Independence Day
Labor Day
Thanksgiving Day
Christmas Day
Additional day in conjunction with Christmas or New Years

To be eligible for holiday or shift pay, you must have worked both your last scheduled work day or shift prior to the holiday and your first scheduled work day or shift following the holiday. If a holiday falls on Sunday, warehouses will be closed on the following Monday. For holidays falling on Saturday, warehouses will be closed on the preceding Friday. You will not receive holiday premium pay unless you actually work on the calendar day on which the holiday falls. Holiday premium pay or additional holidays shall be stipulated in the collective bargaining agreement at each location.

Vacations—All full-time employees earn paid vacations on the anniversary date of hire as a full-time employee in accordance with the following schedule or as indicated in the local collective bargaining agreement.

After the 1st year—
  5 days at regular rate (1 week)
After the 2nd year—
  10 days at regular rate (2 weeks)
After the 10th year—
  15 days at regular rate (3 weeks)

A list will be posted each April for you to request your preferred vacation period. Your supervisor will make every effort to allow you to take your vacation at the time requested. However, the company reserves the right to limit the number of employees on vacation at any one time to insure adequate service to our customers. Eligible employees must take at least one week vacation each year. If a holiday occurs during your vacation, you will be paid for it or your vacation may be extended one day.

Managers must take their full vacation each year. Since managers, supervisors and the corporate clerical staff often work more than a 40 hour week, they receive 2 weeks vacation for one to seven years' service, and 3 weeks after seven years.

Unused vacation may not be accumulated from one anniversary year to the next, except by written approval of the general manager.

Call-in Pay—Hourly employees will receive at least 4 hours pay whenever called to report to work, unless the plant is shut down by an emergency (storm, power failure, etc.) which is beyond the control of management.

Death in Family—Any employee having a death in the immediate family will be given up to a 3 day paid leave to attend the funeral. The immediate family is considered to be wife or husband, child, parent, brother, sister, or as stipulated in the local collective bargaining agreement.

Some Training Examples

There was a time when the average warehouse manager hired employees by seeking people who already had the desired skills. If forklift operators were needed, an advertisement was placed for experienced lift-truck operators. The warehouse manager sometimes evaluated applicants by what they said they knew about operating lift trucks, not how much skill they actually might have. In more recent years, some of the best warehouse managers have come to realize that a well-motivated person who wants to learn will be a better warehouse employee than the applicant who brags that "I can drive anything!"

Training aids have now been developed by warehouse operators, lift-truck manufacturers, and trade associations. They are designed to help inexperienced people become skilled warehousemen.

Additional job training following first-day orientation should be conducted by an experienced employee, a group leader, a supervisor, a member of higher management, or a professional trainer.

Because training is a neverending process, it should not be restricted to newly hired people. New procedures, different equipment, and new customers create conditions where retraining is needed. Most people appreciate an opportunity to be involved in a continuing training and education program, since it gives them a chance to learn to do the job better. In contrast, serious job dissatisfaction can arise from lack of training opportunities.

Training Lift-Truck Operators

OSHA regulations require that every lift-truck operator attend a training course. These courses are therefore offered by every company that sells such equipment.

One course is held in a dealer training center, but it is also offered at customers' facilities, using the customer's materials handling equipment. The training sessions are run as a profit center, and the trainer/dealer recognizes a primary mission to teach safe and productive operation rather than to "push" its brand of equipment.

The training center is equipped with a classroom, a training truck, and an obstacle course. Trainees spend the morning in classroom sessions. After a lunch break, classroom learning is tested with a tough written test followed by a practical test in inspecting a lift truck and then operating it through an obstacle course. Successful graduates are awarded a certificate and an operator's license. Just twelve people participated in one course, and given the difficulty of the operator's test, a larger number could not have been properly trained.

The trainer showed that she is a skilled lift-truck operator, and she understands lift-truck operation and safety. Her competence and dedication quickly earned the respect of every trainee. She spiced her presentation with real anecdotes of actual lift-truck accidents. In fact, one of these accidents inspired her to ask the management to allow her to work as a trainer.

The course opens with a description of the benefits of training: improved attitude, higher productivity, improved safety, better maintenance, lower insurance rates for the company, and most importantly, the fact that training could save your life.

Daydreaming is cited as the primary cause of lift-truck accidents. A second is failure of the operator to look up and see potential collisions high in the air.

The point is made that people on lift trucks are "operators, not drivers." Driving has only a partial role in the successful operation of a truck, and operating an industrial truck is nothing at all like driving a car. Both lectures and videos point out the substantial differences between the handling characteristics of an industrial truck and an automobile. The fact that the truck has front wheel drive and rear wheel steering makes it vastly different. A lift truck can turn in its own length. This maneuverability has advantages and disadvantages, the prime disadvantage being the ability to upset the truck.

Trainees learn causes and effects of lateral tipovers, as well as accidents caused by improper use of industrial equipment. This includes failure to park the truck properly when unattended (forks down, controls neutral, parking brake on and power off). Operators learn how to read and understand the rated capacity of the vehicle and to operate the vehicle within its stated limitations. Accidents have happened when operators substitute a disabled truck with another vehicle that was not specified to do the same job, such as a conventional truck used in a high hazards storage area.

Trainees are taught how to inspect the industrial truck each day at the beginning of the shift, making sure that the vehicle is not damaged or suffering from excessive wear. A most important safety check is to listen for strange noises. If the truck is making noises that it never made before, something is wrong. Using the horn is emphasized and required whenever the vehicle moves into reverse. While seatbelts are not an OSHA requirement, the instructor recommends both seatbelts and a safety seat with side protectors. An excellent video demonstrates that the best way to survive a lateral upset is to ride the truck down rather than to attempt to jump away from the falling vehicle.

Speed is discouraged. The top speed of the average industrial truck is 8.7 miles per hour, but the instructor points out that this is nearly 13 feet per second, a fast pace in a busy warehouse.

Dock accidents are emphasized, and a frequent cause is a confused truck driver who pulls his rig away from the dock before loading is completed, sometimes with a lift operator still aboard his trailer. Chocked wheels and dock board locks are recommended to prevent this kind of accident. Bad trailer floors are another frequent cause of lift-truck accidents, and the operator is instructed that it is sensible to refuse to load a trailer that is in poor condition.

Chain slack in the mast can be a cause of accidents, since it indicates that the lifting cylinder has jammed. A jammed cylinder could snap and cause the lift mechanism to come crashing down.

Supplemental training should be provided for operators of anything other than "conventional" lift trucks. Figure 24-2 shows material developed for trucks with push/pull attachments.

Why Training Courses Fail

Before you consider a training course, take a realistic look at your goals. If it is your goal that the trainee will return from the course with at least one new idea that your company can use, what will you do to implement that goal when the trainee returns? If the culture in your corporation discourages the use of skills or innovation, don't expect the trainee to start a revolution.

Sometimes the teaching method is inappropriate. Not everybody learns in the same way. The best courses provide a balance between instruction, practice, and feedback. Attendees may enjoy being entertained by a jovial and glib lecturer, but will they learn anything?

Some training tasks and situations are not like the "real world" situation. Some warehouse training courses are taught by people who have never managed a warehouse or may not have been in one for a long time. A distressing number of courses downplay the instruction staff in order to emphasize course content or a romantic course location. This tends to reinforce the seminar as a "perk" rather than a real training situation.

When the course is over, the trainees return to their companies. Because nobody else from the company was at the course, the trainee may find that the really tough time begins when he or she tries to implement some of the new ideas received from the management seminar. If people are not reinforced and rewarded for the skills and new ideas they have acquired, the course is unlikely to be worth much to your organization.

Mentoring

Somewhere along our career path, most of us have been influenced by the mentoring activities of one or more senior individuals. If that was a positive circumstance, it is most often remembered as a good learning experience that helped shape our careers.

Finding the perfect fit for that mentoring job in your warehouse may be a difficult task. First of all, you should determine whether your company endorses mentoring, either formally or informally. Consider a mentor with superior general management skills and not just the functional or technical skills in which you are interested. Sometimes it is more difficult to find generalists than specialists. It helps if your mentor has earned the respect of peers both within and outside of your organization.

Load Push/Pull Operating Instructions —Pickup from the Floor

1 Line up the platens squarely with the load.

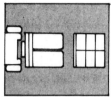

2 Raise the platens about 3″ above the floor.

3 Tilt the mast forward until the tips of the platens touch the floor. *DO NOT* try to pick up a load with the platens flat on the floor. Doing so may rip the slip sheet and/or gouge the bottom layers of cartons.

4 Drive the truck forward until the platen tips are under the slip sheet lip.

5 Extend the pusher plate so that the slip sheet lip fits into the gripper channel opening.

6 Retract the pusher plate. The gripper bar will automatically clamp the slip sheet lip.

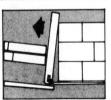

7 Move forward slowly as the load is being pulled onto the platens.

8 As the weight of the load is transferred to the platens, they will deflect downward slightly. Raise the carriage about 1″ to prevent the platen from digging into the floor. Slowly tilt the mast to a vertical position as you scoop up the load.

9 Tilt the mast back, and raise the load 3″—4″ above the floor. You are now ready to transport the load.

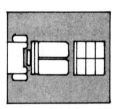

Load Push/Pull Pickup from a Stack

1 Line up the platens squarely with the load.

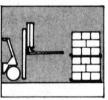

2 Raise the platens to just above the slip sheet of the load you are picking up.

3 Tilt the mast forward 3° to 4°.

Figure 24-2

4 Drive the truck forward until the platen tips are under the slip sheet lip.

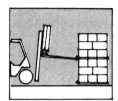

5 Extend the pusher plate so the lip of the slip sheet fits into the gripper channel opening.

6 Apply the truck brakes and retract the pusher plate. The gripper bar will automatically clamp the slip sheet lip.

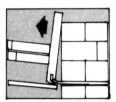

7 Pull the load onto the platens. As the load reaches the platens they will deflect downward. Raise the carriage slightly to compensate for this. DO NOT allow the truck to be pulled forward since the platen tips may gouge the load underneath.

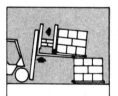

NOTE: Some operations use two slip sheets between the loads. In this case, insert the platens into the opening between the slip sheets and drive the truck forward as the load is being pulled onto the platens.

8 Tilt the mast back and lower the load to 3"—4" above the floor. You are now ready to transport the load.

Load Push/Pull Discharge

1 Carefully position the load exactly above where it is to be discharged.

2 Tilt the mast forward 3° to 4°.

3 Lower the carriage until the platen tips are about 1" above the discharging area.

4 Place the truck's transmission in neutral and start the pushing operation. The gripper bar will automatically release the slip sheet lip. As the pusher plate moves the load off, the truck will move backwards, (with light loads drive backwards slowly). As the weight is withdrawn from the platens they will deflect upward slightly. Compensate for this by lowering the carriage slightly.

NOTE: If your truck does not operate in neutral, place the transmission in reverse and drive backwards slowly to coordinate the pushing motion with the reverse movement.

5 Back away, tilt the mast back and raise or lower the platens to 3"—4" above the floor for traveling.

Don't be embarrassed to ask a potential mentor to help. Be sensitive to the need to accommodate the mentor's time constraints. And remember that you don't learn when you talk, only when you listen. The art of good listening is becoming less and less prevalent in our society.

Mentoring is more than just training. It is the willingness to share experiences and knowledge. There can be a payoff to your company when improved individual performance leads to better services or reduced costs. You're helping individuals to better achieve their aspirations or to feel better about themselves and to be more productive in their environment.

Some individuals crave group acceptance, and mentoring can assist them in achieving that objective. The dignity of an individual may be strongly linked to acceptance by a group. Other individuals need help in recognizing and in coping with the situation in which they find themselves.

Another benefit of mentoring is the reinforcement of your company's value system. If your organization believes in principles such as integrity, customer commitment, ethics, quality, superior performance, and the dignity of the individual, a mentoring effort can be the way in which you reinforce positive results.

Mentoring can also foster leadership. Any emphasis on improving relationships between members of peer groups must be orchestrated by a leader. Sometimes a task force from different divisions or departments is assembled to solve a problem. The leader can help team members work together by recognizing the contribution and capabilities of individual members.

In this context the role of the mentor is almost the same as that of the leader. The ability of any individual to contribute to the group can be strongly influenced by mentors, and it is not unusual to find the mentor operating as a leader and even merging the two roles into one.[12]

Mentoring in the Warehouse

In the warehouse, the mentoring process can involve both hourly workers and managers.

In the case of hourly workers, mentoring usually involves pairing a new employee with someone who serves as tutor. This process takes place in warehouses whether or not management orders it. If management does not assign the mentor, there will probably be an informal leader of the workforce who assumes the task. The problem is that that informal mentor may or may not be the kind of person you would want to provide training and orientation. The mentor is a person familiar with the warehouse routine, but he should also be a people-oriented individual.

One warehouse uses a formal mentoring program with a twenty-week schedule. The mentor works with employees on a scheduled basis, though there are informal visits throughout the program. The mentor provides advice on warehousing procedures as well as company policy. He will also provide personal advice, but only when asked.

Sometimes a warehouse may seek an outside mentor. One manufacturing company had appointed a production foreman to manage a warehouse. While the foreman was skilled in this area, he had no knowledge of warehousing. As a result, there were significant problems with shipping and inventory control during the first rush season.

The mentor, provided by a management consulting firm, evaluated the warehouse supervisor to determine whether that individual had the raw material to do the job. As a second step, he surveyed the warehouse building and equipment. That survey included the people, flow of product, and paperwork. The mentor recommended ways to meet the planned increase in shipping. The mentor gradually reduced frequency of visits to the warehouse and some advice was provided by telephone rather than face-to-face meetings. Eventually, all of the work formerly done by the mentor was now done by the warehouse supervisor. That supervisor has been successful ever since.

There are several sources of mentors in a warehouse environment. Often the best source is another person working in the plant. If an outsider is needed, a recent retiree could provide temporary services as a mentor. If your company has more than one warehouse, the mentor might be a person in another warehouse. When mentoring takes place across locations, both parties may benefit. The mentor may learn about some better warehousing practices at the new location. Successful application of mentoring can prevent warehousing service failures which could be very costly.

Goals of Training

As you create or revise your training program, a final check of it should be a review to be certain that everything in it is related to the goals of the training program. While these goals will differ from one company to another, most warehousing organizations would include at least these six training goals:

- To improve safety, sanitation and security
- To help every employee make decisions that improve productivity and quality
- To reduce damage to warehouse equipment and stored merchandise
- To give trainees a feeling of accomplishment
- To motivate trainees to do a better job
- To ensure that all new employees understand company rules and procedures

Training builds teamwork. It can and should be a morale builder. At times when qualified workers are scarce, an effective training program can be a vital part of the growth of your organization.

25

LABOR RELATIONS

The risk of strikes or other labor disputes is a key consideration in warehousing. In some industries, work stoppages or other disputes among production workers are a likely occurrence, and therefore it's important to have a strategy for continuous distribution of finished products, even when one or more production plants is on strike.

This strategy may be carried out in many ways. Some companies separate distribution facilities from manufacturing locations specifically to cushion the effect of work stoppages. In some cases, the distribution center is located in a rural area with a favorable labor climate so that its non-union status is easier to maintain.

Although such distribution centers can legally be picketed by manufacturing workers engaged in a lawful strike, factors of cost and distance may make such picketing impractical, allowing orderly distribution to take place despite the strike. And if the distribution centers are not unionized, there is little danger of a strike by their employees, either out of sympathy for striking manufacturing employees or in support of economic demands of their own.

Creating a Union-Free Environment in the Warehouse

One retailer provides a dignified and relatively luxurious setting for warehouse workers. In that company, every worker, from president to trainee, enters through the same front door and lobby. Further-

The author is indebted to Theodore J. Tierney, a partner in the law firm of Vedder, Price, Kaufman & Kammholz, Chicago, for extensive research and advice in the preparation of this chapter.

more, everyone on the payroll uses the same cafeteria for lunch. There is no "back door," and every hourly worker eats in the same room with the chief executive. This egalitarian environment is even more pronounced in the plant of a non-union manufacturer where everyone, from the president to factory worker, wears a white shop coat when on the production floor. The white uniform hides the pinstripes of the corporate executive or the blue collar of the factory worker. The result is a removal of the "we–they" symbols of business dress.

In contrast to such egalitarianism, unionism usually flourishes when a company has insecure or highly authoritarian line supervisors.

For the hourly worker, the line supervisor is the symbol of management. Yet, the typical line supervisor was promoted not because of any proven ability in managing people, but because he was a good "doer," or an above-average "producer." There is nothing wrong with rewarding productivity, but rewarding the productive worker with supervisory responsibility is not always appropriate.

Unfortunately, most managers are blinded by certain preconceived ideas about the hourly worker. These include the impression that the hourly worker is interested only in money, and the supposition that unionism is a grass-roots movement which, once under way, cannot be stopped.

Surveys of hourly workers suggest that money is relatively low in their priorities. They are more interested in prestige, security, respect, and a feeling of belonging. These desires can be thwarted by insensitive or poorly trained line supervisors.

The professional union movement is in fact anything but a democracy—it is more like a closely held franchise business with involuntary franchises. Any unionized worker has the right to quit the company, but he or she probably will find it difficult to quit the union without also quitting the job.

A labor problem is often at heart a management problem. The franchise operators who solicit union membership are seldom able to market their product in a happy workplace. Workers who are fairly treated will respect both line supervisors and senior management. They are not susceptible to the appeal of the union organizer. While low pay is occasionally the cause, the motivation to unionize is usually not financial. Typically, the worker who supports a union drive is reacting to either the reality or the perception that he or she has been treated unfairly in the workplace.

An Example

One medium-sized retail organization has a chain of small shops operating all over the country. The organization is also vertically integrated and owns plants that manufacture some of the products it sells. Many of its production facilities are unionized, as are some of its retail stores. In spite of this, the company has carefully preserved the independent status of its distribution operation. Some of its products are distributed with private trucking, but a stevedoring company is retained to hire and control the truck drivers. A group of private distribution centers is maintained on a non-union status.

Some time ago, the company endured a long strike to keep its largest distribution center non-union. Later, a regional distribution center suffered the loss of a union election by just a few votes. The company signed a one-year labor agreement, and its employees subsequently petitioned for a decertification election. At the expiration of the contract, the union was rejected by the great majority of the distribution center workers.

Management of this company feels that logistics is the one branch of their business that can and should be kept non-union. Yet they accept the presence of unions at the manufacturing level, and even at the retail level.

Public Warehousing Alternative

Some manufacturers have used public warehouses rather than their own distribution centers to offer insulation against the impact of strikes in the distribution area. In order to use a public warehouse successfully as part of a strike-cushioning plan, the manufacturer must observe three ground rules:

Rule 1. The manufacturer should enter the warehousing relationship well before the strike begins. Any agreement entered into shortly before the strike and terminated just shortly thereafter would be viewed as legally suspect.

Rule 2. A manufacturer should avoid using more than a limited percentage of a public warehouse's total capacity.

Rule 3. The warehouse's management must have full control over the placement of goods in the facility. To avoid lawful picketing, the customer should not rent any particular physical space at the site. The NLRB has held that, under the secondary boycott provisions of the National Labor Relations Act, a public warehouse cannot be picketed by a union having a dispute with one of its customers unless that customer's employees, trucks, or equipment are on the warehouse premises. In another ruling, the court pointed out that primary picketing at site of a neutral employer may be lawful where primary company's employees are on the premises.

Public Warehouse Strikes

What happens if the public warehouse goes on strike? In most states the user can remove goods through the picket line by filing an action for replevin—a legal move to recover goods unlawfully withheld from their rightful owner. Under this action, the court issues an order to the marshal or sheriff compelling removal of the goods from the warehouse and their placement in the owner's hands.

Using Subcontractor Warehouse Services

A warehousing subcontractor arrangement involves the use of an independent warehouse operator's personnel within the manufacturer's facilities. In using the subcontracting option, the user must beware of creating a "joint-employer" relationship with the subcontractor's employees. To avoid this possibility, the warehousing contractor must be in full and sole charge of determining how many employees are needed, who they will be, and the method used to perform the work. A subcontractor must also be in sole control of hiring, training, disciplining, and discharging workers, as well as all other aspects of the employment relationship.

Transportation Strikes

The transportation industry has a long history of union activity and many warehouses are designed to deal with carrier strikes. For example, a distribution center with a railroad spur can be used in the event of trucking strikes. From a practical standpoint, strike activity at a warehouse seldom impedes a rail movement of goods in and out of the facility. Railroad companies will usually provide switching service with supervisory personnel. It is often physically impossible for pickets to block industrial railroad tracks because they are all on private property. Therefore, rail movement is frequently an effective hedge in the event a strike disrupts trucking operations.

Inventory Hedging

Once a manufacturer sets up a distribution system that provides an effective buffer in the event of work stoppages, the creation of a "strike hedge" inventory becomes possible. This can be a powerful bargaining tool in contract negotiations. Unions can be warned that the company has vast stockpiles strategically located around the country, and that it thus is able to absorb a strike for a significant period without losing its position in the marketplace.

Turning from Union to Union-Free

Turning a unionized warehouse into a non-union warehouse is a difficult and delicate project. The most peaceful way to do it is by a decertification election. One such election took place because the chief executive of the warehouse company suggested to several workers that the union really had not done them any good. The group had been organized for about twenty years, and union activity had been virtually unknown during the last ten years. The workers agreed with the executive and a peaceful decertification election was held.

The details of decertification are closely regulated by law, and the most important thing to remember about the process is the need for close consultation with a lawyer who specializes in this field.

Other ways to remove a union are more difficult, such as closing a plant, or subcontracting the warehouse work and closing it. These matters may also be limited by law, though recent court decisions have greatly enlarged management's right to cease unprofitable operations.

Managers should remember that only the company creates warehouse jobs—the union does not create any. In the heat of organizing campaigns, some workers may get the idea that the union provides security—but in actual fact the opposite is often true. Ultimately, job security is based on the customer's satisfaction with the warehouse services, maintained only by a well-coordinated labor–management team. No union ensures that—only good worker–management teamwork does.

Preemployment Screening

Just as the old saying "Do good and avoid evil" serves as a guide we can all live by, so does "Hire good employees and avoid bad ones" give the most effective method by which an employer can avoid legal problems and maximize productivity. While living by either tenet can be rather difficult, the employer has many acceptable approaches in selecting good employees.

Figure 25-1 lists typical prehiring inquiries, with typical instances of when such an inquiry may be lawful. In addition, such inquiries must be job-related and administered in a uniform manner. Use this information only as a guide and always check with local counsel regarding the law in your own state.

Evaluating the applicant's previous employment record can be a useful tool. While previous employers are understandably skittish about giving out too much information about former employees, any information they provide in this area can be helpful. For example, an employee who had attendance problems on one job is quite likely to have them on another. Employers may be more inclined to honor

Acceptable and Non-acceptable Preemployment Questions—What Can and Cannot be Asked In Most Situations

? Education—Acceptable.

? Work Experience—Acceptable.

? Arrest Record—Acceptable in most states: Have you been convicted of a crime within the last two years? Not acceptable: Have you ever been arrested?

? Language—Acceptable: Inquiry into language applicant speaks and writes fluently, if job-related. Not acceptable: What is your native language?

? Photograph—Not acceptable.

? Financial status—Generally not acceptable: *But*: Have your wages ever been garnished? Do you own a car? Do you have a telephone? Have you ever filed bankruptcy? Have you ever been refused credit? (Some of these questions could be asked of applicants for sensitive financial positions.)

? Reliable means of transportation to work—Acceptable.

? Applicant can supply a phone number—Acceptable.

? Marital status—Not acceptable.

? Religious holidays observed by applicant—Not acceptable.

? Days on which applicant is unavailable for work—Acceptable.

? Relatives—Acceptable: Names of applicant's relative, other than spouse, already employed by you. Not acceptable: Names, addresses, ages, number or other information concerning applicant's spouse, children, or other relatives not employed by you.

? Applicant has any religious objection to paying union dues—Not acceptable.

? Age—Generally acceptable, but rejection after inquiry could create age discrimination problems; application should set forth specific disclaimer regarding this inquiry.

? Applicant has ever filed a Worker's Compensation claim—Not acceptable.

? Applicant has a valid driver's license—Not acceptable (unless the position involved includes driving duties.

? Military experience—Acceptable: Applicant's experience in the U.S. armed forces or in a state militia. Not acceptable: What type of discharge did you receive?

Citizenship—Acceptable: Are you a citizen of the U.S.? If not, do you have the legal right to remain permanently in the U.S.? Not acceptable: Of what country are you a citizen?

? Reasons for leaving former employment—Acceptable.

Are you able to perform any or all job functions with or without reasonable accommodation?

Figure 25-1

requests for objective information, such as a former employee's attendance record, than giving subjective appraisals or conclusions. At the very minimum, verification of job duties and length of service should be completed.

Ask the job applicant for an authorization for release of data by former employers. Obviously, such a step must be taken without regard to the race, sex, religion, or national origin of the applicant. Moreover, be aware that when you do check references, you may have an obligation to furnish the applicant with the results of your questioning.

Employee Complaints

Closely related to good communications is the proper handling of employee complaints. Several suggestions concerning this important management function follow:

1. Appreciate that the employee registering the complaint is paying management a compliment.

2. Listen carefully to the complaint and get all of the details.

3. Even if the complaint sounds frivolous, appreciate that it may be a real problem to the employee.

4. Avoid sarcastic responses and do not attempt to poke holes in the complaint.

5. Investigate the complaint carefully. If it is valid, follow through.

6. Where the complaint calls for a decision, do not hesitate to make it.

7. Realize that complaints may be symptomatic of deeper concerns, so be sure to explore them fully.

One of the keys to good supervision is follow-up. After a full and careful investigation, an employee's complaint must either be dealt with or the employee given a full explanation as to why action was not taken. Simply telling an employee, "You're wrong," or "You don't know what you're talking about" will only create further problems.

The cardinal sin in this respect is for a supervisor to fail to make any response. It is equally wrong for the foreman to "pass the buck" or to say simply that "The front office has turned you down." If the employee was concerned enough to bring the complaint to the foreman, he or she is entitled at least to a complete and meaningful response.

The Role of the Supervisor

In many respects an employee's supervisor is the company to that employee. A supervisor often has more impact on morale and perceptions of the company than any other single factor. So the selection and continued training of supervisors is vital.

Selecting supervisors should involve more analysis than merely promoting the most efficient workers. While a worker's productivity certainly should be considered, it does not guarantee that he or she has the qualities necessary to be an effective supervisor of other workers.

The credibility of your supervisors is critical. It is essential that you do all you can to enhance and maintain that credibility. Supervisors should be kept well informed and, in turn, should be the origin of most information related to the workforce. Supervisors should also, whenever possible, be given credit for working condition improvements.

Arbitration Standards Pertaining to Discharge

No matter how carefully management screens its applicants, and no matter how effective its employee relations program is, there are some occasions where it simply must remove some workers from its workforce. In those situations, nothing can be much more harmful to employee discipline or supervisory morale than to have the discharged person reinstated by an arbitrator or court. Here are some suggestions as to what management can do to minimize the risk that its discharges will be challenged . . . or at least to increase the likelihood of a successful defense!

1. Establish written rules and communicate them to your workforce. If you decide to crack down on the enforcement of a rule, notify your employees in advance, rather than suddenly invoking stiffer or more frequent penalties.

2. Specify those rules violations that will result in immediate discharge.

3. Enforce the rules fairly and consistently.

4. Use progressive discipline, including lost-time suspension prior to discharge, except for the most serious offenses.

5. Conduct a thorough investigation that includes, wherever possible, written statements from all witnesses, including the disciplined employees, prior to discharge.

6. Carefully communicate the full reason for the discharge to the employee. If his past employment record was a factor, say so.

7. Maintain careful, complete records documenting all disciplinary action.

Two of the most aggravating offenses are chronic absenteeism and chronic tardiness. Because arbitrators and government agencies often fail to appreciate the seriousness of chronic absenteeism and tardiness, "building record" is essential. Offenses must be documented, and penalties must be progressive—including at least one suspension prior to discharge. And management must be consistent.

One of the most common mistakes management makes in this area is to be too lenient by merely making oral reprimands— possibly because manpower needs often make that the easiest route, until management's patience is exhausted and the employee is suddenly discharged. Arbitrators often overrule these discharges on the grounds that the employer has lulled the employee into a false sense of security. Furthermore, because such sudden discipline often is not uniform, government agencies often find it to be discriminatory.

From Vol. 15 No. 4, Warehousing and Physical Distribution Productivity Report. © Alexander Research and Communications, New York.

The Essence of Union-Free Management

Union-free management is really a state of mind. It comes out of a shared confidence that teamwork builds a company, and that everyone in the organization is on the same team. It is nurtured by elimination of traditional "perks" which created corporate royalty and corporate commoners. This includes eliminating reserved parking places as well as private restrooms and lunchrooms. In the union-free company, it is common to find everyone on the payroll referred to as an "associate" rather than to have managers and subordinates, officers and employees. In some warehouses, supervisors are called "team leaders" and hourly workers are team members.

In effect, union-free management is not a power play, not a strategy for management to defeat a union. Instead, it is the maintenance of an environment in which nobody in the workplace needs a union to feel secure.

Many warehouse users have gone to great lengths to design a distribution system that is practically immune from labor problems. But there is an element of risk in anything connected with labor relations. The practice of good employee relations within your warehouse is one of the best ways to ensure against labor problems.

26

MOTIVATION

An incentive is defined by Webster as "something that stimulates one to take action, work harder, and is encouraging and motivating." The use of incentives in American industry goes back to the early years of the century, when Frederick Taylor developed a 1911 book called *Scientific Management*. In that book, he described a system in which workers were paid by piecework, receiving a given fee for each wheelbarrow of coal delivered. The steel and textile industries were early users of Taylor's management theory. In many of these companies, the objective of the incentive system was to stabilize costs. Management could precisely predict the cost of each delivered wheelbarrow because it was predetermined in the piecework pay method.

Application of incentives became a fad in the 1950s, but by the 1960s a high percentage of the incentive plans that had been installed in the earlier decade had failed. We need to understand the reasons for both success and failure.

Barriers to Productivity

For warehousing purposes, any standards program typically develops a 100% pace, that which can be maintained by a skilled person working a full day under normal working conditions. An incentive then, is something that motivates a worker to produce more than 100%.

When incorrect methods are used, a worker may be unable to earn the incentive because the deck is stacked against him. People on the floor are not stupid. When they can find short cuts they will.

When other conditions prevent them from getting the job done, they then have a barrier to productivity.

The ideal program is simple and documented in language that is understood by everyone. Every person involved in the system should know how to calculate it. A regular audit will eliminate the possibility of creeping methods changes.

In warehouses, the two greatest barriers to productivity are lack of training in proper methods and poor stock locations.

Who Gets the Incentive?

A good incentive program has to cover everyone. Some warehouse operators have offered incentives only to order pickers, primarily because the selection process is one that is both labor intensive and easily tracked. When this happens, the people who work in replenishment or receiving have every right to be unhappy. They have no motivation to work harder because they are not covered by the incentive program. If your incentive system fails to include all levels of the workforce, productivity will rise in certain areas but may drop in others. Sometimes a group incentive works better for loaders than for checkers because their jobs are interactive and depend upon teamwork. Therefore, consider a program in which there is a group incentive or possibly a combination of individual and group bonuses.

Union Attitudes

Some managers fear that an incentive system will cause problems with a union. In fact, there is ample evidence to show the opposite. Union leaders usually find fewer grievances in a shop that has incentives, because the dissidents know that they can earn more pay by working harder. A grievance free workplace means less work for the labor organizers. The union leader naturally hopes that the standard will be loose enough so that everyone can earn a bonus. The unions know that they will probably have fewer complaints from members when an incentive plan is installed.

If you are installing an incentive plan in a unionized warehouse, don't forget to involve your labor attorney. Union leadership typically has far more experience with incentive systems than most management people, and the information is abundantly shared between union locals.

Never let the incentive be part of the collective bargaining process. In a union shop, the base pay is subject to negotiation, but any incentive bonuses can and should be excluded from the contract.

Once you reach the final step of writing a procedure, it is important to keep the documentation simple. Remember that this documentation need not, and should not, be included in any labor contract. When the system has been written, present it first to supervisors, then to the union, if there is one. Finally it should be presented to the workers in small groups.

Installing and Designing the System

There are four major steps to the installation of an incentive system in your warehouse. The first step is to define your objectives, the second is to define the system design, the third is to check your own readiness for the system, and the last is to write and implement the procedure.

As you look at objectives, consider whether your company wants this system primarily to increase productivity, to offer an incentive opportunity instead of a pay raise, to motivate people, or to reduce turnover. While you might say you want all of the above, consider which of these four points is of greatest importance to you. As you do this, be sure that you are committed to the "long haul" in living with an incentive system.

There are options in system design. First, decide whether you want an individual or group incentive, or a combination of both. Will everyone in your workforce be eligible for the incentive bonus? Do you want that bonus paid in dollars, extra time off, or a combination of both? Do you need an earnings cap to control the maximum amount that will be paid in the incentive? Decide how often you want that bonus paid, since payouts will range from weekly to quarterly. Be sure you know how to deal with quality, safety, damage, and housekeeping issues. Such issues must be controlled so that increased productivity is not made at the expense of warehousing quality.

Part of the process is to create complete reporting and tracking procedures for all warehouse activities. The line supervisor and the warehouse manager must have a clear idea of exactly what each employee does with his or her time.

As the tracking system is designed, it should include ways to receive worker feedback. Employees should be encouraged to comment on their work pace and environment and to point out any roadblocks to further improvement. The existence of a good reporting and tracking system together with well-designed feedback will typically improve productivity by at least 10%, even if nothing else is done with methods.

As you check your readiness for a new system, consider what your productivity level is today. Then decide whether or not to use engineered standards. The unit of measure could be pieces per hour or sales dollars per labor hour. Deciding which unit of measure to use may depend on the quantity and quality of your historical data. Is your current labor reporting system accurate enough to be used in calculating incentive pay? Do you already have the quality checks in place to ensure that a reach for incentive does not get in the way of maintaining accuracy and quality in storage, shipping, and receiving? Finally, do you now have an auditing procedure that can be used to check the incentive program?

The crucial step in preparing for an incentive program is to establish an accurate work measurement system. If you can't measure work, you can't manage it. In a warehouse with a wide variety of products, a fair measurement system is not easily devised. For example, in a warehouse handling major appliances and appliance parts, the work content for finished goods will be substantially different from the job of handling and storing small parts.

Based on the results of the work measurement program, you should then establish productivity standards. Those standards must be obviously reasonable, easily understood, and attainable by those who produce at 100% effort.

Pitfalls

Any incentive system will fail if it does not have continuing support from senior management. Management must have a sincere commitment not just to install the program, but also to see that it works in the future.

Since warehouse work varies constantly, a poorly documented system or one that is not regularly audited will soon become unfair either to the company or to its workers. When this happens, either the workers or the managers will lose confidence in the system.

If the standard is too loose, bonuses will be attained that are not truly earned. If it is too tight, hardworking warehousing people will feel that they have been robbed because of an inability to earn a bonus in spite of a strong effort to do so.

Perhaps the biggest pitfall in developing standards for warehousing is the changing nature of warehouse work. When you design an incentive system, be sure that you have also designed the means of altering it to meet anticipated new developments in work content.

The Role of the Supervisor

Some feel that an incentive program is a cure for poor supervision. In fact, the opposite is true. Excellent supervision is needed to make an incentive program successful, since workers will need to be coached on using the best methods.

Good supervision is also necessary for quality control. When people try to handle merchandise faster to earn more money, they may also create more errors. Both supervision and inspection are needed to identify and control the error rate.

Designing and implementing a warehouse incentive is a tough job, and it is not easily described. Making it work for the long haul requires constant dedication and effort by supervisors as well as senior management.

Incentives Checklist

Before starting an incentive program make sure you:

_____ Have the support of top management

_____ Have the correct methods

_____ Have removed barriers to productivity

_____ Have a good reporting system

_____ Have designed a program that will cover everyone

_____ Have written the program so everyone can understand it

_____ Have designed a procedure to audit the system on a regular basis

_____ Have had the program checked by your labor attorney

SOURCE: Eugene Gagnon, Vol. 7, No. 1, Warehousing Forum, © Ackerman Company, Columbus, Ohio.

Figure 26-1

Shown above is a checklist that you should use when you consider an incentive program for your warehouse.[13]

The Importance of Listening

Successful warehouse managers communicate clearly. The clear communicator has a genuine understanding of what he or she is trying to convey, uses appropriate words to phrase the ideas, chooses an effective way to put them across—and listens effectively. Many managers enjoy talking, telling, informing, teaching, and judging. Listening—active attending to what another is saying—is frequently an undeveloped skill, particularly in those who talk a lot. Successful managers who learn to listen effectively often achieve new insights. Sometimes "generous listening" is used as a way to reward others.

In a warehouse, the continual movement of material in an efficient, profitable manner requires close cooperation and teamwork. Teamwork is based on good listening skills. Choose a place that is relatively quiet and comfortable to give people time to share information. The effective listener minimizes the probability of interruptions from telephone calls and secretaries. You should try to listen in physically comfortable surroundings.

One new warehouse manager was promoted to a new facility from another location in the company for which he worked. He was also promoted to general manager of the operation after serving as administrative assistant to the general manager in his old location. He was nervous and highly self-conscious of his age and minimal experience. The people who worked for him—foremen, customer service representatives, and hourly workers—knew nothing about him or his reputation previous to his arrival. When he made his first appearance, he called everybody together and lectured for forty-five minutes to prove his knowledge and competence. His audience reacted, however, in a very noncommittal, reserved way. They had not been given any opportunity to raise questions, voice concerns, or make their observations about the way the new location was operating. The new general manager's intent was to prove his competence by talking about his experience. In fact, he alienated the people he was trying so hard to please.

Had he taken the opportunity to spend time with individuals, listened to their experiences, heard their observation of how well the warehouse was operating, and then commented on the things he saw they were doing right, he would have been in a much better position to elicit their confidence, loyalty and motivation.

Effective listening is only one part of the complex task of communicating. A manager needs to share information, teach, train, coach, talk at, and inform people—and practice the art of listening. The art of listening is a fundamental management tool but it requires practice. Becoming a better listener will make you become a better warehouse manager.

Feedback

The results of performance must be relayed back to the person in the job, if motivation is to be maintained. In one warehouse, supervisors designed and implemented a major performance improvement project. Each warehouseman was provided with feedback on his or her productivity in terms of "cost/cases handled" in relationship to a budgeted goal. The warehousemen could calculate their own productivity, as at the end of the shift the foremen provided them with their results for the day. Thereafter, "cost/case" feedback was summarized into shift performance. The "cost/case" productivity for all three shifts was continuously posted on a large scoreboard. The foremen insisted that the feedback be on a three-shift, 24-hour basis so that intershift cooperation would be encouraged (they did not want each shift's productivity posted on the scoreboard since they felt that this would cause destructive intershift competition).

Without feedback, personnel in work organizations tend to be self-correcting, which means that on- or off-course deviations from a goal cannot be detected. Feedback on results or outputs leads to performance improvements and behavior, especially if the feedback is positive. Best of all, feedback systems are relatively inexpensive in comparison with the performance improvement payoffs.

Too often, feedback on performance results goes to executives, managers, and sometimes supervisors. Rarely is a system set up to give warehouse employees feedback on their results and accomplishments. Yet, it is the warehouse workers who must change their behavior to improve results. Without feedback there is no way they can improve their own on-the-job performance in a systematic way.

Failing to give feedback to warehousemen is similar to asking them to bowl blindfolded. Even though we provide them with the best job situation (bowling alley, bowling balls, lighting, shoes, etc.), and know exactly what we want in terms of performance results, the most skilled bowlers will not do well. If we go a step further, and provide the bowlers with a supervisor to direct their activities, scores will be improved, but first-line supervision will be very difficult. If we take the blindfolds off, however, the bowlers get immediate feedback on results, can keep their own scores, and modify their behavior to improve those scores.

Feedback makes the employees more independent of the supervisor and sets up a situation where they can manage their behavior in a positive direction. This makes the job of warehouse supervision easier, since it is usually impossible to monitor the activities of workers in storage areas where they cannot be observed. Everyone is interested in improving warehouse productivity. In the final analysis, however, we are not improving warehouse performance, but rather improving the performance of people who work in warehousing. If this distinction is made, then motivation becomes the key element in a performance system.

27

IMPROVING PEOPLE PERFORMANCE

A business theory known as the Peter Principle says that managers in many organizations are promoted steadily to their level of incompetence. In the warehouse there are times when a lift-truck driver is promoted to foreman primarily because of job performance. But most of the skills in operating a lift-truck are not transferable to a supervisory situation.

A management development system will at least reduce the number of times when the Peter Principle applies. Management seminars represent one way in which individuals can be encouraged to grow in personal skills. Individual efforts to gain outside education should be encouraged. When a promotion error cannot be corrected with skill development, the honest but difficult solution is to promptly demote the person.

Maintaining Warehouse Discipline

Because housekeeping, precise performance, and dependability are the hallmarks of a good warehouse, quality is closely related to warehouse discipline. When we see a warehouse that has products scattered throughout aisles and working areas, we quickly form the opinion that there is a lack of discipline in the warehouse.

Just what do we mean by warehouse discipline? A well-disciplined warehouse has these features:

- Managers and workers cooperate in achieving housekeeping excellence.
- Errors and damage are minimized.

- When things are done wrong, workers receive feedback and encouragement to improve the quality.
- When new workers or outsiders drag down the operation, the rest of the warehouse crew pulls together to improve quality.

The best supervisors do not rely only on punishment to manage their workers. They know that positive reinforcement for *following* the rules is more effective than a negative response for *breaking* them. In a warehouse with good discipline, workers know what is expected of them. While rigid work standards may not be employed, there is a formal or informal expectation of what level of productivity is normal in that operation. Neither supervisors nor fellow workers in a well-disciplined warehouse will tolerate an individual who produces substandard work. Finally, the warehouse procedures used in the operation are clear, and everybody follows them. A warehouse procedure that is so complex that nobody can understand it is as bad as no procedure at all.

Warehouse discipline is not easy to define yet experienced operators know it when they see it. When it is lacking, the signs are equally obvious.

Running a warehouse is a people management business. A well-managed warehouse is also a well-disciplined warehouse, one in which people play by the rules because they know that to do otherwise will downgrade the warehouse operation and ultimately their own jobs. Enforcing warehouse discipline requires leadership as well as management. There is a difference between hardness and toughness. If an object made of granite is struck with a hammer, it will shatter into pieces because it is hard but also brittle. An object made of leather may not even be dented by the hammer because it is flexible. Leather is tough without being hard.

The analogy can well be applied to warehouse management. Housekeeping is the hallmark of good warehouse management, but how do you persuade people who have worked in a sloppy warehouse to change it to a neat and well kept facility? The hard manager might threaten suspension to any employee who violates good house-keeping practices, but dissident workers might see how much they could get away with without being caught. A tough-minded manager may enlist the help of all hourly workers and supervisors in devising ways to improve housekeeping. By enlisting the ideas and support of workers, the manager must be flexible. Not all of the ideas will be outstanding, but to achieve real teamwork, all or nearly all of them must be favorably considered.

In today's competitive business climate, your boss expects you to be a tougher manager than ever before. As you do this, remember the difference between toughness and hardness.

Peer Review—A New Approach to Discipline

Traditional methods of disciplining employees sometimes foster an "us against them" mentality that can create the perception that the company is not committed to resolving workplace disputes fairly.[14] Some employers deal with this by using a dispute resolution process known as "peer review."

Peer review systems replace the traditional grievance arbitration process by allowing an employee to appeal disciplinary actions to a committee comprised, at least in part, of co-workers. Normally, all disciplinary actions are subject to peer review. In addition, the employee may challenge either the discipline itself or its severity, although some companies have chosen to narrow the scope of their peer review system to cover only specified actions, such as involuntary terminations, overtime, or the proper application of a company's layoff procedure. Where peer review systems are limited in their coverage, companies usually maintain an "open door" for other review issues.

Generally, the peer review process commences with the lodging of an appeal by the aggrieved employee. An employee typically is afforded five to seven days within which to lodge the appeal.

The next step in the appeal process often is the employee's supervisor, who reviews the discipline; hopefully, the matter can be resolved at this stage. However, if the employee remains dissatisfied following the supervisor's review, he or she may take the appeal to the peer review board.

Although the composition of the peer review board varies from one company to the next, a typical board is comprised of five members: the aggrieved employee selects two members; the plant manager or personnel director also selects two members; and the fifth member is selected by the other four members.

Within a few days following its formation, the board is convened. The matter is presented to the board in an informal, nonlegal setting. Each party—the aggrieved employee and his or her supervisor—presents his or her side of the dispute to the board. The board may ask questions or call additional witnesses and, in addition, may review the employee's personnel records. Once the board has collected all the necessary information, the supervisor and the employee are asked to leave the room while the board deliberates.

In considering the employee's appeal, the board only decides whether there has been a violation of company policy. The board is not empowered to change company policy. The board may uphold the discipline, reduce or increase its severity, or overturn the discipline entirely.

Usually majority rules and the board's decision is final. Some companies, however, require a unanimous decision. If such a decision cannot be agreed upon, a majority and minority report is given to the plant manager or personnel director who then renders a final decision.

Upon arriving at a decision, the board immediately informs the supervisor and the employee. All aspects of the decision-making process remain confidential.

In companies that use some form of peer review, both employees and management comment upon it favorably. Employees are satisfied with peer review systems largely because such systems allow employees to participate in management decisions that have a direct effect on them. Employers have found that board members take their responsibility seriously and do not seek to abuse or undermine the process. In addition, employers recognize that peer review reduces employee discontent. Generally speaking, where discipline is imposed properly in the first place, employees hesitate to invoke their appeal rights under the peer review system. In fact, a complaining employee's peers on the board often are harder on him or her for certain rules violations, such as those related to absenteeism, than is the company.

Summing up, peer review programs:

1. Increase employee morale by providing employees with a voice in disciplinary matters directly affecting them, thus building a bridge between management and employees.

2. Thwart attempts at unionization by removing a common source of employee discontent (i.e., the absence of employee participation in discipline).

3. Help to avoid expensive litigation and arbitration.

4. Help to support the company's position in any subsequent, related litigation, such as EEOC charges.

5. Help to maintain company control over employee relations by keeping workplace disputes in-house.

The Interviewing Process

Can you really afford to leave your hiring decisions to chance? In altogether too many warehousing organizations, the preferred candidates for hiring are people who have worked in other similar warehouses. These may or may not be the best people you could find to work in your warehouse.[15]

One of the biggest mistakes in the hiring process is the lack of a systematic approach. Picking up the resume or application two minutes before the candidate enters your office is not a systematic approach. A system is defined as an established way of doing something, using a set of rules arranged in an orderly form, to show a logical plan of linking the various parts.

A system is important because it gives you a structure that you can follow to get the best possible information. Also, your skills as an interviewer will increase by practicing the same techniques over again. It is very difficult to think of a basketball player getting very good at shooting free throws when he uses a different style each time and only practices once or twice a month.

The same can be said for interviewing. The process you use

does not have to be difficult. It just needs to be clearly defined, repeatable, and effective. Figure 27-1 shows a simple yet very effective process.

Always begin by defining what it is you are looking for, a profile of the successful person. This appears simple but is the area that causes the most problems. Generally people list requirements simply as education and experience, but these are not indicators of success on the job. If they were, any person with those qualifications should be successful and we know this is not true. A person's knowledge, skill, and abilities will determine whether or not he or she will be successful and these should be the keys to our search.

It is easiest to think of a job as a series of tasks. And, successful

Figure 27-1. The Interview Process

people accomplish the tasks in an appropriate manner. Ask yourself the question, "What tasks must this person perform to be successful?" And then ask, "What knowledge, skills, and abilities does it take to perform those tasks?" Finally, ask, "How must the person go about accomplishing the task to be successful?"

The first step in developing your list of requirements is to think about the technical skills necessary to accomplish the tasks. Think about the tasks and try to determine what a person must know and do to accomplish the tasks. Does he or she operate any special equipment? What systems does he or she need to use? Is any special knowledge required? The answers to these questions should give you a list of technical requirements.

Next, we need to determine the behavioral requirements so think about how the tasks are performed. Analyze the environment. What are the specific behavioral traits necessary to accomplish the tasks? Do you work in teams or individually? Does the person solve problems, come in contact with the public? Does he or she have to influence people? How does the person need to act to be successful on the job? The answers to these questions should generate a list of behavioral requirements such as problem-solving skills, initiative, and flexibility.

One of the best methods to determine the list of requirements is to profile a successful person. Think of someone doing the job now or who has recently done the job and been successful. Why is he or she successful? What does he or she do, or not do, that causes him or her to be successful? How does he or she react to situations? How does he or she deal with people? The answers to these questions will help you identify the requirements necessary to build a profile of the successful person.

Once you have developed your list of requirements, a qualified candidate pool needs to be developed. This is the recruitment process. In a warehouse, your first question is whether to recruit from other warehouse companies or from people outside the industry. In times of high unemployment, recruitment may be primarily a screening process. When qualified people are scarce, recruitment may become the toughest part of the process. Assume that you now have a group of applicants. From this group you must screen to determine whom to interview. This process moves you from a group of applicants to a group of candidates.

This screening process is very difficult. The decisions made at this point are made with the least amount of data and are usually irreversible. We very seldom go back to the pile of "No's" for a second look. It is important to keep in mind the ultimate objective when screening applications, which is to hire a person who will be successful.

With this in mind, try to "screen in" not "screen out." Don't look for a reason to throw people out; look for a reason to screen them in. This is consistent with our ultimate objective and will also help eliminate the mistake of losing good candidates because there is one thing we don't like on the application.

Examine each resume and look for indicators that the person will be successful on your job. Review your list of requirements and determine which might be indicated on a resume. Management skills, knowledge of a specific piece of equipment, and sales skills are examples of skills that should be indicated on a resume. Initiative, flexibility, and problem-solving are not as evident and will need to be researched in an interview. Review the resume with the intention of finding as many indicators as possible that the person has the necessary knowledge, skills, and abilities. To get in this habit, try counting the things you like on each resume you read.

Once you have narrowed your list, it is time to prepare for the face-to-face interview. Part of a systematic approach is to prepare for every interview the same way. The first step is to prepare an interview guide which is your list of questions, in the order you will ask them. It is important to prepare your questions in advance for two reasons. First, you can be certain to generate questions that cover all the important information. Second, with your questions available, you can concentrate on listening as opposed to worrying about what to ask next.

Start with the application and your list of requirements and develop your list of questions in correct chronological order. Interviewing in correct chronological order, from the beginning to the present, allows you to follow a person through his or her career. You can see patterns of growth or lack of growth. This method will help you ascertain whether a person has plateaued, whether he or she repeat mistakes, and how quickly and how often he or she learns new skills.

Choose the first time period you want to explore, and with your requirements in mind, develop questions designed to generate information that will help determine the candidate's knowledge, skill, and ability levels. Create enough questions during each time period to ensure you get enough information to assess his or her skills. Continue to advance in correct chronological order toward the present. What are you doing is simply taking a journey through the person's background and determining what he or she accomplished along the way.

It is necessary to pay particular attention to the types of questions you ask in an interview. Asking the right types of questions is the single biggest contributor to a successful interview. Much of the conventional wisdom is misleading. Try to avoid open-ended, nonspecific questions. An example is, "Tell me about managing a facility." And, "What is the best way to control expenses?" Although these questions do tend to get the candidate talking, they do not get specific information that will help predict if the person will be successful.

Also, try to avoid theoretical questions such as, "What would you do if . . .?" These force the candidate to manufacture an answer he or she thinks you want to hear. This only measures their ability to figure out what he or she thinks you want to hear. Instead of asking "What would you do if . . ." ask "What did you do . . ." which takes the question out of the future and generates specific factual information.

The two most effective types of questions to ask are factual and action questions. Factual questions require the person to respond with a simple discreet fact. "How many people did you supervise?" "What type of equipment did you use?" "What inventory management system did you use?" These simple, straightforward questions require the person to answer with a factual piece of information that helps you determine his or her knowledge, skill, and ability level.

Action questions require the candidate to describe some action he took. "Tell me how you dealt with the last problem employee you had." "Step me through how you implemented your quality control program." "Explain how you closed your last sale." Each of these questions requires the person to explain his actions. The answers will contain information that helps determine how he will act on your job. This helps predict whether he will be successful.

Now that your interview guide is complete, the next step is to implement your interview. The face-to-face interview is the easiest part of the process. Begin with an opening designed to put the candidate at ease. The goal is to establish rapport and reduce stress and anxiety. Begin with some small talk to break the ice. Talk about a current event or something interesting you saw on his or her resume. You might lead with an issue or event that is hot in the industry.

Next, explain the process you will use and ask if he or she has any questions about the process. The goal here is to take the mystery out of the process. The person should be told exactly what you are going to do, in the exact order, and why. Explaining these items to the person has the effect of relaxing him or her and removing some of the anxiety.

The opening has put the person at ease, established rapport, and reduced the anxiety. You are now ready for the body of the interview. Refer to your interview guide and proceed to ask your prepared questions in the order you wrote them. Be sure to probe for details. Don't assume you know what they mean. Ask for clarifications and explanations when you need more information to assess his or her abilities.

It is difficult to plan for every eventuality in an interview. If you get off on a tangent, determine if you are getting information that will help you predict success. If not, get back on track with one of your prepared questions.

Once you have asked all your questions, you can begin the closing phase of the interview. This begins with your explanation of the job and company. This is the time to sell the applicant on your company. Don't spend time in the beginning of the process marketing the job because (1) you don't know if you need to market the job and (2) you won't know how to market until you understand the candidate's background.

After you have described the job and company, outline the remaining steps in the selection process. Make sure the candidate knows what will happen next and when. Finally, end the interview on a positive note.

You now have a tremendous amount of information about the person's knowledge, skill, and ability level. The next step is evaluation of the data. This phase is not difficult if you break the process down to small manageable parts.

First, go back to your list of requirements and create a chart similar to the one in Figure 27-2. List the requirements down the left side of the chart and the candidates along the top. Rate each candidate against each requirement working from the top of the list.

Requirement	Candidates			Comments
	A	B	C	

KEY ✓ Meets Requirement
– Does Not Meet Requirement
+ Exceeds Requirement

Figure 27-2. Matrix Evaluation

For each requirement, go back to your notes from the interview and determine if the person meets the requirement. Then proceed to the next requirement until you have a rating for each.

The grades should be as simple as possible. One system has three grades: meets requirement, exceeds requirement, and does not meet requirement.

Candidates are always measured against your list of requirements which is the standard. Evaluation against a standard has two major benefits. First, it satisfies the major legal requirement of the equal employment opportunity laws which say that all candidates should have an equal opportunity to compete. Second, it will prevent you from hiring someone who is the best of the worst (a person who looks good compared to the other candidates but still doesn't meet the requirement).

The best candidate will meet or exceed all the requirements. A candidate who is missing an important requirement will probably not succeed on the job and should not be hired.

The last step, hiring a successful candidate, is the reward. The process is simple but very effective. Over time your skill level will increase as will your odds of hiring a successful person.

Substance Abuse in the Warehouse

Substance abuse problems probably exist in your warehouse if you have more than ten people in the crew. In times of relatively full employment, up to 50% of those people who are looking for a job have a substance abuse problem. Substance abuse includes the legal use of alcohol and even the abuse of prescription drugs. Ignoring the problem will not help it go away—it only aggravates it. Most of the writings on this subject are either by law enforcement people, healthcare experts, or social workers. Each brings the viewpoint of that particular profession, which is not necessarily a managerial approach. We look at the subject from the standpoint of business managers, and one goal of every manager is to attract and retain a safe and healthy workforce.

Coupled with the recurrence of substance abuse, the standards for quality work in today's warehouse are higher now than they were when the problem was first identified. A growing number of today's warehouse workers must be computer literate. If they are impaired by mood altering substances, they could put bad information into your warehouse information system as well as cause accidents, tardiness, or excess absenteeism. Therefore the potential loss from tolerating substance abuse is higher than ever before.

The starting point for controlling substance abuse in your warehouse is to develop and publish a company policy on this subject. If your company has never done this, your people will not understand your position. The majority of people in our society do not abuse substances and would prefer not to work with those who do. They need to know where you stand and what standards of conduct you expect. The policy statement is the first step in creating this understanding.

Here is a model policy statement, developed by us after reading and distilling the statements issued from many different sources:

Our company is committed to providing our employees with a safe workplace and an atmosphere that allows people to protect merchandise placed in their care. Our employees are expected to be in suitable mental and physical condition while at work to allow them to perform their jobs effectively and safely. Whenever use or abuse of any mood altering substance interferes with a safe workplace, appropriate action must be taken. Our company has no desire to intrude into its employees' personal lives. However, both on the job and off the job involvement with any mood altering substances (alcohol and drugs) can have an impact on the workplace and on our company's ability to achieve its objective of safety and security. Therefore, employees are expected to report to the workplace with no mood altering substances in their bodies. While employees may make their own lifestyle choices, the company cannot accept the risk in the workplace that substance use may create. The possession, sale, or use of mood altering substances at the workplace, or coming to work under the influence of such substances, shall be a violation of safe work practices. This violation may result in termination of employment.

This policy does several important things. It relates your position to your company's mission as a warehousing operation. It relates substance abuse to health, safety, property protection, and quality control. Everyone understands that these are essentials to any quality warehouse operation. The policy sets standards of conduct.

Supplements to this policy may recognize the availability of a rehabilitation program and the ways in which an employee can get help through that program. You may also wish to state the penalties for violating the policy.

But what if your company already has a policy? First, is everybody aware of the policy? Second, is the policy actually enforced? Unless you can answer yes, you may have a policy that is not really working.

Consider the fact that there are three violation levels that differ markedly in their impact. The first level is the employee who reports to work still under the influence of a substance consumed while off the job. That worker is unsafe in the workplace and therefore both a threat and a nuisance to fellow workers.

A more serious offender is the employee who comes to work in possession of mood altering substances. This suggests that the worker is consuming those substances in the warehouse and is therefore clearly working under their influence.

Most serious of all is the worker who brings mood altering substances to your warehouse for the purpose of selling them to others who work there. This individual may or may not be consuming these substances, but your workplace is being used as a market. The third level deserves the strongest discipline, and the second is far more serious than the first.

The most difficult challenge involving testing is the means of intervening with a warehouse employee who exhibits symptoms on the job that might be related to substance abuse. A proven answer is to use the services of your company physician, or any competent M.D. who has an established relationship with your organization.

You have the right to question the health of any employee who appears to be impaired while on the job during working hours. That right includes the requirement that the employee immediately visit a doctor for the purpose of having a physical checkup to find out what is wrong. If you send the employee to the doctor, you have an obligation to pay for the time spent during this process, and you should transport that employee to and from the doctor's office. Any competent physician can tell whether the symptoms are the result of substance abuse, and this is the most potent weapon available for dealing with the problem. If the worker refuses to visit the doctor, dismissal for insubordination is the available solution.

Four steps are necessary for successful implementation of a substance abuse prevention program. The first step is the publication and dissemination of a substance abuse policy. The second step is a detailed screening program for all new hires. A third step is the training and retraining of supervisors and managers to recognize and intervene when there are behavior problems that could be related to substance abuse. The fourth and most critical step is a rehabilitation and discipline program when abuse problems are uncovered.

Job Performance Appraisal

A properly conducted job performance appraisal increases motivation and allows constructive criticism in both directions. The interview may also provide a needed pat on the back for good performance. There are many forms and procedures for appraisal, but the core of the program is the interview between subordinate and immediate supervisor.

Like wage reviews, appraisal interviews should be handled on a regularly scheduled basis. Such interviews should be objective, emphasizing actions and things rather than individuals. At the same time, the interview should clearly point out needed improvements and provide a plan for accomplishing them. If the employee participates, the plan is more likely to be successful. Figure 27-3 shows a performance evaluation form.

ASSOCIATE PERFORMANCE EVALUATION

ASSOCIATE: .   POSITION .

D.C. LOCATION

DATE OF LAST EVALUATION   REVIEW DATE .

PREPARED BY   TITLE   DATE

FOR EACH "CHARACTERISTIC" BELOW, CHECK THE MOST APPROPRIATE RATING CATEGORY. USE THE "COMMENTS" SECTION FOR ANY SIGNIFICANT REMARKS DESCRIPTIVE OF THE ASSOCIATE.

CODE "A" FACTORS RELATE PRIMARILY TO WAREHOUSE AND CLERICAL POSITIONS, CODE "B" PRIMARILY TO SUPERVISORY POSITIONS, AND CODE "C" RELATES TO MANAGEMENT POSITIONS

BELOW EXPECTED FACTORS ARE THOSE WHICH REQUIRE IMPROVEMENT IN ORDER TO MEET EXPECTED REQUIREMENTS. EXPECTED FACTORS ARE THOSE WHICH ARE ADDITIONAL AREAS FOR IMPROVEMENT POTENTIAL. ABOVE EXPECTED FACTORS ARE THOSE WHICH ARE POSITIVE CHARACTERISTICS.

CODE	JOB PERFORM-ANCE FACTORS	ABOVE EXPECTED	EXPECTED	BELOW EXPECTED	COMMENTS	CODE	JOB PERFORM-ANCE FACTORS	ABOVE EXPECTED	EXPECTED	BELOW EXPECTED	COMMENTS
A	WORK PERFORMANCE: Quantity or Volume					BC	JUDGMENT:				
AB	DAMAGE MINIMIZATION AND CONTROL:					BC	DEALING WITH CUSTOMERS:				
AB	WORK PERFORMANCE: Quality or Accuracy					BC	COURTESY:				
ABC	DEPENDABILITY: Follow up reliability, Meeting work deadlines, Attendance & Tardiness					BC	SUPERVISING:				
						BC	DISCIPLINE: Firmness				
ABC	PERSONAL CHARAC-TERISTICS: Disposition, Poise, Appearance & Sincerity					BC	DELEGATING:				
						BC	SELF-IMPROVEMENT:				
ABC	JOB KNOWLEDGE:					BC	FOLLOW UP:				
ABC	SAFETY:					BC	DEVELOPING SUBORDINATES:				
ABC	EQUIPMENT & FACIL-ITIES: Care & Conser-vation					BC	LEADERSHIP:				
						C	SALES & COMPANY IMAGE BUILDING:				
ABC	HOUSEKEEPING:					C	WRITTEN COMMUNICATIONS:				
ABC	EFFICIENT USE OF STORAGE SPACE:					C	SELECTING COMPETETENT SUBORDINATES:				
ABC	COMMUNICATION: With Peers, Subordinates & Superiors					C	EVALUATING SUBORDINATES:				
ABC	CONSISTENCY:					C	TERMINATING INCOMPE-TENT SUBORDINATES:				
ABC	SELF IMPROVEMENT: Acceptance of Criticism					C	SETTING FAST PACE:				
ABC	DRIVE & INITIATIVE:					C	MAKES DEMANDS ON SUBORDINATES:				
ABC	COOPERATION:					C	FOCUSING ATTENTION ON GETTING JOB DONE:				
ABC	INNOVATION:					ABC	OVERALL EVALUATION:				
ABC	STABILITY:					REMARKS:					
ABC	ORGANIZATION:										
BC	PLANNING:										
BC•	OVERTIME CONTROL:										

Figure 27-3

Promotion from Within

When seeking individuals for higher positions, management should look inside the company first. In spite of the dangers of the Peter Principle mentioned earlier, upward movement is an important morale builder. If no opportunities for promotion exist, talented employees will soon leave the company.

Pride in the Company

A hallmark of dedicated employees is their pride in the company. The task of creating such pride is more difficult today than it was a generation ago. Nevertheless, company identification can be stimulated through athletic teams, service pins, and, sometimes, uniforms. A company newsletter also can be a fine means of building pride in the organization.

Amazing things can happen when a warehouse is staffed with dedicated personnel. An unbroken chain of dedicated warehouse workers will provide the best possible security against cargo theft. Most importantly, highly motivated people will suggest ideas for improving methods that might never be discovered by top management. An unbroken chain of dedicated warehouse workers will provide the best possible security against cargo theft.

Part VII

PRODUCTIVITY AND QUALITY CONTROL

28

MAKING WAREHOUSING MORE EFFICIENT

Warehousing people, by the nature of their jobs, deal in quantity-based activity. They calculate pounds-per-man-hour and other quantitative measures to show how well they are doing.*

It is important to avoid confusing quantity with quality. A highly productive warehouse operation that has high error rates, poor housekeeping, and poor service could deteriorate to a point where quantity is no longer important.

In manufacturing, product quality results from eliminating defects. In warehousing, product quality is measured by the loyalty of customers and workers.

Only your imagination will limit the ways you can improve productivity in a warehouse. Of the eight productivity improvements that will be considered, five are critical:

1. Establish targets for improvement.
2. Reduce distances traveled.
3. Increase the average size of each load handled.
4. Seek round-trip movements within the warehouse.
5. Improve cube utilization.

Establish Targets for Improvement

Since warehousing involves more random operations than manufacturing, the development of work standards is more difficult. Yet

---

* Adapted from an article by B.J. LaLonde and K.B. Ackerman, Harvard Business Review, May–June, 1980.

the use of productivity goals, based on predetermined engineered standards, is feasible in a warehouse operation.

Forecasting

Warehousing is often a hedge against uncertainties. Therefore, an accurate forecast may eliminate the need to store materials. A good forecast will prevent the deployment of items, for example, in Chicago when they are needed only in Miami. Better forecasting also reduces two prime sources of waste: the cost of *reserving* space the user thinks will be needed and the *hoarding* of workers who may be needed when volume increases.

A few companies give the estimating and planning responsibilities to logistics managers. This means that logistics people are responsible for market forecasts and the scheduling that follows them. Consider the advantage of concentrating this responsibility in a department that must both forecast the future and fulfill its own forecasts.

Forecasting should include a prediction of variation in flow. Most companies have rush seasons based on demand peaks or responses to sales incentives. Such peaks can waste storage space and increase labor costs.

There is a way to control such waste. Smoothing the flow usually involves cooperation with marketing personnel. Can we give customers an incentive to do their own warehousing during the off-season? If a sales contest is involved, must the goods be shipped immediately? Entering the order at the warehouse might permit it to be counted for incentive purposes, but the shipment could actually be made later to smooth the work flow.

Reduce Distance Traveled

In controlling the distances traveled in moving material between storage bays and shipping or receiving docks, the warehouse manager finds that one of the easiest ways to cut costs is to examine the layout. Pareto's Law, or the 80–20 rule, states that in most enterprises, 80% of the demand is satisfied with only 20% of the stockkeeping units. Clearly, if management identifies this 20% of the items and locates them near the shipping and receiving doors the effect on distance traveled will be dramatic. Yet in most warehouse operations, goods are stored by product family. A grocery product operation, for example, may have all canned goods in one section, all paper items in another, housewares in another, and so on. In some cases, storage characteristics require such a separation, but often they do not.

The planner who desires to change the layout to conform to demand must first determine the demand pattern. In doing so, recognize that your target is constantly moving. In any dynamic business, a study of Pareto's Law must be repeated to reflect changes.

Are your order pick lists designed for maximum efficiency in the warehouse? A document that lists items in the same order they are found in an order-picking list allows the stockpicker to start at the top of the sheet and at the head of the aisle, and move down the aisle and down the document at the same time.

Reduction of distances traveled is a function of planning, data processing, and materials handling. The planner can use Pareto's Law or some procedure to design aisles and staging areas to reduce unnecessary movement. Then data processing personnel can design a locator system that functions effectively with random locations and order-pick lists that conform to the physical layout.

In materials handling, inbound movements offer good opportunities for travel economy. In some warehouses, the responsibility for finding a location for inbound loads rests with the inbound materials handlers. In the absence of instructions, they will put the incoming load in the first empty slot. The result is a needless proliferation of stock locations for the same item.

Look at the location and use of dock doors. Many warehouses have dock areas dedicated strictly to receiving, while other docks are used only for shipping. But time can be saved through use of the same doors for both shipping and receiving.

If you allow any dock door to serve either purpose, you reduce travel by assigning inbound loads to the door closest to the area where the items are to be stored. Do the same with empty vehicles arriving for outbound loads.

Unlike our national highway network, the warehouse "highway system" of aisles can be changed without substantial cost. The layout of aisles should be constantly fine-tuned with a goal of reducing travel distance.

Increase Unit Load Size

The warehouse that supplies convenience stores frequently may open cases of tomato soup and ship individual cans because the small store cannot justify ordering in full case quantities. While the cost of breaking cases is high, the distributor is meeting the needs of the convenience store. Contrast that situation to the harbor scene where a 40-foot marine container carrying bulk whiskey from Scotland is unloaded. Handling large units greatly reduces the opportunity for breakage and pilferage, as well as the cost-per-ton of handling.

As you examine your own warehouse, ask whether the average size of units handled could be increased. Doing so may require changes in marketing policy. Some grocery product companies offer customers an incentive to buy in pallet-load quantities, with the discount offered only when the order quantity is an exact pallet load.

Seek Round-Trip Opportunities

Because most fork-lifts carry a payload in just one direction, the truck travels empty half of the time. In contrast, high-rise facilities equipped with stacker cranes include a computer control with a memory unit. This memory is used to maximize round-trip travel opportunities. When the crane is putting away an inbound truckload containing item F, the computer already has determined the best storage address for that item. Meanwhile, the computer remembers that item J, stored in a nearby slot, is wanted for an outbound order being staged. Under instructions from the computer, the crane takes item F to its storage address, then moves to the address for item J and returns with this merchandise for the outbound order.

While the opportunities for reducing one-way travel are not limited to cranes, moving payloads in two directions does require effective communication. Equipping each lift-truck operator with a radio helps. If the operator advises the dispatcher that item C is being moved to a given storage address, the dispatcher can instruct the driver to return with a pallet of item N.

Admittedly, maximizing round-trip hauling adds some complications. For one thing, it requires precise planning by supervisors. Moreover, involving the lift-truck driver in both shipping and receiving creates new opportunities for errors. Also, the shipping and receiving docks may require larger staging areas to accumulate a bank of work to support roundtrip movements.

While the process is complex, your supervisors can control the complications if they are given reasonable lead times to plan and control round-trip travel.

Improve Cube Utilization

The cheapest space in any warehouse is that closest to the ceiling. If you were constructing a new 22-foot clear warehouse, you might develop a cost figure of $20 per square foot, or $.91 per cubic foot ($20 divided by 22). If you raise the clearance from 22 feet to 24 feet, the total cost of the building will increase only slightly, to about $20.30 per square foot in the example. This means a drop in the cost per cubic foot from $.91 to $.85.

While packaging strength limits the practical pile height, management can do something about the packaging problem. The most obvious solution is to improve the package. If you can't change the package, consider high-density storage racks, such as drive-in racks to permit higher stacking of products with weak packages.

As cube utilization improves, travel distance in the building can be reduced. But there is always a tradeoff in using overhead space. Elevating the product to the roof takes time. The most economical storage plan is influenced by the speed of turnover. The fastest moving product should be stored not only close to the door but also close to the floor.

There are limits to cube utilization. To allow effective functioning of sprinkler systems, fire insurance underwriters impose a maximum pile height, particularly of hazardous materials. They also require a buffer between the sprinkler heads and the storage pile. This distance depends on the capacities of the sprinkler system, the pile height, and the type of merchandise stored.

It is important not to overemphasize space utilization. Consider the plight of a grocery warehouse manager who had inherited a system that was the ultimate in narrow-aisle design. As he gloomily surveyed the cramped environment, he made this observation: "One thing about aisles is they cost a little less each year. Aisles don't bargain for pay increases or increased benefits. If we save a few aisles and add a lot of people, we'll never be ahead of the game."

Though maximum use of overhead space may seem to be the cheapest way to get more product into your building, the key to improving warehousing efficiency is to seek the best tradeoff between storage costs and handling costs.

The next three areas for improvement may be less critical, yet they will significantly change warehouse productivity.

Free Labor Bottlenecks

Bottlenecks are always at the top of the bottle. Since management is also at the top, it's your responsibility to correct the situation.

A typical bottleneck occurs in unloading a floor-loaded boxcar. Two laborers may be assigned to the boxcar to palletize cases for a fork-lift truck driver who removes loaded pallets. If the laborers' speed exceeds the driver's, they wait until he returns to remove a load. Changing the crew size will clear the bottleneck.

Another approach is the *one-man, one-machine* technique, which gives each worker a lift truck, and means that worker has no need to wait for anyone else. As you tour your warehouse, look for workers who are waiting and find out what has caused the bottleneck.

Reduce Item Handling

In a typical factory warehouse, each product is handled 16 times. Many of these handlings can be eliminated and they should be. Each movement of a product is an opportunity to damage it. Each lifting further fatigues the package.

As you examine your warehousing operation, ask why it is necessary to stage every inbound load on the dock before moving it to a storage bay. Ask the same thing about the staging of outbound loads. In trying to reduce the number of product handlings, pay particular attention to temporary storage locations and see if they can be eliminated.

Improve the Packaging

From the warehouse operator's point of view, the perfect package is made of cast iron and filled with feathers; it is indestructible and can be stacked 50 feet high with no artificial support. Such perfection is nonexistent, of course, but the warehouseman is justified in objecting to packages that won't permit use of the available space in the warehouse. Warehouses built in the last few decades have ceiling heights that exceed the freestanding stacking capabilities of packages going into them.

The packaging engineer wants a container that is light and cheap and just strong enough to get the item to the consumer.

The manufacturer must consider the tradeoffs. The product manager should ask whether it would pay to spend more money on packaging, and achieve savings through avoidance of product damage in warehousing and distribution.

Forces for Gain in Productivity

The turbulent business era of the 1990s has created four major change factors for warehousing. They will have a significant influence on people who work in warehouses, people who manage them, and those who use them. Here are the four propositions:

1. *Time* is one of the most important ingredients in effective warehousing. Therefore, the best warehouse operations are those that are designed to reduce every aspect of order cycle time.

2. *Quality* is just as important as time, and users of warehouse services now expect performance that is very close to perfection.

3. The emphasis in using warehouses is to improve *asset productivity*. Three critical functions are to reduce total cost, reuse and recycle.

4. To enter the 21st century, warehouse managers must develop a *new kind of workforce,* and requirements for both management and labor will change significantly.

Improved capabilities in computer modeling and communications have enabled today's manager to substitute information for inventory, and to move that information faster and more accurately than ever before. The best examples are found in a relationship between retailers and their sources. Just a few decades ago, it was common for appliance manufacturers to stockpile huge quantities of finished goods because nobody knew which colors or models would sell, and it was important to have inventory available in case somebody was ready to buy. Today, information systems at point of sale allow nearly all retailers to report the sale of a given refrigerator or clothes dryer within minutes after the transaction is concluded. This information moves to the retailer's headquarters and back to the manufacturing resource within minutes. Production schedules can be changed frequently to reflect current transaction activity. The same kind of "real time" reporting exists in many other commodities, even for a nondurable such as soap.

Improving quality is just as important as reducing cycle time. Through the use of bar code control on outbound shipping, some warehouses report an error rate of less than one in ten thousand. With this capability has come a growing intolerance of errors and damage by users of warehousing services. What would happen to baseball if the rules changed so that a batter was out after one strike? Something similar is happening in warehousing, and those who don't recognize it are out of touch with today's expectations and capabilities. In warehousing, quality is measured by the operator's ability to deliver product on time, in good condition, and precisely as ordered without overages, shortages, or any other discrepancy— "on spec., on time, and on budget" as one logistics executive said it.

Use of bar coding is part of the quality revolution in warehousing. While only 10% of warehouse transactions are controlled by bar code scanning today, by the end of the decade over half of them will be.

Today's user of warehouse services is forced to place primary emphasis on asset productivity. This drive to improve asset productivity comes from the highest level of senior management, as a growing number of corporations rate their success by measuring return on assets rather than return on sales.

The drive to improve asset productivity is manifested in three areas:

1. How can we reduce assets or otherwise improve turnover of capital?
2. Can we reuse any materials?
3. Can the material we discard be recycled?

There are two ways to improve asset productivity. One is to improve the operations to the extent that the same asset investment can be used to handle a significantly greater volume. For example, if you could double the throughput of your warehouse without adding space or lift trucks, you would improve asset productivity. The second way to improve asset productivity is to transfer the investment in assets to a third party.

Senior management must develop a new workforce for warehouses, and this includes managers as well as workers. The logistics executive of the 1990s is younger and more diverse than the traffic manager who filled this function a generation ago. He or she also has a higher rank in the organization. Researchers asked corporate executives when the first senior logistics executive position was created in their organization. Forty-six percent reported that this happened in the 1980s. More than 41% of logistics executives today have a graduate degree, and over half of these have some kind of professional certification.

The challenge for this younger and better educated management team is to deal with a changing hourly workforce. Customers are more concerned than ever before about quality of labor. In some cases, the decline of unions in warehouse workforces is accompanied by a decline of warehouse discipline. The presence of a union was once a stabilizing influence on the workforce, but this situation has changed. When asked to compare private with third-party warehousing, our respondents felt that the third-party providers have fewer labor problems than their corporate warehouses.[16]

Until recently many United States manufacturers treated warehousing as a necessary evil, rather than an integral part of the business system. Commitments of capital, technology, and engineering emphasized manufacturing and marketing and gave little attention to warehousing. As a result, warehousing has not kept pace with other corporate functions. Today, however, this situation is changing.*

---

* From an article by Bernard J. LaLonde, The Ohio State University.

29

MONITORING PRODUCTIVITY

Some executives are convinced that warehousing is a dead expense, a function that adds no value and therefore should be minimized if not eliminated. Others believe that warehousing adds value and they can show the contribution of warehousing to the enterprise. How do you calculate the total costs and value of your warehouse operation? The job is not simple, but we can give you a few clues.

Twelve options are available to most warehouse managers. Figure 29-1 illustrates these options, each a combination of four real estate and three operations choices.

There are three different options in managing warehouse operations:

1. A private operation is one that is done entirely by your own people.

2. A third-party operation is managed entirely by people who work for another organization.

3. Some companies combine the two, with third parties operating some warehouse locations and others remaining private. In other situations, the supervisors may be your own people but the hourly labor may be provided by others. Another combination might have the office staff provided by the owner and the warehouse crew provided by a third party.

There are four options in use of warehouse property:

1. The simplest option is to own the property.

		Own	Lease	Third Party Ownership	Combination
Operation	Private				
	Combination				
	Third Party				
				Real Estate	

Figure 29-1

2. A lease of varying length can be obtained on many warehouse properties.

3. If the property is in the hands of a third-party warehouse operator, it could be either leased or utilized under a month-to-month public warehouse agreement.

4. Some users work out a combination of the first two or three options, with a mix of ownership, long-term lease, and thirty-day agreements.

Your operation can be any of the twelve options.

In looking at property, your selection will be governed by your view of the market. If you believe that the price of warehouse facilities will increase, you may choose to own warehouses as a business investment and to anticipate a profitable future resale. If you are concerned that an unscrupulous landlord might refuse to renew a lease at a reasonable rate, you may feel that property ownership is necessary for security.

Some corporations avoid investment in real estate. Capital is conserved so that it can be invested in other areas of the business, particularly when there is no shortage of real property.

Tax considerations can be an important issue, particularly in connection with real estate. Owning property will mean that taxes on the facility are generated. When a new warehouse is involved, sometimes an eager community will provide tax abatements that benefit only the owner. The owner can also earn depreciation, which is a shelter against taxes on earnings. The owner will pay interest expenses in financing the building, and these may provide a deduction from taxation. If your company owns property with the intent to sell it in the future at a profit, you should factor into the analysis the estimated capital gains taxes at sale. When the property is leased or taken from some other third party, the owner bears the expenses and some of the potential advantages of taxation on depreciation. When you negotiate a lease, it is important to be aware of the tax impact on the owner. In some cases, a favorable impact can be used as a negotiating tool to achieve concessions on the lease or rental agreement.

Is Your Warehouse Economical?

Numbers do not provide the only measure by which to justify warehousing. The decision may be driven by long-range plans and corporate policy. Here are some of the questions that could influence a decision on choice of the twelve options shown in Figure 29-1.

- *Is your company moving toward operation of more or fewer warehouses?* If there is a clear trend to increase or decrease the number of warehouses used, you may choose an option that provides maximum flexibility for change in the future.

- *How stable are your product lines?* If your company is rapidly changing its products and procurement sources, the warehouse configuration that you have today may be less than ideal a few years from now. A dynamic product situation would again drive you towards the most flexible warehousing resource.

- *How stable are the markets for your product?* A changing marketplace could cause the size or locations of your current warehouses to be unsuitable long before the buildings are depreciated.

- *If you are planning to expand production, will it be moved into existing warehouse space?* A warehouse located adjacent to existing production facilities can serve two purposes, both its present purpose and as a buffer to allow easy expansion of production in the future.

- *Are you willing to hire and retain the experienced people needed to operate warehouses?* Some decide that they cannot or should not make the management commitment to warehousing.

Figure 29-2 is an outline that will allow you to track all costs of property management, when the facility is owned, leased, or run by a third party.

Figure 29-3 is a similar listing of costs for the materials handling operation.

Figure 29-4 is a summary that provides a grand total of costs developed in Figures 29-2 and 29-3. The purpose of these exhibits is to provide a checklist to avoid overlooking any of the expenses involved in warehousing.

1. Challenge and justify your company policies. Past practices are often an overwhelming influence on warehousing decisions. Sometimes these are actual formal policies, but frequently they are just old assumptions. If your company has always done it that way, don't be afraid to ask why. Sometimes a view that is believed to be company policy is actually a never challenged past practice.

2. Analyze every alternative. Figure 29-1 listed twelve different options. Be sure that you consider each. Your goal is to find the very best option for your company.

3. Include all costs. Figure 29-2, 29-3, and 29-4 provide a checklist for every cost item that might apply.

4. Quantify the "service effects" of each option. It is not easy to put a number on customer service enhancements.

5. Compare warehouse ownership options on the basis of return on investment. After assessing full cost of each option, calculate a rate of return.

6. Develop decision rules to compare warehousing options that yield unequal levels of service. How do you compare

Section A: Real Estate and Operation

1. **Owned Facilities**
   Land:
   Book or Market Value if Owned .... _____
   Market Value if Newly Purchased .... _____
   Related Acquisition Costs .... _____
   Other .... _____
   Building:
   Construction of Shell and Flooring .... _____
   Sprinkler, Safety Systems .... _____
   Electrical Systems .... _____
   Design Fees .... _____
   Office Area Costs .... _____
   Water Supply/Sanitation .... _____
   Special Equipment Required .... _____
   Other .... _____
   Grounds:
   Site Preparation (tests, surveys, grading and fill,
   pilings, etc.) .... _____
   Parking Lot (paving, painting, repairs) .... _____
   Installation of Drives, Rail Sidings .... _____
   Outdoor Storage Facilitiees .... _____
   Other .... _____
   Taxes:
   Property Taxes .... _____
   Depreciation .... _____
   Interest Tax Shields for Loans .... _____
   Other .... _____
   Total Costs of Owned Facilities .... _____
2. **Leased Facilities**
   Annual Lease Payment .... _____
   Building Improvements .... _____
   Grounds Improvements .... _____
   Tax Effects (paymt. deduct. if appl.) .... _____
   Other .... _____
   Total Cost of Leased Facilities .... _____
3. **Third Party Operation**
   Handling Charges .... _____
   Storage Charges .... _____
   Other Charges .... _____
   Site Inspection/Personnel Evaluations .... _____
   Dedicated Resources (contract only) .... _____
   Other .... _____
   Total Costs of Third Party Services .... _____

Figure 29-2

311

Section B: Operational Costs

1. **Operating Equipment**
   Forklifts, and Related Support Items .... _____
   Conveyors .... _____
   Racks and Bins .... _____
   Pallets and Slipsheets .... _____
   Safety/Security Equipment .... _____
   Cleaning Supplies .... _____
   Other .... _____
   Total Cost of Equipment .... _____
2. **Office Equipment**
   Furniture and Accessories .... _____
   Computers & Communications Equip. .... _____
   Office Supplies .... _____
   Other .... _____
   Total Cost of Office Equipment .... _____
3. **Cost of Transportation**
   Cost of Transport. to Storage Fac. .... _____
   Cost of Transport. from Storage Fac. .... _____
   Total Cost of Transportation .... _____
4. **Labor Costs**
   Warehouse Employees .... _____
   Maintenance Employees .... _____
   Office Staff .... _____
   Supervisors .... _____
   Administrative Personnel .... _____
   Overtime .... _____
   Seasonal Employees .... _____
   Other .... _____
   Total Cost of Labor .... _____
5. **Employee Benefits** (these may be included in the labor
   estimates above) .... _____
   Unemployment Insurance .... _____
   Social Security Contributions .... _____

Figure 29-3

warehouse option A, which provides a 10% fill rate, with option B, which provides a higher return on investment? Dealing with this decision will require a consensus on acceptable customer service levels.

7. Do not confuse accounting rules with cash flow effects. The focus should be on cash flow. Some accounting methods can

Pension, Company Savings Plans .... ——————
Health and Life Insurance .... ——————
Paid Vacations .... ——————
Sick, Emergency Leave .... ——————
Other .... ——————
Total Cost of Employee Benefits .... ——————
6. Maintenance, Repairs & Replacement
Building, Grounds .... ——————
Operating Equipment .... ——————
Pallets .... ——————
Racks, Bins .... ——————
Other .... ——————
Total Maint., Repair & Replace .... ——————
7. **Utilities**
Electric .... ——————
Natural Gas/Fuel Oil .... ——————
Water .... ——————
Sewer .... ——————
Refuse Disposal .... ——————
Total Cost of Utilities .... ——————
8. **Insurance**
General Liability .... ——————
Public Liability .... ——————
Fire and Catastrophe .... ——————
Business Interruption .... ——————
Workman's Compensation .... ——————
Other .... ——————
Total Cost of Insurance .... ——————
9. **Over, Short and Damage** .... ——————
10. **Other Costs as Applicable** .... ——————
Total of Operating Expenses .... ——————

be a distraction. For example, a capitalized lease may change your balance sheet, but the lease payment is a cash expense.

8. Audit to evaluate asset management performance. When you make a warehouse investment, compare actual performance with forecasted performance.

9. *Periodically review current warehousing approaches.* It is

313

healthy to periodically reexamine the way you are handling warehousing now to see if there is a better way to do it in the future.

10. *Compare the operation of the warehouse with costs of distribution if the warehouse did not exist.* Suppose you had no warehouse? What would be the cost and service penalties? How would production scheduling and marketing be changed?

Many companies put warehousing projects into one of two categories:

1. Capacity maintenance projects
2. Improvement or profit enhancing projects

The first category is those investments needed to maintain or modify the firm's capacity. If the warehouse is a site for future production expansion, the control of space adjacent to an existing manufacturing plant makes sense.

If your goal is to improve profitability, the warehousing investment is considered differently. Many companies have a hurdle rate, and these rates will range from 9% to 20%. If the return on your warehousing investment is not higher than your company's hurdle rate, the investment cannot be made.

There are many methods of calculation, and the best of these will account for the time value of money. It is obviously impossible for this chapter to produce a calculation for your company. However, the guidelines provided should allow you to make the financial analyses to justify your present warehouse operation and evaluate the alternatives.

The three figures shown below provide a checklist and a place to list each of the component costs of a warehouse operation. Not

---

Section C: Total Option Costs

Owned, Leased or Third Party Costs .... _____
Total of All Operating Expenses .... _____
Grand Total for This Option .... _____

Figure 29-4

every item is appropriate for every operation. Users of third-party operations will fill in some blanks, and private warehousemen will omit those items.

Many warehouse operators fail to include all of the costs that are part of the operation, and this list will help you to be sure that no cost item is overlooked.[17]

Monitoring Public Warehouses

One company uses eleven categories to monitor warehouse performance, and they are the following:

1. Prompt receipt of product into inventory
2. Prompt submission of documents
3. Annual DC Inspection
4. Batch control error rate
5. Damage and shortages
6. Freight savings
7. Product receiving and storage
8. Order picking and staging
9. Order processing
10. Customer service
11. Management rating

Consider each category in the performance matrix, since each has a significant role in the overall performance.

Prompt Receipt

All product receipts should be entered into inventory within 24 hours after the product has been received. It is very costly to handle back orders, and you should avoid back orders that occurred because actual receipts were not promptly recorded. The scoring matrix is shown in Figure 29-5, with a maximum of 10 points for perfect performance.

Prompt Submission of Documents

All documentation should be received by us within 10 days after product has been received, including customer returns. It is

Score	Points
98.0 – 100. =	10
96.0 – 97.9 =	9
94.0 – 95.9 =	8
92.0 – 93.9 =	7
90.0 – 91.9 =	6
88.0 – 89.9 =	5
86.0 – 87.9 =	4
84.0 – 85.9 =	3
82.0 – 83.9 =	2
80.0 – 81.9 =	1
< 80.0 =	0
Receipt Matrix	

SOURCE: From an article by C. Alan McCarrell published in Warehousing Forum, Vol. 8, No. 4, © Ackerman Co.

Figure 29-5

essential for us to issue proper credit to a customer for returned product as quickly as possible.

Reconditioning product must also be accomplished during this 10-day time period. The scoring in this category is based on the number of documents received within 10 days divided by the total number of documents received for the month. Points are awarded based on this percentage. The maximum number of points awarded in this category is 10 points. The scoring matrix is the same as the one used to measure product receipt.

Annual DC Inspection

One company inspects all DCs on a scheduled basis. A grade is issued as a result of this comprehensive inspection which takes into account all aspects of handling our business. The warehouse location will carry this grade until the next inspection. Points are assigned based on 20 points for a very good rating, 10 points for an average rating, and 0 points for a below standard rating. The success of the measurement program has allowed the company to reduce the number of inspections for warehouses that are running smoothly.

Batch Control Error Rate

Because of the critical quality control in a health care industry, batch integrity is absolutely essential. The maximum batch error rate shown on the matrix in Figure 29-6 is 1%, and the highest score is achieved by having an error rate of less than 0.09%.

Damage and Shortages

The goal is to achieve the lowest possible damage and inventory shrinkage rates. This category is reviewed monthly by location to identify trends. The damage/shrinkage rate is first calculated separately, then added together to determine overall performance. Again, begin with a zero shrinkage rate and award points based on the shrinkage percentage experienced by location. The maximum number of points awarded in this category is 10 points. As shown in Figure 29-7, a zero score is given if damage and shrinkage rates exceed 0.135% of product moved through the warehouse. One company has a damage and shrinkage allowance of 0.1% of inventory.

% of Error			Points
.0	–	.09% =	10
.10	–	.19% =	9
.20	–	.29% =	8
.30	–	.39% =	7
.40	–	.49% =	6
.50	–	.59% =	5
.60	–	.69% =	4
.70	–	.79% =	3
.80	–	.89% =	2
.90	–	.99% =	1
	≥	1.00% =	0
Error Rate Matrix			

SOURCE: From an article by C. Alan McCarrell published in Warehousing Forum, Vol. 8, No. 4, © Ackerman Co.

Figure 29-6

Percent			Points
	.0	=	10
(.001)	– (.015)	=	9
(.016)	– (.030)	=	8
(.031)	– (.045)	=	7
(.046)	– (.060)	=	6
(.061)	– (.075)	=	5
(.076)	– (.090)	=	4
(.091)	– (.105)	=	3
(.106)	– (.120)	=	2
(.121)	– (.135)	=	1
	< (.135)	=	0
Damage and Shortages Matrix			

SOURCE: From an article by C. Alan McCarrell published in Warehousing Forum, Vol. 8, No. 4, © Ackerman Co.

Figure 29-7

Freight Savings

Today's transportation environment offers opportunities for transportation savings through negotiations with individual carriers. Warehouse suppliers should be active in negotiating freight reductions for their clients. Measure each location's effectiveness in negotiating discounts.

Many public warehouses ship large volumes with carriers by combining freight from several customers in their warehouse operation. Therefore, they have more leverage with the transportation carriers in negotiating discounts. Points are awarded based on the percentage of discounts experienced. The maximum number of points awarded in this category is 10 points. Figure 29-8 shows how the scoring is developed.

Product Receiving and Storage

A standard for receiving and putting away product lets you measure each location's productivity. Measurements are based on the number of pounds of product received and placed in storage per

% Saved			Points
	≥ 45.0%	=	10
40.%	− 45.0%	=	9
35.%	− 35.9%	=	8
30.%	− 34.9%	=	7
25.%	− 29.9%	=	6
20.%	− 24.9%	=	5
15.%	− 19.9%	=	4
10.%	− 14.9%	=	3
5.%	− 9.9%	=	2
.1%	− 4.9%	=	1
	0.%	=	0
Freight Savings Matrix			

SOURCE: From an article by C. Alan McCarrell published in Warehousing Forum, Vol. 8, No. 4, © Ackerman Co.

Figure 29-8

hour. Points are awarded based on the actual pounds per hour handled compared with the standard. The maximum number of points awarded in this category is 10 points. Each location reports the actual number of pounds received per hour. Points are assigned for warehouses that handle more than 11,000 pounds per hour as shown in Figure 29-9.

Ordering Picking and Staging

A standard for picking and staging individual orders has been developed to measure each location's performance. Points are awarded based on the number of cases picked and staged which is compared with the standard. Note that this is based on cases per hour, while receiving is based on pounds per hour.

Provide your warehouses with a recommended product layout which is developed by showing actual usage and inventory turnover by product. Use of the layout is optional, but it should enhance picking productivity. Each warehouse should provide a quarterly report indicating the actual number of cases picked and staged per

Pounds Per Hour			Points
>	20,000	=	10
19,000 –	19,999	=	9
18,000 –	18,999	=	8
17,000 –	17,999	=	7
16,000 –	16,999	=	6
15,000 –	15,999	=	5
14,000 –	14,999	=	4
13,000 –	13,999	=	3
12,000 –	12,999	=	2
11,000 –	11,999	=	1
<	11,000	=	0
Receiving and Storage Matrix			

SOURCE: From an article by C. Alan McCarrell published in Warehousing Forum, Vol. 8, No. 4, © Ackerman Co.

Figure 29-9

hour. The scoring is shown in Figure 29-10. Each location reports quarterly the actual number of cases picked and staged per hour.

Order Processing

This standard measures office activities. Assume that every warehouse document should be completed in fifteen minutes or less. The information is reported quarterly by each warehouse.

Customer Service

There are several ways to measure customer service. Compare line items ordered with the number of line items shipped with the original order. The goal is to reduce the number of back orders generated. This measurement is closely tied to the prompt receipt of product as described earlier.

Management Rating

This is the scoring area which allows you to measure the intangibles involved in handling your business (Figure 29-11). Scoring

Cases Per Hour			Points
	>	260 =	20
253	–	259 =	18
246	–	252 =	16
239	–	245 =	14
232	–	238 =	12
225	–	231 =	10
218	–	224 =	8
211	–	217 =	6
204	–	210 =	4
197	–	203 =	2
	<	197 =	0
Picking and Staging Matrix			

SOURCE: From an article by C. Alan McCarrell published in Warehousing Forum, Vol. 8, No. 4, © Ackerman Co.

Figure 29-10

Cust. Srv. Level			Points
98.0%	–	100.% =	10
97.8%	–	97.9% =	9
97.6%	–	97.7% =	8
97.4%	–	97.5% =	7
97.2%	–	97.3% =	6
97.0%	–	97.1% =	5
96.8%	–	96.9% =	4
96.6%	–	96.7% =	3
96.4%	–	96.5% =	2
96.2%	–	96.3% =	1
	<	96.2% =	0
Customer Service Matrix			

SOURCE: From an article by C. Alan McCarrell published in Warehousing Forum, Vol. 8, No. 4, © Ackerman Co.

Figure 29-11

will include the overall relationship with your company, including responsiveness, cooperation, flexibility, and the attitude of the people at the warehouse.

Implementation

Develop a monthly report to show points awarded by category for each location, and this is compared with scoring for the most recent month and year-to-date totals. Each warehouse is ranked, based on the most recent month's performance, with a second ranking for year-to-date performance. Share this information with each warehouse on a quarterly basis and solicit the comments of each warehouse manager.

Manage the performance rating on an exception basis, with major emphasis placed on the ten lowest ranked warehouses. Try to determine the reasons for the low ranking and work with each warehouse manager to improve performance. This effort obviously improves the overall efficiency of the warehouse network.

As the business changes, you can add, delete, or modify each of the individual scoring categories. Figure 29-12 illustrates how to use the performance matrix in working with your warehouse suppliers.[18]

Guidelines for Measurement Systems

Many warehouse measurement systems have not been successful. Those that have succeeded generally follow certain principles, which are listed below:

- Keep it simple. The best systems are those every warehouse employee can understand. They not only know how the system works, but also what the system is and why it is there. Systems designed with participation from warehouse employees have an excellent chance of success.

- Avoid making frequent changes in standards. With warehouse standards, accuracy may be less important than consistency. Warehouse work changes frequently, and constant

Ross Laboratories Performance Matrix
Quarterly Results

DC Name: Minneapolis Public Warehouse   Quarter: 2
                                           Year: 1991
                                           DC No.: 56

Category	YTD Actual	YTD Goal
Prompt receipt of product into inventory ..	60	60
Prompt submission of documentation	50	60
Annual DC inspection	120	120
Batch control error rate	50	60
Product damage/Inventory shrinkage	41	60
Freight savings	18	60
Product receiving and storage	54	60
Order picking and staging	102	120
Order processing	45	60
Customer service	60	60
Management rating	54	60
Totals	**654**	**780**

Year-To-Date Efficiency Score = 83.85%
Year-To-Date Overall Rank =   3 out of 32

SOURCE: From an article by C. Alan McCarrell published in Warehousing Forum, Vol. 8, No. 4, © Ackerman Co.

Figure 29-12

changes in standards may make measurements meaningless. Furthermore, frequent standard changes may be the source of employee distrust.

- Create measurement systems to cope with predictable changes in the product line. Every warehouse inventory has certain predetermined variations in product mix that can be anticipated when establishing a standard.

- Never use performance measurement systems for worker discipline. Maintenance of such systems depends on worker cooperation. If they are seen as a way for management to spy on the workers for disciplinary purposes, the entire reporting system will break down.

30

SCHEDULING WAREHOUSE OPERATIONS

Perhaps the biggest drain on productivity in most warehouses is the "hurry up and wait" syndrome popularly associated with the military. One example is the unloading of a truckload of floor-loaded bags of product at a warehouse. There are at least three alternative loading methods. First, two men may work in the truck to palletize the freight for a lift-truck driver who removes the loaded pallets. Second, a single man may load the pallets, receiving assistance as needed from the fork-lift driver who is transporting the pallets. Third, a single man could work the truck alone, equipped with either a lift truck or a pallet jack to remove loaded pallets. The third method eliminates lost time because there is never a time when anybody is waiting for anyone else. With the first method it is likely that the two workers will sometimes be waiting for the lift-truck driver. With the second method the man in the truck will have no work to do if the lift-truck driver is delayed in returning.

Scheduling to reduce waiting is possible once the manager knows the time required for each element of each job, as well as the likelihood of delay or interruption.

By measuring elapsed times, the manager can estimate when one task will be completed and another started. With this knowledge, truck appointment times can be set to reduce waiting time and congestion. If carrier drivers know they will receive prompt service, they are more inclined to cooperate with the appointment demanded by warehouse management.

Scheduling for Peak Demand

In some businesses the scheduling problem of the warehouse is complicated by uneven demand. For example, in the retail grocery business there is a typical peak on Thursday, with abnormally low demand on the first three days of the week. As a result, the food chain warehouse will probably have too much work for Monday and Tuesday to supply the Thursday peak and not enough for Friday. Such unbalanced workloads can be partially balanced by delaying some work having minimal penalty, such as unloading of inbound box cars.

One force pushing against the 40-hour week is the need to improve asset productivity. When capital is not available to purchase more facilities and equipment, the only way to extend the life of the existing assets is to use them on a multishift basis. When you lack either the time or the money to acquire additional capital assets, moving to multishifts may be the only answer to increasing the output in the warehouse.

Why 40 Hours Is No Longer Enough

A growing number of managers recognize that customer expectations today cannot be met with a 40-hour work week. With a continuing emphasis on shorter order cycle times and better service, 40 hours is simply not enough. At the same time, recognize that it is more difficult to manage people on a second or third shift. It is on these late shifts where you find the greatest frequency of accidents, substance abuse, and other disciplinary problems. Three of the worst accidents in recent history—the nuclear accidents at Three Mile Island and Chernobyl and the explosion of the Challenger space capsule—were traceable to fatigue of workers functioning on a night work schedule. A major question for warehouse managers today is how to maintain motivation and quality while keeping the night shifts running.

A New Kind of Work Week

As you consider alternates to the 5/8 week, notice the change in the description of many warehouse jobs during the past few decades. There was a time when warehouse work was physically demanding and fatiguing, but many warehouse jobs today are easier than ever before. Increased unitization, better use of power equipment, and improved ergonomics have all combined to reduce physical effort. While yesterday's order picker may have been exhausted at the end of an 8-hour shift, today's worker can work 10 or even 12 hours without undue fatigue. Another change is the similarity of work in the warehouse office to that done on the warehouse floor. Today's warehouse worker is likely to operate a computer terminal as well as a lift truck, and the line between clerical and materials handling tasks is less distinct than ever before.

Given these conditions, you can be creative in introducing alternates to the 5/8 week. One option with proven popularity is the 4/10 week. This schedule is very popular with hourly workers, because everybody has a three-day weekend which eliminates one day of commuting between home and work. But what if customer service demands will not permit your warehouse to operate on a schedule of only four days per week? How do you offer the 4/10 schedule while keeping the warehouse open five or six days per week?

One answer is a two-shift schedule with an overlap to allow the warehouse to be open five or six days per week.

Four Options

Under the schedule shown in Figure 30-1, the first shift works a 4/10 week Wednesday through Saturday. A second shift starts on Sunday and works through Wednesday. On Wednesday only, both shifts function, which means that the second shift moves from day to night on that day only. This schedule allows the warehouse to run for 80 hours per week with night work performed on only one day of the week. The burden of working on the weekend is shared by both first and second shift, with one group working Saturday and the other on Sunday. A third shift could be added as shown to increase open hours per week to 75% of the available hours.

Some warehouse operations utilize the availability of people

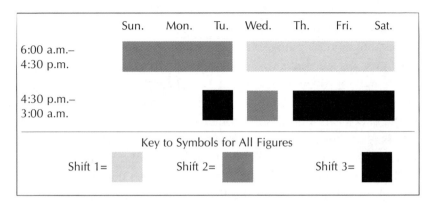

Figure 30-1

who are seeking part-time work to supplement another job or to finance college. These part-time workers may be anxious to have work on nights or weekends when they have available time for a second job. Such people might be accommodated by a second shift which is concentrated on the weekend, with 8-hour shifts on Friday night and Monday night and 12-hour shifts on Saturday and Sunday. The first shift works a 4/10 schedule as shown in Figure 30-2.

Another accommodation for the part-time worker would be a 5/6 schedule, with a second group working a five-night work week at shifts of just 6 hours in length. Working mothers or students would welcome this kind of schedule which leaves the daytime hours free for other activities. Figure 30-3 shows how the 4/10

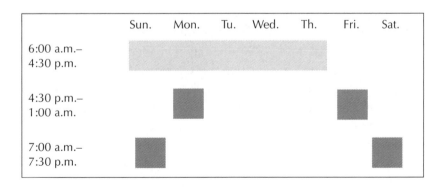

Figure 30-2

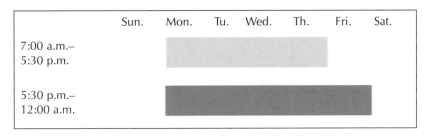

Figure 30-3

schedule described earlier could be supplemented by a late shift designed specifically for part-time people.

There are countless variations to these schedules that can be developed once you recognize that the traditional 5/8 work week is not a necessity. The question of whether shifts of varying length will be popular with your workforce is one that should be tested by asking people if they would like to try it.

Many managers worry that changes such as those described will be unpopular and even disruptive. Yet there is ample evidence that the majority of hourly workers would welcome a job that doesn't call for five 8-hour days. If you doubt this, ask the people working in your warehouse!

31

CUSTOMER SATISFACTION—
THE ROLE OF
THE WAREHOUSE

Traditionally, the orientation of most private and third-party warehousing has been toward the need of the producers (manufacturers). This mind-set is typical of a developing market in what is sometimes called an "If I make it they will come" mentality. We know this as "push distribution" with the manufacturers calling the shots, "pushing" inventory into the market.

Who is YOUR Customer?

You don't like change? Well, if you are not changing, reinventing yourself, you are not just standing still, but are going backward!

Next time you are sitting down with one of your customers, ask yourself if you are in a commodity business. Is price the only thing that you are competing on? If this is the case, you definitely are in a commodity business. You can market and sell "value-added differentiation" all you want, but if your customers are buying price . . . you had better be the low-cost producer!

Have you asked your customers what they really think of the services you are providing them? Are you selling one thing and your customer is buying something else?

If you will not or cannot be that low-cost producer the challenge returns to: "How can we differentiate ourselves?" We can start by asking ourselves and our customers those questions that identify our core competencies.

Essentially, warehousing service is a commodity that is usually bought at the cheapest price. It is the relationship that you have with your customer that can change the game into a vital service that can demand a premium value. The ideal warehousing information system needs to provide the best in operations support, but it has to do that "next best thing." Quality, value, and the timeliness of service have to be a given.

Why Do Something Well That You Should Not Be Doing At All?

Your company's information system reflects your operation's culture and carries a tremendous amount of momentum. Change is difficult. Most warehousing information systems are oriented toward the production aspect of distribution, not the pull aspect of distribution.

A few years ago most warehousing information systems demonstrated Pareto's 80/20 rule—80% of what the system was used for was completed with 20% of the total effort. The remaining 20%—if it would only stand still—would require the remaining 80% of the work. And just what was that work? It has been adding the "sizzle" that warehouse operators want to sell today; things like bar coding and RF (radio frequency) terminal capability. Some companies are continuing to pour big money into that last bit of technology. But will this be that "magic bullet" that you are looking for? Or, in the end, is it the low-cost warehouse operator who will win the new business anyway?[19]

The Importance of the Customer's Customer

Years ago, there was a time when the responsibility of the warehouse manager ended when shipments were loaded in a boxcar or placed on the tailgate of a truck. Once the bill of lading was signed by a carrier representative, the warehouseman's responsibility ended. If the delivery was late or in poor condition, the fault lay with the carrier, not with the warehouse. Those days of buck passing are long gone, and nearly every warehouse manager today recognizes that the responsibility for quality does not end until goods are received on time and in excellent condition by the ultimate consignee. And if you are a third-party warehouse operator, that consignee is not the customer, but rather the customer's customer. If you are a private warehouse operator, the consignee might be your customer, another division of your company, or the customer of another division.

In recent years, progressive warehouse managers, in both third-party warehousing and private warehousing, began to pay attention to the customer's customer or the ultimate consignee. They recognize this as an opportunity to improve quality of service to the primary customer by listening to the suggestion of the most critical party, the person who receives the shipment.

Visits to consignees have both long-range and short-range goals.

For years, both truckers and warehouse people had a short-range goal in contacting the receiving manager at the consignee. That short-range goal was simply to expedite the unloading of their trucks. Particularly in the food industry, the receiving dock is typically a serious bottleneck, aggravated by excess volume, inadequate dock capacity, and poor scheduling. Every party delivering freight to those docks has been keenly aware of the value of a friendly relationship with the receiving manager. Unfortunately, establishing this relationship has occasionally involved actions that could best be described as bribery. Furthermore, the relationship with the receiving people was generally negative. The goal of the shipper was to prevent complaints and freight claims and to avoid lengthy delays in unloading. It was rare that visits to the receiving people involved any exploration of ways to improve service.

In contrast, other visits to consignees have taken a long-range view. The objective of these visits is to achieve a thorough understanding of the goals and the problems of the consignee, as well as the steps that the warehouse and carrier can take to achieve those goals. These visits are not made to grease skids or solve complaints. Furthermore, such visits should not become the "bitch session" that sometimes occurs between receiver and shipper.

The long-range goal of both warehouse and consignee should

be to improve utilization of assets. Obviously asset utilization is improved by reducing detention of delivery vehicles, but it is also improved by moving inventory to the right place at the right time and in good condition. Improving utilization of assets is a win/win goal, one that helps both the consignee and the service producer.

How Will You Use What You Learn?

Consignee visits can produce mountains of data, and this must be sorted into useable information. As a guideline for future visits, a format for a call report can be developed that captures the essential information about that visit. Table 31-1 shows the format for a call report developed by DSC Logistics.

Table 31.1

ABC GROCERS 1100 West Adams Chicago, IL 60622	Midwest Ringmaster Program Chicago - 43rd Street; Des Plaines; Melrose Distribution Pick.N.Pack; Revel
Participants:	Jane Doe—Distribution Mgr. John Doe—Traffic Coordinator
Call Date: Next Anticipated Call:	01/28/91 07/91
Primary Accounts:	Vendor A Vendor E Vendor B Vendor F Vendor C Vendor G
Number/Type of Customers:	325 independently owned grocery stores
Service Area:	Illinois, Indiana, Michigan Iowa and Wisconsin
Total Receipts:	75,000 cases per week
Receipts from DSC:	33,758,701 pounds; 1,438,337 cases; 1,522 orders

---

An executive summary is maintained to record key statistics regarding each consignee. Typically this information will be a comparison of two different years for the same consignee, and it will reflect the following statistics:

- Total pounds shipped to consignee
- Total pounds picked up by consignee
- Average pounds per order
- Percent of pounds delivered
- Percent of pounds picked up

You may wish to rank consignee activity for comparison purposes. A consignee activity summary would list the following items:

- Rank of consignee among all
- Percent of pounds compared to total
- Total number of orders
- Average pounds per order
- Average cases per order
- Percent of orders delivered
- Percent of orders picked up
- Percent of pounds delivered
- Percent of pounds picked up

These visits help the consignee as well as your company. The contact allows the consignee to make suggestions that will improve operations, and those suggestions should be carefully noted and recorded. Furthermore, you need to track your ability to implement those suggestions so that the process is not perceived as a waste of time.

One warehouse lists action items that should be achieved when contact is made with the customer's customer:

- Identify improvement in warehousing efficiencies.
- Identify improvement in transportation efficiencies.
- Is on-time delivery service up to expectations?

- Is completeness of order fill within expectations?
- Is the method of ordering as efficient as it could be?
- Have we created the proper network for contacts?
- Can we divert freight from distribution centers directly to stores?

While it is useful to note changes in activity and relative importance of each consignee, certainly the most important value of consignee visits is the creation of a cooperative relationship with the people who receive merchandise from your warehouse.

What you learn from them can and should enable you to make your operations more service oriented as well as more efficient. Some of the things you learn will save money both for the consignee and for you. Your goal is not just to win additional business, but rather to improve the quality of business being handled today. Improving relationships with the ultimate consignee may be one of the best ways to raise quality in your warehouse.

32

IMPROVING ASSET UTILIZATION

Most productivity improvement programs concentrate on improving utilization of labor. Yet there are four other major cost areas in warehousing where making better use of assets can boost productivity: space, energy, equipment, and inventory.

Space Utilization

The warehouse space most frequently wasted is that closest to the ceiling. While there are limitations to the pile height for most commodities, it is not unusual to see cube space wasted because merchandise is not piled as effectively as it might be. Racks or stacking platforms will improve cube utilization as will good planning.

Reducing the size of working aisles also may improve utilization of cube. However, reducing aisles can involve a tradeoff between space and labor. Since better labor productivity is achieved in wider aisles, it is necessary to balance the cost of the space against the cost of the time.

Energy Utilization

Perhaps no commodity has undergone such violent and unpredictable price changes within the past several decades as energy. With these sudden changes of pricing have also come great changes in our attitude toward energy as a resource. Buildings constructed when energy was cheap will typically show enormous heat losses through insulation which, by today's standards, is grossly inadequate. In contrast, buildings built more recently are governed by building codes that require substantial protection from heat loss. Changes in the price of energy have changed many procedures in construction. Extensive use of skylighting in older buildings was justified by a reduction in the price of illumination. More recently, some types of skylighting have been found to lose large amounts of heat energy, which is more costly than the electricity for lighting that they might save.

Warehousing practices of a few years ago were based on the assumption that energy was a cheap resource. Many storage buildings were designed to allow workers to be comfortable in shirt sleeves, even in mid-winter. More recently, warehouse operators have questioned some of the reasons for heating a building to such levels.

Essentially, there are three reasons to control warehouse temperatures:

1. *Product environment:* Many products are damaged by freezing, and others will deteriorate if the temperature is too cold. Other products must be kept frozen or cool in order to be preserved.

2. *Fire protection:* Most sprinkler systems are wet-pipe systems, which require protection from freezing in order to operate.

3. *Labor productivity:* Warehouse workers are likely to work more slowly or require warm-up breaks if they work in a totally unheated area in which temperatures fall below freezing. On the other hand, a properly dressed warehouse worker is likely to be nearly as productive in a 40°F building as in a 70°F building. The only exception might be precision warehousing jobs such as order picking or packing. Comfort levels in these special areas can be maintained with spot-heating units, such as infrared fixtures.

With the emphasis in recent years on conservation, many practical means have been discovered to use energy wisely. High-intensity discharge (HID) lighting fixtures save substantial costs in electricity. However, most HID lighting systems require a high-voltage wiring which is expensive to relocate. HID lighting systems should be specifically designed for relocation, with long cords to allow fixtures to be moved.

Any warehouse heating system designed before the price of energy started to climb is obsolescent today. Perhaps the greatest advance in warehouse heating is a realization that circulation is as important as heat. For this reason, heaters with fans are a great deal more effective than radiant heating devices.

In addition to insulation, great progress has been made in developing better sealers for all kinds of warehouse openings. This includes the use of flexible vestibule-type doors that swing open when a lift truck or any other kind of transporting mechanism moves through the doorway opening. The same principle can be used to provide flexible doors for craneways or other openings used by more complex materials handling equipment.

Many warehouse operators purchase energy surveys of their buildings by professionals, but a simple checklist, such as that shown in Figure 32-1, will help.

Increasing use of semiautomated warehouse equipment is an-

Energy Checklist

1. Are order pick-areas in the warmest part of the warehouse?
2. Are lighting levels high in order pick-areas?
3. Are picking vehicles equipped with headlights which could replace warehouse lights?
4. Are thermostats locked so that heat level is controlled only by management?
5. Are heat losses through doors and automatic dock boards controlled with weather stripping and flexible vestibule doors?
6. Is grounds lighting controlled by timers or daylight sensors?
7. Are all heat circulators functioning without blockage?
8. Are walls and ceilings painted white to maximize light reflection?
9. Are warehouse lights designed so that they can easily be relocated when aisles are changed?

Figure 32-1

other means of saving energy costs. One warehouse has equipped its stock-picking trucks to operate in darkness. Each truck has headlights that illuminate the stock being picked. Because the truck has its own lights, lighting in the warehouse can be kept to a minimum.

Saving energy in the warehouse is largely a matter of common sense. Remember that warehouse workers do not need to be in shirt sleeves to be effective. While personnel comfort is obviously desirable, in the warehouse a degree of comfort can be achieved through clothing rather than temperature control. Because most warehouse jobs are active, a reasonably dressed warehouse worker can function effectively in temperatures that are somewhat more variable than those found in an office.

With reasonable vigilance and good equipment maintenance, any warehouse operator can achieve significant savings in cost of light and heat.

Equipment Utilization

Every lift truck should be equipped with an hour meter that can be used not only to show engine running time, but also to indicate when to perform regular preventive maintenance. Hour meters do not measure the amount of time a lift truck is actually in use. In-use time must include time when the engine may not be running, but the lift truck cannot be made available to another operation. To illustrate, a driver may be unloading a truck and at the same time must hand-stack cartons or process paperwork. The engine may be turned off for that portion of the time, which stops the hour meter. Yet the truck cannot be used for another operation during this brief waiting period.

One warehouse operator uses a rule of thumb that lift trucks are actually in use for 75% of each shift. This means that an hour meter that shows 5.5 hours for a 7.5-hour shift is actually at 100% of normal capacity.

The Equipment Utilization Ratio

There are three basic elements in calculating an equipment utilization ratio:

1. The number of days in the work month (less weekends and holidays)
2. The number of active lift trucks in the system
3. The total hour meter readings from the active lift trucks

The equipment utilization ratio is calculated by dividing the total hours from all hour meters by the available-hours base number. The base number is determined by multiplying the number of active lift trucks times the number of worked days in the month times 5.5 available hours per lift. Figure 32-2 is an example of how to calculate the ratio.

A monthly utilization report shows monthly equipment utilization figures for each operating location. Managers can see their own utilization, compare themselves with others, and begin to inquire into the whys and hows.

Improving Performance

There are several uses for this program. At the corporate level, utilization ratios may be used to evaluate and plan equipment allocation among various operating locations. The combined utilization ratios also will provide an indication of whether there might be too much equipment at a particular location.

On the operating level, local managers should use this report to evaluate their own performance. When regularly confronted by a low utilization ratio, they should ask themselves if they have a surplus of equipment. Because demand for equipment can fluctuate depending on the seasonality of distribution activity, managers should consider the alternative of short-term rentals to complement their base lift-truck fleet during periods of high activity.

Although an equipment utilization ratio of 80% is certainly acceptable, managers can achieve ratios greater than 100%. This is accomplished by adding more shifts to the operation. The available hour base (see Figure 32-3) is determined by the number of lifts required to operate one shift. By spreading the workload over two or three shifts, less equipment would be needed to accomplish the same amount of work.

Calculation of Utilization Ratio
At Location A

I. **Data**

Number of days worked	23 days
Number of active lift trucks	10 trucks
Total hour meter readings	1,199 hours

II. **Base Calculation**

Base = (# days) × (# trucks) × (5.5 hrs./truck)

= 23 × 10 × 5.5

= 1,265 hours

III. **Utilization Ratio Calculation**

$$\text{Ratio} = \frac{\text{Total hour meter readings}}{\text{Available base hours}}$$

$$= \frac{1,199 \text{ hours}}{1,265 \text{ hours}}$$

= 94.8%

SOURCE: Warehousing and Physical Distribution Productivity Report, Vol. 17, No. 12. .

Figure 32-2

There are additional costs in multiple shifts, such as supervision, shift premiums, and utilities. These extra costs must be weighed against the savings of decreasing the equipment asset base. Additional shifts are one way to absorb some periodic high activity without resorting to short-term equipment rental.

Lift-Truck Rebuild Program

Reducing the investment required for a fleet of lift trucks is of major importance, but it's equally important to extend the life of that investment. A lift-truck rebuild program is an integral part of increased utilization planning.

Calculation of Utilization
With Different Shifts

I.	**One Shift Operation:**	
	Usage hours	1,199 hours
	Available base (10 trucks)	1,265 hours
	Utilization ratio	94.8%
II.	**Two Shift Operation:**	
	Usage hours	1,199 hours
	Available base (8 trucks)	1,012 hours
	Utilization ratio	118.5%
III.	**Three Shift Operation:**	
	Usage hours	1,199 hours
	Available base (6 trucks)	759 hours
	Utilization ratio	158.0%

SOURCE: Warehousing and Physical Distribution Productivity Report, Vol. 17, No. 12.

Figure 32-3

Overall lift-truck utilization is calculated at 75% efficiency, or approximately 1,500 hours per year. From a controlled maintenance program you should expect to get about 10,000 hours' operation prior to requiring a major overhaul. Depending on usage, this would be six to ten years. If you rebuild in your own workshop, you can expect the average cost to be between 30% and 35% of a new truck. If you have the rebuilding done by one of the major truck dealers, you can expect the cost to be about half that of a new truck.

Key Points of the Rebuild Program

A good rebuild program requires the complete dismantling of the lift truck and inspection of every operating part.

The engines should be completely rebuilt, including the reboring of cylinders, or installation of a new block. It is normally more economical to have this job done by a contractor.

Transmissions and differentials must be inspected and worn parts replaced. Steering mechanisms, brakes, and electrical portions (including the electrical harness) also should be inspected. Most chains, cylinders, and hydraulics should be repaired and/or rebuilt as necessary.

Every part should be inspected and any one with discernible wear replaced. This includes seats, pedals, meters, tires, etc.

The truck should be completely repainted and equipped with new anti-skid strips, knobs, handles, and decals. The vehicle should look, feel, and run like a new machine.

To make sound decisions on rebuilding, the warehouse operator must have a meticulous maintenance program. Without accurate repair costs, it is difficult to determine whether it is time to rebuild, or whether the rebuild program will be economically feasible.

For example, a schedule might call for rebuilding at 10,000 hours. However, if one truck had a major failure at 8,000 hours, it would be sensible to schedule rebuilding at that point, rather than fix the failure and then rebuild just 2,000 hours later.

If hourly maintenance costs remain relatively low, a second rebuild may be justified.

The Value of the Program

A rebuild program has several advantages. For example, at a cost of a third to half that of a new lift truck, the rebuilt truck's service life is extended by 80%. Even with the extension, the resale or trade-in value of the truck will increase.

Materials handling equipment is a major budget item for any distribution facility. A study of the grocery products industry showed that equipment costs represented an average of 4% of total distribution center costs, nearly as much as the percentage spent for warehouse supervisors. Yet few organizations carefully control equipment costs.

When controls are lacking, equipment is purchased that is not really needed. Frequently, managers will "squirrel" spare equipment that is underutilized and should be sold or traded for newer models. In other cases excessive maintenance is expended on a piece of equipment not worth repairing.

Good controls based on meter utilization will show when there are too many lift trucks in a fleet. When you find excess equipment you might move some of the excess to another location where more equipment is needed.

Records of repairs expressed as a cost per operating hour will show which units are worth repairing, as well as which ones have been unusually good or bad from a maintenance standpoint. Judgment is an important part of the rebuild program, since there may be cases where an older truck has been made obsolete by technological advances.

With a combination of fleet utilization records and judicious use of a rebuild program, you can bring down the costs of a materials handling fleet as a percentage of total operating costs. And you can do it with no sacrifice in warehouse productivity.

Inventory Performance

Not every warehouse manager is responsible for inventory performance and very few public warehouse operators have any input at all into inventory planning.[20] Some private warehouse operators are not held responsible for this phase of the operation either. Yet failure to achieve satisfactory inventory performance has caused the closing of many a warehouse, both public and private; so every warehouse operator should understand the essentials of effective inventory performance.

After personnel, inventory is the most important asset in most companies. Inventory is volatile, dynamic, and complex. It is affected by both external and internal forces. External forces that influence inventory include general business conditions, inflation, customers, market conditions, technology, vendors, seasonality, and competition. Internal forces include management awareness, company policy, engineering development, inventory control techniques, forecasts, availability of capital, and storage capability at multiple levels (raw materials, work-in-process, finished goods).

Realistically, a company will never have a completely accurate inventory. It is the degree of imbalance that is critical, since there will always be some excess inventory or shortages. This can be managed, however, since inventory problems do respond to an organized, focused effort.

The Management Factor

The most influential factor affecting inventory and service level—more important than the computer system or forecasting methodology—is senior management's interest or awareness. Observe the effect on inventories when a president or senior officer isn't happy with present inventory levels (whether they are high or low). Frequently, the source of bungled decisions is senior management's sudden realization that inventory levels are too high, or service levels too low. Management, simply by expressing concern, can have a significant impact on inventory levels and their profit contribution to the business.

Internal forces are more significant than external forces in determining inventory and service levels. Decisions to reduce inventory, when not part of a planned program, generally result in shortages of fast-moving items while the slow-movers sit in the warehouse.

Why Measure?

Why should you measure inventory performance? Why should you circulate the results of your measurements when, after all, they could be embarrassing? Measurement highlights many tough issues—many that some managers would prefer not to face. Yet measuring inventory levels is necessary to evaluate and monitor the effectiveness of your inventory management system. It identifies where you can plan improvements, track progress, and highlight success. It provides a forum for communicating and working with other managers on a topic of critical importance to the profitability of the business.

By measuring inventory performance you can determine the most desirable inventory levels for the top 20% of the high sales/usage items, develop programs to use (or scrap) obsolete items, and organize the warehouse so the fast-movers are located nearest the shipping doors.

The performance measurement system can be used as a base from which to measure improvement in inventory turns. Too often, inventory management and control policies or systems are implemented without the means to measure their effect.

Danger Signals

Here are some of the red flags to watch for:

Your inventory turns are less than industry experience. Is the majority of your inventory turning markedly less except for a few items, or are a few items responsible for the majority of your turns?

Sales volume shows significant growth, with little increase in profitability. Could the cause be pricing structure, product mix, product cost, or, perhaps, excessive inventories?

Inventory is growing at a faster rate than sales. Should production be slowed or purchases delayed? Perhaps it is time to consider reducing the labor force—a layoff may be in order.

You are hanging on to surplus or obsolete inventory with the hope of selling it one day. Do the costs associated with warehouse space, handling, and administrative time exceed the potential that can be realized from selling the inventory at a reduced price?

Your inventory and on-order records are not 95%-to-98% accurate. Inaccurate records generally result in additional safety (or fudge) stocks, with a corresponding increase in expenses for inventory carrying cost.

Your deliveries are consistently too late or your lead times are too long. In a manufacturing company late deliveries or a longer than necessary lead time most often means excessive work-in-process inventories. In a distribution company, you may be holding or storing items while waiting for an order to be shipped complete, or you may be taking longer than necessary to process, pack, and ship.

There are material shortages, or production is delayed due to lack of material. Perhaps order reviews should be more frequent, or the inventory management methodology revised.

There is a reluctance to mark down stock. Often, inventory carrying costs for an item over weeks or months will exceed the markdown of a special promotion.

Inventory is growing as a percentage of assets. Instead of investing in productivity improvements, is capital going to unnecessary inventory?

Cost systems are inadequate. Perhaps your inventory is overvalued or undervalued (particularly if scrap is not properly accounted for).

Sales forecasts are used to manage inventory. Are the forecasts consistently too optimistic, resulting in excess stock?

Sales or usage is not analyzed by customer, product line, or profitability. If not, do you really know the inventory or service level—or what is desirable?

Senior management does not review inventory performance. It should review inventory performance, obsolescence, and service level periodically. Is senior management attentive, interested, and aware of inventory performance?

Your warehouse or storeroom is running out of space. This can be a sign of serious trouble, particularly if sales (adjusted for inflation) are not increasing significantly, or lead times are lengthening.

Your service level is increasing significantly with no apparent reason, or backlogs are declining. Excessive inventories may be just over the horizon.

Your company has had a large inventory adjustment or write-off recently. This can indicate recordkeeping (receipt/issue control) that frequently results in simultaneous excessive inventory and stock-outs.

If any of these red flags come up in your company, you should take a hard look at your inventory performance on an item-by-item basis.

Inventory Turnover and Distribution Analysis

Every company with inventory can and should apply straightforward, tested methods to measure its inventory performance. Two basic reports are the key: a turnover analysis and a distribution analysis. Both should include a one-page age summary by major inventory category. Properly prepared and used, these analyses can provide an objective and quantitative diagnosis of your inventory.

It is important to measure performance on an item-by-item basis as customers buy individual items, not groups or product lines. Then, summarize your individual measurements by category to help identify particular problem categories or situations that are not obvious from either the detail or the total.

Often, managers judge inventory turns by dividing annual sales or usage dollars by the present or average inventory level, thus failing to recognize the fluctuations—which can be significant—or status of the individual items making up the totals.

For example, take the case of the firm whose inventory turned an average of five times a year. This was acceptable to management and in line with other firms in the industry. Yet an item analysis showed that 15% of the items were moving 25 times annually, while 85% of the items were turning only twice a year. On an item-by-item basis, the firm was in trouble. The average turnover of five turns simply did not reflect the true condition of the inventory.

Always use a common denominator in performance measurement. Dollars are generally accepted, although for some companies pounds or gallons are more relevant.

A turnover analysis shows the number of months' supply on hand, calculated at the current or forecasted rate of usage. Inventory balances are divided by the average usage per month, and arrayed in descending order. Stated another way, turnover is computed as the ratio of annual usage divided by the inventory.

Identifying low-stock or no-stock items in the inventory with demand during the year helps you spot a potential service-level problem. Conversely, you can age the inventory to determine the value that is potentially excess by identifying that portion of inventory that has had little or no turnover during the past year.

Case History

Figure 32-4 is a turnover analysis report, using actual data from a small company that manufactures products sold primarily through distributors. Thirteen hundred finished goods products, with sales or inventory, have a manufacturing lead time of two to six weeks.

The company has received occasional complaints from customers on product availability. The firm has had a computer for six years, used primarily for financial reporting. Production scheduling was done by plant management. A tour of its crowded warehouse indicated that some stock was inactive.

The red flags clearly indicated a potential problem. A turnover and distribution analysis was run to quantify the problem and to aid in developing solutions.

The charts shown here are "snapshots" of the actual reports. Data are shown on only 10 out of the 1,300 line items that were actually analyzed. The item/part number has been disguised to maintain confidentiality, but the balance of the data is real.

In this illustration only those columns critical to a turnover analysis are shown. Other data pertinent to an item-by-item analysis within a company (such as description, unit of measure, category codes, new item identification, and on-order status) were included on the actual report but are omitted from the illustration.

The full report shows that 33% of the total number of items (430 out of 1,300) are stock out. This is an indicator of a potential customer service level problem. Sixty percent of the items have on-hand inventory equal to or greater than a four months' supply with 33% of these items having a greater than 12 months' supply. Many of these are surplus or potentially obsolete and should be reviewed for discontinuance and disposition.

The turnover analysis report summarized by stocking category code is shown in Figure 32-5. Note that 77% of the excess inventory is in Category 03, which has only 49% of the part numbers. This summary exposes areas that require more or less emphasis.

Figure 34-4. Turnover Analysis (Abbreviated format)
(Snapshot of 10 items out of 1,300 in the full report)

Item/Part No.	Cumulative % Items	Inventory Units	Inventory $	# Months of Coverage**	$ Coverage More Than 12 months	24 months	Comments
1001	1 (13th item)*	0	0	0	-	-	
4050	10	0	0	0	-	-	Probable service
4123	33 (#430)	0	0	0	-	-	level/lost sales!
9701	40	3	136	.7	-	-	
5945	50	391	586	1.9	-	-	
6173	60	4	20	4.0	-	-	4 to 6 weeks
1711	66 (#858)	50	2,000	12.0	400	-	mfg. lead time!
2105	80	30	891	24.0	446	-	At 30% per year!
1796	90	617	23,175	57.4	18,337	13,500	4.7 yr. supply
8731	100	28	1,498	999.9	1,498	1,498	All stock inactive!
Totals			$1,262,000		$402,811	$243,877	
				32%		19% of total inventory	

*of 1,300
**At a forecasted rate of usage

Figure 34-5. Turnover Analysis Summary (Abbreviated format)

Inventory Category Code	# of items	% of items	$ Coverage more than 12 months	24 months
01	332	26	38,782	9,094
02	201	15	55,341	30,247
03	640	49	288,439	189,060
04	38	3	4,514	4,232
05	89	7	15,735	11,244
	1,300	100	$402,811	$243,877

SOURCE: Warehousing and Physical Distribution Productivity Report, Vol. 17, No. 11., © Marketing Publications, Inc., Silver Spring, MD.

A distribution analysis (or ABC analysis) is illustrated in Figure 32-6. This company's inventory situation follows Pareto's Law (the 80–20 rule) very closely. The careful control of a few items will provide a high degree of overall control. Vilfredo Pareto (1842–1923) was an Italian philosopher who determined that income distribution patterns were basically the same in different countries. That is, a small percentage of the population had the majority of the income. Translated more broadly, a relatively small number of items account for a large proportion of the activity. This phenomenon is one of the most applicable and effective, yet least used, of the basic principles of inventory management.

The items in the distribution analysis are arrayed in descending value of annual sales (or usage) dollars at inventory cost. In Figure 32-6, through item number 2004 (the 13th item on the total report of 1,300 items), the forecasted annual costed sales are $58,289, with a cumulative sales usage of 19.9% (of the total sales). These can be considered the "A" items; therefore, they need to be tightly controlled.

Figure 32-6 also shows that 20% (through item 4004) or 260 of the items account for 80.8% of the total sales, but only 55.6% of the inventory.

What was learned from the turnover analysis and the distribution analysis? At first management was not concerned with the total number of 4.8 turns a year. However, when the inventory was detailed on an item-by-item basis, an alarm was raised.

Figure 32-4 shows that 33% of the items (below the third line) have an inventory greater than a 12-month supply. Another 33% of the items (above the second line) with demand during the most recent 12 months have no inventory. Clearly, this inventory is out of control. The company was losing sales to competition because of lack of inventory, yet its inventory investment was excessive, and there was a disproportionate amount of slow-moving or inactive items.

What was done? After an in-depth management analysis of the situation, the company developed and implemented an action plan that included:

Distribution Analysis
(Abbreviated form — snapshot of 10 items out of 1300 in the full report)

Item Code	Cumulative % Items	Descending Order Annual Sales/Usage $	Cumulative % Usage $	Inventory Units	Cumulative $	Cumulative % Inventory
2004	1 (13th item)*	58,289	19.9	0	0	9.8
6041	10	10,692	65.0	0	0	44.4
4004	20 (260th item)	4,981	80.8	19	830	55.6
5321	30	2,841	88.8	0	0	65.6
4013	40	1,824	93.7	45	1,026	72.5
4348	50	1,008	96.6	13	222	77.8
5632	56	775	97.7	8	304	79.3
5721	70	342	100.0	46	2,097	87.3
7051	91	1	100.0	12	3	93.5
8017	100	0	100.0	102	2,703	100.0
* of 1300	TOTALS	$6,057,922			$1,262,300	

Average turnover rate: 4.8 times

SOURCE: Warehousing and Physical Distribution Productivity Report, Vol 17, No. 11. © Marketing Publications, Inc., Silver Spring, MD.

Figure 32-6. Abbreviated Distribution Analysis

- A weekly senior management review of schedules and inventories of the top 10% (A items) of the sales items.

- Hiring a production planner to schedule production based on needs versus ease of manufacture.

- A rework disposition program for inactive items.

- A product line review that resulted in dropping many slow-moving, unprofitable products.

- Installation of a forecasting system to help the company better anticipate its requirements.

- Rearrangement of warehouse stocking locations to make it easier to pick high-volume items.

- An ongoing inventory performance measurement system to provide the base from which to measure improvement.

An Action Plan—And Potential Benefits

To implement a comprehensive inventory performance measurement system within the company, you must have a specific action plan. The major steps in the plan include: designing a turnover and distribution analysis, with appropriate summaries; reviewing requirements with the data processing staff; establishing targets or goals for service and inventory levels; analyzing the reports generated to identify opportunities for improvement; summarizing the analysis and recommendation for a management presentation; formulating a strategy for the best approach; and making specific corrective action recommendations.

Corrective action could include developing an obsolete/surplus utilization (or disposition) program; improving the inventory management or forecasting system; emphasizing the need for better sales forecasts; improving inventory recordkeeping; designing and implementing an improved cost system; or looking at warehousing operations or facilities. Specific corrective action recommendations can include anything that improves your inventory or service level.

What benefits can you realize as the result of an aggressive inventory performance program? Improved cash flow with less inventory; recovered cash from the sale of surplus or obsolete items; deferred manufacturing cost; recovery of space—perhaps avoiding the cost of additional storage or racking.

This results in improved warehouse productivity from reorganizing the warehouse with high-activity items stored more effectively; better product quality by allowing optional engineering change sooner; and increased sales with improved product availability. All of these potential benefits are directly related to profitability.*

Summary

Of the four assets considered inventory represents the greatest potential saving, although warehouse managers may have a greater degree of control over space, energy, and equipment. However, a good manager should exert influence, if not control, over every asset that could affect warehousing performance.

---

* From an article by John A. Tetz, published in Warehousing and Physical Distribution Productivity Report, Vol. 17, No. 11. © Marketing Publications, Inc., Silver Springs, MD.

33

"JUST-IN-TIME" AND ITS VARIATIONS

The last few decades have seen the popularizing of an old distribution strategy referred to as "just-in-time" or alternatively as "kan ban." *Kan ban* means signboard in Japanese, and it was so named because suppliers to a Japanese auto manufacturer were instructed to place inbound materials beneath posted signs that indicated the appropriate part number. Like many popular business developments, just-in-time or JIT is a product of both fact and myth.

The first myth is that it is a Japanese idea. It is something that the Japanese learned from others. In fact, JIT is a variant of the postponement strategy (see Chapter 18) which has been practiced in business for many decades.

A second myth is that JIT is a production strategy. In fact, JIT was used by wholesale and retail distributors at least as early as it was in production. Its use in merchandising is probably more widespread than in manufacturing, and that merchandising use nearly always involves warehousing.

A third fable is that JIT requires that a supplier's plant be close to the user. This in fact is the case in Japan, a compact country where suppliers and customers are pushed close together by both geography and culture. In the United States and other countries accustomed to larger distances, use of overnight delivery systems will permit JIT to work effectively with a supply line that stretches over hundreds or even thousands of miles.

JIT in Manufacturing

A Japanese auto builder some years ago ordered that tires of a certain size be delivered to a specific spot in the plant at 8:00 a.m., with another delivery at 11:00 a.m., another at 2:00 p.m., and another at 5:00 p.m. Instead of having a tire warehouse adjacent to the assembly plant, the manufacturer relied upon his supplier to have the right tires in a spot that was convenient to the assembly process. Also, instead of the normal rehandling and quality checking, the auto assembler relied upon his supplier to deliver that product in perfect condition at a precise time in front of a specific sign board (*Kanban* in Japanese).

The previous system of storage, rehandling, and checking had generated excessive costs, and the Japanese auto maker demonstrated that the *kan ban* concept was one way in which such costs could be eliminated.

Because the term *kan ban* was associated with one company (Toyota) the phrase "just-in-time" was adopted as a more universal method of describing a timed delivery program.

The practice spread from automotive to other industries. A computer manufacturer uses air freight transportation to provide reliable timed delivery to its manufacturing lines. Parts to supply the assembly lines are classified as A, B, or C. Many of the "A" parts are supplied just-in-time; the remainder are warehoused at the assembly plants. Not all of the "A" parts come by air, particularly those moving only a short distance. Quality charts are maintained to measure every carrier under contract. A delivery record of 98% or lower is not acceptable. The manufacturer provides routing instructions to all suppliers to maintain reliability. Quality charts are updated frequently to be certain that delivery reliability is under control.

JIT in Service Support

Sometimes JIT is used to eliminate parts and service centers in many locations around the country. One health-care supplier, for example, produces an appliance used in hospitals. If it breaks down, a premium air service is used to fly in a new unit and to return the one that has malfunctioned. Then the repair of the malfunctioning unit can be done at one central location rather than in scattered repair depots. Furthermore, the service is usually better than it would be in field repair stations.

Similarly, a computer service company maintains detailed records on the equipment used by each customer. If a computer breaks down, the service company isolates the problem through telephone consultation. In about one-third of the cases, a technician is able to solve the problem by providing advice over the telephone. When this cannot be done, the failed component is replaced rather than repaired, again using air freight to deliver the replacement unit and return the one that has malfunctioned.

Quick Response

Like JIT, quick response (QR) is as much a process as it is a concept. It depends on other tools such as bar coding, point of sale registers, electronic scanners, and electronic data interchange. Like JIT, QR promises benefits to all parties who are involved. While the concept as described was originally designed by and for distributors, the expectation is that it will provide benefits to the supplier as well. By practicing QR, the distributor enjoys lower inventories, better stock availability on critical items, reduced operating costs, and reduced markdowns. At the same time, the firms supplying that distributor should have fewer returns and higher productivity of their own inventories.

Some of the early reporting of quick response involved the use of bar codes and electronic data interchange (EDI) to cut textile manufacturing cycles. Milliken and Company reduced its order turn around time from six weeks to one week. Some predict that an element of QR will be flexible manufacturing in small lots.

One of the fundamental influences on the development of QR is an article published in 1988 in *Harvard Business Review* by George Stalk, Jr., "Time—The Next Source of Competitive Advantage." While this article describes manufacturing, the illustrations are related to merchandising. For example, Stalk describes a group of companies engaged in this strategy and lists companies such as The Limited, Federal Express, Domino's Pizza, and McDonalds. Not one of these is primarily a manufacturer, and all of them are engaged in providing retail services.

Quick response is a customer-driven system that pulls merchandise through the supply chain rather than pushing it. When a customer purchases a retail item, the sales clerk scans the bar code at the cash register. The stock information contained in that code is compared with existing inventory, and when inventory is low, the same bar code is immediately transmitted to the supplier and the carrier. Once informed, the carrier arranges to move replenishment stock to the distributor. Since the carrier was previously notified, he is able to rapidly move the replenishment stock to the store where it is needed. As a result, the retailer replaces sold stock within days instead of months.

QR changes warehousing as much as it changes retailing. Some manufacturers may choose not to handle a QR program from the factory. Time and cost considerations may make it difficult or impossible to provide the timed response without a staged inventory. Just as new warehouses were established to support JIT, other new warehouses are being established to support QR.

In some cases, the warehouse may be oriented to provide a consolidation point for the retailer. For example, the ABC retail company might persuade twenty different suppliers to position inventories at a convenient public warehouse. Rather than receive LTL (less than truckload) shipments from each of these suppliers, ABC retailing can order a consolidated truckload containing a mix of all of the products just at the time it is needed. Furthermore, because the merchandise is staged at a public warehouse, title to that inventory does not pass until it is delivered to the distributor. In effect, by combining QR with the strategic use of warehousing, the distributor can avoid taking title to inbound shipments until just the time when they are needed and can be sold. As a result, the distributor's inventory investment is dramatically reduced. At the same time, as the distributor becomes more efficient, he is less likely

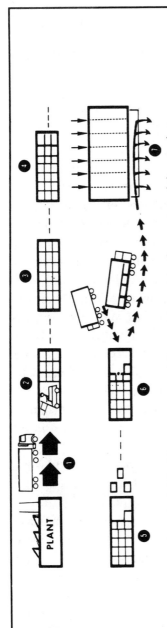

1. Product moves in bulk shipment to a public warehouse in the center of the market area.

2. Product is put into storage at the manufacturer's expense, until it is sold by the producer to wholesalers or retailers in the area.

3. Full pallet loads of product are used for order selection.

4. The warehouseman provides continued communication to the processor about the number of cases which remain in stock. Sometimes product is transfered to the new owner without being moved from one storage location to another.

5. Local buyers place orders on the warehouse, and these are picked and either shipped or held for pick-up. Records are adjusted to show withdrawals.

6. The buyer's truck arrives at the warehosue to trade empty pallets for full pallet loads of selected merchandise.

7. At the buyer's grocery warehouse, containers with store orders are merged with the merchandise which was just received from the public warehouse. Because of the speed of this merger, frozen foods can be treated the same way as dry groceries.

Figure 33-1

Diagram courtesy of Customer Service Institute

to be forced to engage in reduced price sales to unload merchandise that never should have been purchased in the first place.

QR changes warehousing as much as it changed retailing. Some manufacturers may choose not to handle a QR program from the factory. Time and cost considerations may make it difficult or impossible to provide the time response without a staged inventory. Just a new warehouses were established to support JIT, other new warehouses are being established to support QR.

In some cases, the warehouses may be oriented to provide a consolidation point for the retailer. For example, the ABC retail company might persuade twenty different suppliers to position inventories at a convenient public warehouse. Rather than receive LTL shipments from each of these suppliers, ABC can order a consolidated truckload containing a mix of all of the products just at the time it is needed. Furthermore, because the merchandise is staged at a public warehouse, title to that inventory does not pass until it is delivered to the distributor. In effect, by combining QR with the strategic use of warehousing, the distributor can avoid taking title to merchandise until a few days before it can be sold. As a result, the distributor's inventory investment is dramatically reduced. At the same time, as the distributor becomes more efficient, he is less likely to be forced to engage in reduced price sales to unload merchandise that never should have been purchased in the first place.

For those who believe that we have moved from the industrial age to the age of information, it seems clear that quick response is part of a massive trend to use time as a competitive weapon. Buyers are no longer interested in just quantity and price; the emphasis today is on the ability to move products quickly and effectively. Regional brands and brand proliferation could disappear. The successful company will be the one that is able to respond the fastest, and warehousing will be an integral part of that rapid response.

Efficient Customer Response and Other Variations

As the JIT concept moved to other industries, the term *efficient customer response* (ECR) was used to describe a similar process in the grocery industry. Other JIT variations are likely to occur as the concept spreads from one industry to another.

While early writings about JIT suggested that this was a strategy to eliminate warehousing, it is probable that JIT has created more new warehouses. The difference is that these newer JIT centers are flow-through terminals as much as they are warehouses. JIT is a method of increasing the velocity of inventory turns, but that increased velocity is not accomplished by eliminating warehousing, but simply by changing the way in which the warehousing function is performed.

34

WAREHOUSING COSTS

As you consider the true cost of controlling space and operating your warehouse, look both at costs related to function and at the concept of *opportunity costs*.

If you own a building that your company constructed in the 1960s, your occupancy cost is probably low because the construction cost was low compared to today's rates and the mortgage debt on the building has likely been paid in full.

But what would it cost to replace this building if it were destroyed? Alternatively, what would a third party pay in rent if you elected to close the building and put it on the market?

Real estate, like any other product, has an opportunity cost as well as an accounting cost. If the value of your warehousing facility is substantially higher or lower on the open market than the rent that is currently charged for the space, you should anticipate a change as you adjust to the realities of today's market.

Costs Related to Function

The first function of warehousing is storage, and one group of costs are those associated with "goods at rest," or those expenses that would occur even if the stored products were never moved. The second group of costs are those related to "goods in motion." These are the handling costs that occur only when materials are moved.

To understand and control overall warehousing costs, it is best to make the distinction between goods at rest or goods in motion. Because the allocation of these costs is always a matter of judgment, there can be reasonable debate about whether any cost item should be applied to one function or another. Some warehouse operators assign all fixed costs to goods at rest because these costs cannot be reduced, even when materials-handling activity decreases.

Developing a Cost for Warehouse Labor

What is a realistic cost for each hour of work in the warehouse? Figure 34-1 shows how a base wage rate of $7.50 per hour is converted into a total cost per employee of $11.00 per hour. Note that allowances are estimated for time lost to vacations, paid holidays, and paid sick days. Fringe benefits are shown as a lump sum. Another allowance is for the cost of supervision.

As you examine labor budgeting, you should look critically at several factors. Is the supervision burden realistic? If you use part-time workers who do not receive fringe benefits, has this saving been calculated in developing a total cost?

Sometimes the accounting people who develop a standard labor cost have scant familiarity with warehouse operations. As a manager, you should question the standard costs in the light of current experience. In this situation, the standard cost is 147% of the base labor cost. Does this ratio seem realistic as compared to those used in other operations?

If your internal warehouse labor cost is higher than hourly prices charged by third-party providers of contract warehousing services, you might ask why your company is charged warehousing costs that are higher than those of services available from outside suppliers.[21]

Justifying Purchase of Equipment

For most equipment purchases, the price is only a part of the total cost of ownership. In the case of a lift truck, maintenance over the life of the equipment may be greater than the cost of the vehicle. Also, be sure to include the cost of training people to use a piece of equipment, particularly one that is substantially different from anything now used in your facility. While people are learning, there

Figure 34-1
Total Cost of an Employee

Cost area	Days	Hours/ Day	Wage Rate	Annual Cost
Days on job	223	8	$7.50	$13,380
Plus: Paid vacations	10	8	7.50	600
Paid holidays	10	8	7.50	600
Paid sick days (avg)	7	8	7.50	420
Total days paid	250	8	$7.50	$15,000
Fringes				$2,500
Supervision				
Cost	$17,000 per year			
Span of control	8 people			
Total supervision per employee				$2,125
Total Annual Cost				$19,625
Actual Rate of Pay				
Effective cost	Total Annual Cost/Hours on Job			
	$19,625/1784 hours = $11.00 per hour			

Courtesy of Joseph L. Cavinato

will be waste or damage. In many companies, management overhead must be factored into the cost of any new asset.

Against these costs, you must measure the benefit.

- What benefit will you realize from purchasing another lift truck? Additional equipment may reduce overtime, and the overtime savings should be calculated.

- Perhaps the new truck will operate in narrower aisles than an old one, allowing you to change the warehouse layout to place an additional number of pallets in the same amount of space. What is the value of this increased storage as compared to the cost of the new truck?

- If the new truck is faster than an old one and will therefore save time, how much time? What is the value of that time?

As you perform this analysis, look at the danger points. If the proposed new lift truck breaks down, how will you operate the warehouse until it is repaired? Does the selling dealer have the ability to keep the new lift truck running? What is the financial stability and service reputation of the dealer?

As you review an appropriation request, beware of common pitfalls. If you omit costs, you may delay the project. Try to think of every cost related to owning the new asset. Be sure the cause-and-effect relationships that you describe are accurate. Avoid creating detailed tables that represent a degree of measurement that does not exist in your warehouse operation.

In any capital commitment, there are at least three common risks. These are the chances that:

1. The business environment might change.
2. Operating situations might change.
3. Project estimates might be wrong.

If any one of these should occur, the need for the equipment could disappear. What is the resale value of the equipment on the open market if this happens?

Establishing a Value of Inventory

While some warehouse managers have no control over the inventory they store, others have some degree of influence—possibly even total control. Even if you do not have full control of the inventory, you can be a better warehouse manager if you understand the real and hidden costs of inventory.

The purpose of a finished-goods inventory is to afford an opportunity to make a sale by having product in the right place at the right time. When inventories are in your warehouse and are not moving, it seems evident that they are not filling their purpose. There are four general costs of inventory: capital costs, opportunity costs, risk costs, and administrative costs.

As a warehouse manager, how will knowledge of inventory costs affect your operation? First, you will get a better understanding of when the use of high-speed transportation is a good investment. Consider the situation illustrated in Figure 34-2. In this case, management had the option of moving a shipment of machinery from the United States to Europe by air freight or by ship. Even though the line-haul costs more than doubled, the *total cost* by air is less.

The air freight cost is less primarily because the shipment has a value of $50,000, and the inventory has a value of 13% per year, or nearly $18 per day. Saving 19 days by air, as opposed to marine transportation, represents a saving of $198 in the cost of inventory. This would be lost if the product cannot be used because it is sitting in the hold of a ship.

This cost analysis presumes the merchandise will be put to use as soon as it arrives in Europe. If the inventory will be stored while awaiting a future sale, the assumptions about inventory value would be different.

As you make decisions about transportation modes, as well as other warehousing decisions, your knowledge of the value of your inventory can and should play an important part in the process.

Figure 34-2
Total Cost Analysis

Task: Transport machinery from Marietta, Georgia, to Rome, Italy

Activity	Air	Ship
Transit time	2 days	21 days
Transportation		
Linehaul	$2,200	$900
Plant to terminal	0	260
Terminal to customer	0	180
Packaging/crating	100	350
Insurance	220	650
Documentation	110	150
Inventory cost*	36	374
Total	**$2,666**	**$2,864**
*$50,000 at 13% times days in transit		**Savings by air: $198**

Courtesy of Joseph L. Cavinato

Cost of Goods At Rest

Most warehouse operators measure storage by using the cube of the item. In some operations, product cube is not the best way to make the allocation. Space required by the product is not always a function of the cube, but instead may be defined by the size of the storage rack space it occupies, or by the amount of floor space taken up by floor-stacked product.[22]

When product is stored in pallet rack, floor rack, shelving, or cantilever rack, the rack fixture itself is a constraint because you should not mix product in a fixture. Once the product occupies any part of the fixture, the entire space is unavailable for any other use until that product is gone.

Because of this, the fixture opening is the space for which storage is charged. How the product fits into the fixture opening determines how efficiently the space can be used. The more units stored in the rack fixture, the lower the cost per individual unit stored.

For floor-stacked product, the space is calculated based on the product's footprint—the amount of floor space used by a stack of product. Space is charged based on the number of square feet occupied. The higher you can stack the product, the lower the cost per individual unit stored.

Storage might be thought of as charging rent for a square foot of floor space, or the use of a fixture. Then the cost allocated per unit of product is a measurement of how efficiently the product fills fixture or the floor space.

The cost is also influenced by how much time the product spends in the fixture or on the floor. Therefore, product turn must be calculated. The faster the product turns, the less time it spends in the fixture or on the floor, and the lower its share of the annual rent.

To start the calculation process, identify the type of storage fixture you will use for each item. Then calculate the number of units that can be stored in each fixture. In the case of floor-stacked product, determine the number of square feet that the stack will require and then how many units high the product can be stacked. Divide the total number of units in storage into the occupancy costs to find the cost for each item presuming it remains in storage for an entire year.

The figure you just calculated would be the storage cost if the inventory turns once or less per year. Use the actual turn rate of the item to calculate that portion of annual occupancy cost you should charge to each item.

The Influence of Inventory Turns

Storage costs may be calculated as a unit cost per month, or possibly per year. However, within that storage period, the same spot in a warehouse may be occupied by more than one unit of a very fast-turning inventory. Because of this, calculating the number of turns involved requires relating throughput inventory to average inventory.

An inventory that turns faster is less costly to store than one that turns slowly.

Throughput is generally defined as the total of units received and units shipped divided by 2. The average inventory is defined as the total of beginning inventory plus ending inventory divided by 2. The relationship of throughput to average inventory indicates the speed of product turn.

The Influence of Warehouse Layout

Warehousing costs are influenced by factors that affect the ability to store and handle a product. These factors include stacking height, which is always limited by the clearance available in the building, and also by the strength of the product or the package being stored. If you have a 20-foot high building and you are storing product that can be stacked only 5 feet high, you must purchase storage racks to use all the space. In these cases, an amount must be allowed for the cost of pallet rack or other storage equipment.[23]

The computation shown in Figure 34-3 illustrates a method of determining a cost per hundred weight for an inventory consisting of 115 SKUs, 50 of which are fast movers. The inventory is stored on pallets, with 100 small cases on each pallet. The warehouse operator has selected a layout with deep rows on one side of a 12-foot aisle, and shallow rows on the opposite side.

The total square feet needed in this layout is increased to allow for a honeycomb factor, the lost space in front of partial stacks. There is an additional 20% added as an efficiency factor, which is actually the allowance for docks and staging area.

Line 21 of Figure 34-3 shows the rate calculation that multiples square feet used by cost to develop a dollar cost for storage. Total storage cost is divided by tonnage throughput, which is shown in the computation as billing units. Note that product received is reduced by half to reflect the fact that goods flow in at a steady rate throughout the month. On average the inventory is in the warehouse only for half a month.

This calculation is used by a third-party warehouse in developing costs that are converted into storage fees. The same kind of calculation is valid for any warehouse operator who wishes to convert storage into a cost per unit.

Cost of Goods in Motion

Per-unit handling rates are simply an expression of item devoted to handling per unit multiplied by total labor cost. Observation of handling procedures will allow reasonable calculations of the time involved. Here are the results of some typical observations:

- In unloading, the time required depends on whether the product arrives floor-loaded, palletized, or slipsheeted. For each of these shipping methods, a different unloading time per pallet was created. For this and subsequent full-pallet activities, the time required to handle a unit is calculated by dividing the handling time per pallet by the number of units stored on a pallet.

- The time required for checking, identifying the product, and validating the quantity against the vendor's packing list is a flat rate per pallet.

- The time for put-away and picking line replenishment will vary, depending upon the location of the product. This time

---

Figure 34-3
Storage Computation

Volume/Criteria

1. Average inventory: 6 million pounds   300,000 Cs   3,000 Pallets
2. Monthly throughput: 4 million pounds   200,000 Cs   2,000 Pallets
3. 16 SKUs are fast movers   65 SKUs are slow movers

Layout

4. 3,000 pallets divided by 3-high stacking = 1,000 spots.
5. 1,000 spots divided by 80% honeycomb factor = 1,250 storage spots.
6. 1,250 storage spots times 80% fast movers = 1,000 deep-row spots.
   1,250 storage spots times 20% slow movers = 250 short-row spots.
7. 1,000 deep spots = 3,000 pallets with 50 SKUs.
8. 1,000 spots divided by 50 SKUs = 20 spots per SKU. (Fast movers.)
9. 250 spots divided by 65 SKUs = 3.8 spots per SKU. (Slow movers.)
10. 45 feet–12-foot aisle = 33 feet storage
11. 33 feet divided by 4 feet per pallet = 8 pallet rows per bay

Space calculations

12. Deep facings = 167
13. (6 pallets times 4 feet deep) times 4 feet wide = 96, plus
    (4 times 12-foot aisle divided by 2) = 120 square feet per facing
14. 120 square feet per facing times 167 facings = 20,040 square feet
15. Short-row facings = 125
16. (2 pallets times 4 feet deep) times 4 feet wide plus
    (4 times 12-foot aisle divided by 2) = 56 square feet per facing
17. 56 square feet per facing times 125 facings = 7,000 square feet
18. 20,040 square feet plus 7,000 square feet = 27,040 square feet
19. 27,040 square feet divided by 80% efficiency = 33,800 square feet

Rate calculations

20. 33,800 square feet times $0.195 per square foot per month = $6,591
    Beginning inventory   = 60,000 cwt
    Product received   = 40,000 cwt times .5 month (20,000 cwt)
    (mid-month)
    Total billing units   = 80,000 cwt
21. $6,591 divided by 80,000 cwt = $0.082 per cwt cost

---

is determined by the travel time required to move the product from one area to another.

- In picking, the total time is a function of both the travel through the picking line and the time to handle the product at the picking station. All products in the warehouse are assigned to a predetermined storage and picking type classi-

Figure 34-4
Root Vent Handling Time

Activity	Driving characteristics	Handling per pallet	Units per pallet	Handling per unit
Unloading	Palletized	2.8		
	Floor-loaded	5.7	+2	2.9 min
	Slipsheeted	3.8		
Checking	Automated	0.4	+2	0.2 min
	Manual	1.2		
Strapping	Unstrapped	0.0		
	Automated	0.0	+2	0.0 min
	Manual	0.7		
Put-away	Bin	5.2		
	Cantilever	3.7	+2	3.1 min
	High bay	0.5		
	Low bay	6.3		
	Tier	3.7		
Picking line replenishment	Bin	2.4		
	Cantilever	6.3	+2	2.0 min
	High bay	0.5		
	Low bay	4.0		
	Tier	2.6		

		Handling time	Cube of product	
Order pick	Fixed travel	.11		0.1 min
	Pick/Rate/Cube	.04	× 16.35	0.7 min
Loading	Handling/cube < 20/cube	.17	× 16.35	2.8 min.
	Handling/cube > 20/cube	.10		

Total handling time for seven activities: 11.8 minutes

fication. The travel cycle is a constant for each classification of picking. The handling for a particular product is related to the product cube and reflects the time required to select and transfer the product onto a picking vehicle.

- Outbound trailer loading is a function of the cube of the product. The bulkier the item, the more time is needed to move it onto a trailer. This is true up to the point at which the product should no longer be hand loaded, but instead would be loaded with a fork truck. At that point in time, a mechanical loading rate is applied.

* If strapping is required, the time is based on the cost of moving the product through an automated or manual strapping process.

Figure 34-4 shows a sample of times calculated for a roof vent. While the times calculated, or characteristics selected, will be different in your warehouse, the approach to develop them is the same.[24]

35

MANAGEMENT PRODUCTIVITY

If managers are not productive, it is unlikely that productivity on the warehouse floor can be improved. In a company where senior managers look down on junior managers, or in a warehouse where supervisors show contempt for workers, even the best external techniques for improving productivity are likely to fail. Effective management is a prerequisite for any warehousing improvement program.

Defining Management

Management has been defined as the process of getting things done through others. A critical difference between worker and manager is that the manager doesn't pick up his tools and do the job—he or she causes that job to be done by *others*. Therefore, a primary requisite to success in management is the ability to identify and hire good "others," the hourly people who will get things done as directed. Some managers are promoted from the hourly workforce into management, and must make the transition from being an "other" to performing as a manager. Everyone finds this transition difficult, and some never quite accomplish it. Signs of failure to manage can be seen in the manager who interrupts the day to do a task that should have been delegated to an hourly worker. Another sign of failure is the inability to find and keep good people. Still another sign is an inability to recognize the difference between having a job skill and having the ability to lead others in performing that same job.

Why Warehousing Is Different

Warehousing creates its own peculiar problems for managers. In most warehouses, a high percentage of supervisors and managers are ex "others," former forklift operators or checkers. Therefore the problem of transition from an "other" to a manager is seen more frequently in warehousing than in most businesses.

Entry level work in warehousing has an undeservedly low image. Some managers believe warehousing is a job for a strong back and a weak mind, and that pay scales can be lower than those for manufacturing. It is a false image. A warehouse employee often needs greater skills than those of the closely supervised assembly line worker. The very nature of warehousing defeats close supervision, so a warehouse worker must be a self-starter, a person who will deliver quality work at a highly productive level without someone looking over his shoulder. The problem is management's failure to build community respect for warehousing as an occupational specialty. In a community with labor shortages, the problem and the opportunity become apparent.

The Effective Manager

Have you taken an inventory of your own strengths and weaknesses as a manager?[25]

A first step in managing is getting to know yourself. It isn't easy for most of us to engage in critical self-appraisal. One way to learn objectively about strengths and weaknesses is by working with a consulting industrial psychologist. The psychologist usually provides no real surprises, but the interview will give you an objective appraisal of your own strengths and shortcomings as a manager.

If your organization uses an industrial psychologist you should welcome the chance for an interview. You'll find out a few things you may not have been aware of before. This can give you a chance to compensate for shortcomings by trying to change your behavior or by hiring others in your organization whose skills complement your own limitations in those areas.

If you have no access to an industrial psychologist, run through the self-appraisal checklist shown in Figure 35-1. And, of course, you probably already have a pretty good idea of the things you do well and where you may have a weakness. The next step is to figure out what to do about it.

Measurement Tools for the Manager

Nothing can be managed well without being measured. In business, the ultimate measurement for every organization is the balance sheet with the profit and loss statement. The expression "the bottom line" is overworked, but its meaning is still important; the last line on a financial statement, the profit or loss of your organization, is still the vital statistic.

Depending on your rank in the managerial organization, and the size of your company, your ability to affect your company's bottom line may be great—or it may be nearly insignificant. Yet there are many things you can do to measure managerial progress.

The first of these is productivity. In the warehouse, you can measure productivity in many different ways. Your measurements must be simple, readily understood, and relevant to the tasks performed. They might include percentage of damage, frequency and severity of customer complaints, frequency and quantity of overage or shortage, or frequency and quantity of customer returns. All can be valid measurements of effectiveness.

What you measure is important, but it is equally important that the measurements be consistent. And if no action is taken, the measurements are meaningless.

Consider the following example of how measurement affects managerial skills in warehousing. As a warehouse supervisor, you are giving out assignments at the beginning of the day. You have given a two-man crew the task of unloading a box car of merchandise. Available work standards indicate that two men will take four hours to complete the job.

When you give the assignment, you describe the job to be done and indicate to the lead man that you will expect to see him at lunch time ready for a second assignment. In making the work assignment

12 Factors to Rate Yourself as a Manager

	Excellent	Average	Needs Development
Administration			
Job knowledge			
Planning			
Innovation			
Communication initiative			
Responsibility			
Team player			
Sales person			
Decision maker			
Leader			
Selector and developer of personnel			

Figure 35-1

this way, you communicate your expectations. You also communicate your knowledge of your operation. If the job can really be done in three hours, for example, your communication of a four-hour expectation can be seen as an invitation to slow down. The opposite is true if the job would normally take five hours.

Workers will typically accept and conform to standards that are perceived as fair, Above all, use of standards such as these shows that you as a manager have measured and will continue to measure the output of your work crew, and that you have communicated your expectations and they are reachable.

The Critical Tasks of a Manager

Since management involves getting things done through others, how do you find those other people who make your job a success? Some managers have little experience or training in attracting and selecting the people they need for their team. A first step is to set the specifications for the kind of people you need to work in your warehouse. If the work is strenuous, you need people in good health, free of ailments such as back injury that could be aggravated by warehouse work. Substance abuse is another problem that must be recognized. You need people who are honest and unlikely to fall prey to temptations to steal. You need people who will be content with the work in your warehouse and who are not anxious to change the existing situation. Some feel that it is desirable to find workers who have past experience in the kind of work done in your warehouse, though others feel that training of an inexperienced person will quickly produce a quality worker who does not need to "unlearn" bad habits from a previous job. The job specifications of the people you are seeking should be placed in priority. Some may be absolutely essential, and others could be subordinated if the essential talents were relatively strong. As you set priorities, remember that it is easier to teach new skills than it is to change attitudes.

The interview process is a key step in selection. Entire books have been written about interviewing, and courses are available to enhance this skill. A successful interview is one in which the applicant does 80% to 90% of the talking. The best questions asked are nondirective ones which typically cannot be answered in few words. The goal of the interview is to learn as much as possible about the attitudes as well as the experience of the applicant. Not all successful managers are good interviewers, but they should appoint someone to do this job who is good at it. The checking of references is equally critical to successful selection. Use of a probationary period for new employees is essential to be sure that any mistakes made in the selection process are corrected during probation.

The development of people begins with employee "indoctrination" that starts with the first day on the job. The impressions of your company that are gained on that first day of work are usually the strongest impressions the individual gains of your company. Every worker will be motivated and oriented by somebody—if you have no program, the job will no doubt be done by a leader of the work crew. If you want the image of the company to be a good one in the minds of its employees, then it is necessary for management to have a well-designed program.

New employees, hourly workers or managers, are "socialized" through several processes. The new worker develops skills and learns the tasks involved in the job. At the same time, that new person learns what behavior is acceptable or unacceptable within the company. Often these things are learned from fellow workers rather than from management. If the socialization of the new worker comes from other workers, the new employee is likely to be exposed to conflicting demands. The supervisor may provide one set of demands, the union steward another, and the informal leader of the work group still another.

Training is a critical part of the development process. Sometimes the training given is not relevant. In other cases training is given only to new employees and is not available to older workers. If this happens, an older employee who is given a new task may fail at the new job because training was not provided. This is one valid reason why many workers resist the chance to change jobs.

Part of the development task is to provide up-to-date procedures manuals. Such a manual is effective only if kept up-to-date.

Appraisal is perhaps the most critical part of a manager's task in developing people. Many managers don't know how well they are doing, because their performance is kept in the dark. Many managers are "too busy" to provide appraisals, which usually means that they are not comfortable in giving them. Others lack a format or training in conducting appraisal interviews. One format is shown as Figure 35-2. A successful appraisal emphasizes performance. It provides both positive and negative feedback, recognizing that nobody is perfect and everyone has a few good traits. The appraisal interview should include discussion of a step-by-step approach for achieving improvement in future performance.

How Effective A Manager Are You?

1. How productive are you? Do you spend your time putting out fires or do you concentrate on managerial tasks?

2. Are your work habits ones you would want your employes to adopt?

3. Do you allow employes room to grow?

4. Do you have any employes you should have fired, but haven't?

5. Do you have a formal training program for new employes?

6. Do you have a continuing education program or does formal training stop at the entry level?

7. Are long-time employes continually challenged, given new tasks, rotated in other jobs?

8. Do you have written procedures? When was the last time they were updated?

9. Are your supervisors trained to manage others? Do they teach employes to think, rather than simply tell them what to do?

10. Do your supervisors regularly recognize performance of workers?

11. Does your company make an effort to recognize superior performance?

12. Do your fellow workers recognize good performance?

13. Is there a performance rating system? Is it fair?

14. Do you have a work measurement standard? Is it understandable? Can you influence the standard? Is it easy to follow? Does it measure in positive terms?

15. Do your managers/supervisors look down their noses at the workers?

16. Do managers/supervisors further their own education?

17. How often do you ask, "What are my firm's distinctive traits? Where are we ahead of—behind—the competition? Where will our corporation be in five years?" Answering these questions—planning—is what management is all about.

18. Do you tackle the tough decisions first?

19. Are you happy to tell outsiders good things about your company?

20. Can you easily defend your company if it is criticized?

21. Do you frequently produce creative new ideas for improving the job?

22. Are you willing to do everything that is necessary to get a job done?

23. Do you challenge others if a job is not being done correctly?

SOURCE: Vol. 17, No. 2, Warehousing and Physical Distribution Productivity Report. © Marketing Publications, Inc., Silver Spring, MD.

Figure 35-2

Coaching and counseling is another aspect of developing the people who report to you. Much of this coaching and counseling is providing feedback, and an equally important part is the role of being a good listener.

The result of your efforts in developing people is the promotion to greater responsibilities and/or the awarding of pay increases for those who have met or surpassed management's expectations. After you have communicated your expectations on job performance, you should reward those who meet or exceed them.

The Manager's Hardest Job

Terminating incompetent subordinates is probably the most emotionally demanding task required of managers. Many hesitate to fire employees because they like them, or feel sorry for them, or are afraid to fire them, or shy away from the unpleasant task. Firing an employee is often considered a personally hostile act. Whatever the reason, many managers have a difficult time terminating an unsatisfactory worker. No matter what the circumstances, firing an employee is never pleasant. The following steps are suggested to make it as easy as possible for both you and the employee.[26]

1. Never surprise the employee. Be deliberate and take the necessary time. Never do it in haste.

2. Never fire an employee in anger. You must establish that the employee cannot be salvaged, that there is no chance for improvement, and that all the responsibilities have been clearly understood.

3. The employee has to know in advance that the company is dissatisfied because the expected standards of performance have not been met. Make it clear that termination will follow if there is no improvement.

4. Conduct the termination interview on neutral territory. Allow the employee to leave without an audience. Be sure to have ready a paycheck for everything due the employee and deliver it at the end of the interview.

5. Don't drag it out or make a lengthy recitation of the employee's mistakes. Just do what has to be done.

6. Later, try to find out what went wrong in the hiring process that eventually led to this individual's dismissal.

The Buck Stops Here

Another problem for the warehouse manger is the intensity and the nature of the business pressure that builds up at the warehouse level. The shipping dock is the place where "the buck stops" in the corporate shell game which begins with salesmen's promises and ends with the question of whether or not the order has been delivered—which must be answered with either a *yes* or a *no*. If the answer is no, senior management often displays scant interest in the reasons why, and this creates a unique pressure seldom found elsewhere. With this pressure come risks of failure that are peculiar to warehousing. A late shipment is a failure, but so is a shipping error. Failure of a common carrier is often blamed on the warehouse, even when warehouse management had no authority to select that carrier. Damage, whether caused by the carrier or warehousing people, is a cause of failure, particularly if the damage is not discovered promptly. Theft, from whatever cause, is generally regarded as something that good management should have been able to control. Even casualty losses from fire or wind storm are sometimes considered as events that might have been prevented with better management. It is neither accurate nor fair to blame these failures on warehouse management, but that's the way it happens in many companies. Managers in other businesses face their risks of failure, but warehouse managers have more risks to cope with.

How Good Are Your Warehouse Supervisors?

One of the hallmarks of any good supervisor is skill and effectiveness in delegating. Those who do not know how to delegate are doomed to failure. In many cases, supervisors receive little training in delegating the job to be done. And too many supervisors show this lack of training. Here are some points you should caution your supervisors about:

1. Be careful to whom you delegate a task. Particularly if your

supervisors are new to delegating, caution them to pick carefully their first choice for the job at hand. It's essential that the first efforts be met with success. Not everyone is right for every job. And not everyone wants added responsibility. So it's important that the right person be chosen for the job.

2. Be sure instructions are given clearly. The effective supervisor gives instructions clearly and explicitly and then checks to be sure those instructions were received and understood. Poor initial communications often cause things to go wrong. It's particularly important that both the supervisor and worker understand precisely *what* is to be done, *how* it is to be done, and *when*.

3. Ask someone to do a task, rather than order that it be done. One of the responsibilities of a supervisor is to motivate workers. Ordering someone to do something is apt to have the opposite effect. Unfortunately, many improperly trained supervisors don't realize the power of asking rather than ordering. Emphasize this point to supervisors in their training sessions.

4. Remember that someone may do a job other than the way you would do it—but, of course, this doesn't necessarily mean it has not been done correctly. An effective supervisor allows for different approaches and methods—and, in fact, may be amazed by the freshness and perception of an employee's approach, even if it is different.

5. Follow up. The supervisor needs to realize that delegating a task doesn't mean delegating responsibility for the success of the task. The old Army saying is true; you can delegate authority but you cannot delegate responsibility. So follow-up is crucial. The supervisor should check the employee's progress at specified intervals—and also on a spotcheck basis. Thus the supervisor is aware of the potential for a problem situation in time to take steps to prevent it.

6. Give credit where it's due. When a task is completed, the supervisor should be sure to give credit to the worker. Similarly, when one of your supervisors delegates a job successfully, be sure to recognize appropriately both the supervisor and the individual who actually did the job.

The Supervisor Promoted from Within

In warehousing it is not uncommon for supervisors to be promoted from the workforce. One of the most difficult tasks of supervision is to forge a new constructive relationship with people who were once peers. Supervision requires a certain detachment and freedom from "cronyism." It requires a certain measure of discipline, and the ability to instill pride in performance. It demands a responsibility to management, and a responsibility to the workforce. And the worker who is promoted into supervision must make the transition without being a crony and at the same time without making enemies.

This is a delicate balancing act, and requires sensitivity and tact on the part of the worker promoted from the ranks. It's important that warehouse management prepare the new supervisor for such problems, and suggest ways to handle them.

The Importance of Clarifying Expectations

Do the workers in your warehouse know what their supervisors expect of them? Expectations are best met when they are clearly stated—when both worker and supervisor know without question what is expected. Chances are good that your supervisor can tell the workers exactly what management expectations are. But it might be an interesting exercise to find out if workers have a clear understanding of management expectations—and if their understanding agrees with their supervisor's. Not knowing what management expects can be a major cause of worker dissatisfaction.

An atmosphere of positive achievement, with praise given when it has been earned, is a big help. Positive discipline—constuctive criticism—involves keeping track of results, as well as acknowledging success. Workers in any field of activity including warehousing will tolerate strict discipline as long as it is accompanied by positive feedback.

In one large operation, a straightforward system of recognition is used. When a supervisor sees a worker doing a superior job, the supervisor gives the worker a card with a comment about the work. Records are kept of the number of cards awarded, and a worker who receives a certain number of cards is awarded a small prize, such as a ticket to a sports event or a dinner out.

While incentive programs have strong supporters, they also have vocal critics. Whatever approach you choose to use, the basic underpinning is your supervisors. Do they understand the value and importance of positive feedback? Do they recognize progress or the opportunity for a better job? Or are they negative news carriers who recognize only failure? In some cases, there's a widespread feeling that it is impossible to please a particular supervisor. Yet, when counseled, the supervisor honestly believes that he or she is fair, and does use positive feedback properly. So part of management's responsibility is to make supervisors aware of the value and proper use of positive reinforcement.

Effective management is much more than symbolic behavior. It is based on meaningful, fair measurement. This means more than simply establishing productivity standards. It means getting management commitment—throughout the department, throughout the company—to a productivity improvement program. It means evaluating the standards, giving workers feedback and giving recognition when a job is well done.

36

REDUCING ERRORS

One of the best ways to improve productivity is to get every job done right the first time.[27] The cost of errors in warehousing is high; the cost of an order-picking error, for example, has been estimated as ranging from $30 to $70 for each occurrence. These costs are derived by considering the writeoffs of undiscovered warehouse shortages caused by picking mistakes, the cost of handling returns, cost of supplemental shipment to replace shortages, cost of reversing errors caused by substitution of the wrong product, correspondence to handle credits and adjustments, and, finally, the almost unmeasurable cost of customer dissatisfaction or loss of confidence.

In many warehouse operations, errors are the worst problem that management faces. Errors not only mean lost customers—in a few situations, an error can mean the loss of a life. Regardless of the consequences, as long as you have human beings in the warehouse, there will be some errors. Still, it is management's job to reduce them to the greatest possible extent. In most warehouse operations, a decrease in errors may be the best way to increase overall productivity.

The Value of Order Checking

There is little evidence that checking orders substantially reduces order-filling errors. No amount of double checking or triple checking eliminates the errors; only a few of them will be caught. How few is evidenced by the fact that customer complaints about order errors continue despite all the checking.

The very fact that there is a checking department (or an order checker) seems to constitute management recognition that mistakes are inevitable. Thus, the checker becomes responsible for the accuracy of order-filling, rather than that being the responsibility of the order filler himself.

The fact that double-checking frequently turns up additional errors not caught by the original checker should signal management that order checking is futile and pointless—but apparently it does not do so.

To illustrate the futility of order checking, in one company five full-time order checkers were required in a particular broken case order filling department. Even with these five order checkers, customer-reported errors averaged 25 per day. When the order checking department was eliminated, management saved five man-years of labor at a cost of (an increase in) only 10 additional reported order-filling errors per day. Since no effort was made to upgrade the caliber of work or worker, it seemed that 10 additional order errors per day was a small price to pay for a saving of five salaries.

Order-Picking Errors

Good order-filling includes not only error-free picking, but also reaching a standard picking speed. High work standards exist in any number of warehouses where management has been motivated to take some action. But it takes a concerted effort on the part of management and the worker to achieve even the minimal concentration needed for reasonably error-free order-filling. If management refuses to accept substandard work performance for standard pay, an acceptable rate of order-filling accuracy is more likely to be reached.

Between 10% and 12% of all errors are not due to poor performance in the warehouse, but occur because the order taker fails to take down completely what the customer wanted or had ordered. The remaining 90% of order errors are equally divided among wrong items, wrong quantities, and items completely missing from the order.

Order-Taking Errors

A misunderstanding when the order is taken by telephone is a common source of errors. Consider the procedure used by a fast food company that makes home deliveries. This firm asks for the telephone number whenever a phone order for its products is taken. Then the operator phones back to the customer to repeat the order. This second phone call prevents a high percentage of errors.

For error-free order-filling, it is important that the order sheet submitted to the warehouse for filling is correct or the customer is going to be dissatisfied—even if the warehouse is right. Clerical order-taking errors are easy to eliminate if the order taker takes the time to get it right.

Error-free order-picking has four principal requirements:

1. A good item location system
2. Clear item identification
3. A clear description of quantity required
4. A good order-picking document

A Good Item Location System

The code system must be clear enough so that any order picker can quickly find any picking location in the warehouse.

A six-digit locator system will accommodate the needs of all but the very largest of warehouses. The first two digits indicate the aisle or rack-row number. The next two digits refer to the shelf or rack "section" (ranging from 3 feet in width to 8 or 10 feet). A section is defined as the space between two adjacent uprights, so the system will work not only for shelving, but pallet racks as well. The fifth digit indicates the elevation of the shelf or the rack beam. The last, or sixth, digit indicates the item location on the shelf, from location one up to nine counting from left to right from the nearest shelf or rack upright on the shelf.

Some warehouses have more than 100 aisles and in others the number of small parts on a shelf is too large for a single-digit "last-number" system. Accordingly, a larger system may have to be considered. In any event the shorter the location number the better.

In the six-digit system, the digits are generally three, two-digit numbers as follows: 12–34–56. This indicates to the order filler that he should first go to aisle 12, section 34. This system normally has even numbers on the right-hand side and odd numbers on the left side of the aisle, as the order filler enters, so section 34 would be found on the right side of the aisle. The order filler now counts five levels of shelving up and six items over from the left-hand side of the shelf (the upright support) and comes up with item location 12–34–56.

Clear Item Identification

Items are normally identified first by trade name, then by size, color, or retail value, as for example, XYZ mouthwash 16 oz., or ABC toothpaste, giant size. Some industries require more precise item identification: "Hex head bolt, $\frac{3}{4} \times 2\frac{1}{2}$ cad. plated" becomes "1- 6 - $\frac{3}{4} \times 2\frac{1}{2}$ C."

Wherever a product can be concisely listed by its name or name and specifications, errors are reduced. Avoid further complicating item identification by the use of in-house, computer, or manufacturer's numbers, unless no other practical means of item identification is possible. Confusion is created by identification systems that include information that does not appear anywhere else but on the order document. Internal or "house numbers" that have been programmed into the computer for precise identification tend to create confusion unless they also appear on the product.

Most important in item identification is that the product on the shelf, the description of the product on the order-picking document, and the description on the shelf label are all identical. If the product on the order says XYZ toothpaste, the label on the package says XYZ dental cream, and the shelf label says XYZ TP, the order filler will not know whether the location was wrong, the order was wrong, or the item on the shelf was wrong.

One of the causes of item identification confusion is the use of manufacturer's catalogs—particularly if the catalog sheet has not been prepared by the advertising or sales department. Make sure the item description on the catalog sheet is the same as the item identification on the package that is going to be picked off a shelf.

The problem of item identification can be eliminated when the computer program used to produce shelf labels also produces the order-filling document. When the order filler arrives at a numbered location and checks the shelf label against the product description on the order document, they must correspond or else he's in trouble.

Clear Description of Quantity Required

Industries such as wholesale hardware, wholesale drugs, auto parts, and toys have a practice of shipping merchandise in shelf packs, overwraps, tray packs, and even reshippable containers within master cases. This creates confusion with respect to quantity, and many order-filling errors result.

An example is a brand of hacksaw blade that comes 10 to a package and 10 envelopes to a carton. The outside cover is marked "10 units saw blades." We must assume that the manufacturer of this product did not deliberately design the package to induce the order filler to ship 10 times the quantity required. The proper way to describe the quantity required would be "one envelope containing 10 hacksaw blades." A conscientious order filler, finding the cartons and hacksaw blades on the shelf and recognizing that the order did not call for a carton but an envelope, would either open the carton to determine the contents, or would check with his supervisor.

An opposite example is wholesale drugs, where ampules are prepackaged in lots of 25 units. While a customer's order for 25 ampules should not be a difficult problem to an order filler, one wholesale drug house described the product as XYZ ampules 25 × 1 cc, and in the quantity column "one each." Hopefully the order filler would know that "one each" meant one carton of 25, but the order did not say so. One package plainly marked in red letters, "Do not break this carton," was broken open and contained 24 ampules. Evidently an unhappy customer had received one instead of 25, with the remaining 24 on the shelf constituting an unsalable package that must go to the return goods room for a lot of handling to get the credit. This could probably have been avoided by indicating clearly in the item description that XYZ ampules come 25 per box, and, in the quantity column, indicate "one box of 25 ampules."

Order Identification

It is easier to control order numbers if they are reduced to the fewest possible number of digits. The order number system has to be workable within the framework of the particular warehouse. The reuse of sequential numbers is best limited to a 10- or 15-day cycle. The order number is best listed on the top of a page on the right-hand side.

Eliminate all unnecessary information on the order-filling document. Such items as "bill to" or "ship to" are not needed by the order filler. Likewise, credit terms, sales data, cost at retail, or cost at wholesale have nothing to do with order-picking efficiency or accuracy.

Order entry systems should be designed to convey information of value to the order filler, rather than to the accounting or sales departments. The order filler's task is to proceed to the proper location, verify that this location is correct and the wanted product is stored, and then select the right number of pieces of that item. Superfluous, nonessential information only creates confusion.

"Good" Order-Picking Document

The order-picking document for selecting items should be specifically designed for and used only by the order filler. Frequently, however, it is designed for so many other purposes that its original function (error-free order-picking) is forgotten, the order itself buried in other important information of little interest to the order filler. The best possible order-filling form is one that has only essential information on it, such as customer identification and an order number or date so orders can be filled sequentially.

In many warehousing operations the order-filling document is one part of a multipart form. Sometimes it is possible to remove the unnecessary information from the document by blocking it out on the order-filling copy only.

The body of the order-filling document should have no more than four columns of information. The information should be supplied to the order filler in proper sequence, as follows:

First column: Product Location
Second column: Product Identification
Third column: Check Marking
Fourth column: Quantity Ordered

The sequence of the columns must appear as indicated above, since the information needed by the order filler is in that order. If the picker can get to the right location, there is an excellent chance the correct product will be selected. A good order filler checks the order-picking document against the shelf label and against the merchandise on the shelf to verify that the correct product is selected. The order filler then counts the desired number of units or pieces of the product ordered. The order-filling document should be prepared in the logical sequence of these actions.

If the quantity filled is not the same as the quantity ordered, the place to make changes is in the checking column—not in an additional column. (See Figure 36-1.)

Importance of Uniform Procedures

Frequently, warehouse management allows each worker to use his or her own method of indicating a line has been filled. One order filler will use a checkmark, another will circle the quantity ordered, and still another might circle the quantity when it was not filled. All order fillers should use the same means of identifying what work has been completed, what changes in quantity were made, or what items were omitted.

Two other important factors in the preparation of a good order-picking document include:

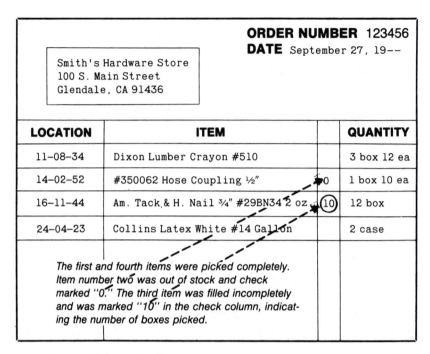

Figure 36-1. Order-Filling Document for Error-Free Broken Case Order Filling

- *Using item location sequence*: The order should list line items in order of location sequencing. Without this, order filling may be accurate, but the rate of order picking will be slower.

- *Clarity in spacing or ruling*: Single-spaced order-picking documents should be avoided—double-spacing is an absolute minimum. It is even better to rule the pages horizontally and print the picking information within the rules.

Illustrated in Figure 36-1 is an example of an order-filling document for error-free broken case (repack) order filling. There is no superfluous information on the form. It was designed specifically for use by the order filler and contains only information needed by that person. The "heading" is limited to basic requirements only, including an identifying order number, the date and the customer's name and address. The customer's identification can be enhanced by the addition of a four-digit matrix-print number (Figure 36-2) indicating the route and stop number.

The check column is placed to the left of the quantity column so that any changes in quantity or any omissions appear next to it.

Other Factors to Consider

Several other factors can help to reduce order-filling errors. These include:

1. *Sufficient illumination*: When illumination in the order-filling aisle drops below 20 foot-candles at 36-inch elevation, there are more errors of the wrong-item-pick type. This applies to both full-case picking and repack picking. When it comes to order-picking broken case or repack quantities, a minimum of 40 to 50 foot-candles of illumination at 36-inch elevation is recommended.

2. *Shelf locations*: Merchandise not easily accessible tends to be "overlooked" or omitted from the order. Frequently, the items that cannot be found are located on the upper and/or lower shelves beyond easy reach. This includes shelves that exceed 6 feet in height, or less than 14 or 16 inches off the floor. The order filler may have to get down on hands and knees. So it's not surprising that he may "forget" to pick items located too high or too low. The answer to this sort of problem is no shelf picking below a 14 to 16-inch shelf, or above a shelf 5 feet 8 inches high.

3. *Item placement*: Naturally, the owner of a trademark wants to make that trademark familiar to the public. But what works well in attracting the buyer works against the order filler. An order filler looking for a particular color of hair dye may find five or six dozen packages on a shelf, all with similar design, color, and decoration. The only difference may be an inconspicuous secondary identification. So the door is open for mistakes, by both the order filler and the stockman who replenishes the shelves. One way to prevent this type of order-filling error is to separate trademarks, so that no trademarked product constitutes a full shelf of merchandise.

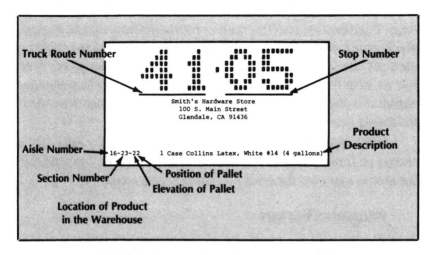

Figure 36-2. Label for Error-Free Case Filling

4. *Document clarity*: If the order filler's document is one part of a multipart form, the warehouse copy should be the first or second copy of it and be clean and legible. In too many instances the order filler gets the copy on the bottom of the stack and sometimes has to guess what's written on it.

Personnel Factors

The way you deal with people can have a great deal to do with error reduction in your warehouse.

Tell your people what you expect. Your expectations may be expressed in a written procedure, or meetings and other sessions dedicated to reduction of errors.

The first and best place to communicate your expectations is with the training of new people. Your people should have a chance to review written procedures and discuss them in training meetings. When new order pickers are retained, there should be supervised dry runs to teach them to do the job right the first time.

Order-filling training is too frequently neglected. Many order fillers graduate from the shipping room to stockroom, and from stockroom to order filler. Without actual training, a worker is likely to start with bad habits. And an order filler who is trained by a worker on the order line may pick up any bad habits that worker may have.

Identification with Work

The worker who is identified by his work usually does it better. One method of doing this is for each order filler to sign each page of an order document as it is completed. If an entire page isn't completed, the picker initials those lines that he or she handled.

Yet another means of encouraging worker identification is to have a picking ticket placed in every order filled. This ticket should include the picker's name, instead of a number (which is anonymous). A particularly effective picker's ticket reads: "This order was picked by Thomas E. Jones. If you find any discrepancy between the packing ticket and your order, please call 123/456/7890. Ask for the Claims Department, and give them my name when reporting the error."

A further refinement is to put the order filler's name and photograph on the picker's ticket. This personalization has resulted in a great improvement in order accuracy.

Posting Error Rates

Using peer pressure to encourage order pickers to be more careful sometimes works by listing each worker's error rate in declining sequence on the warehouse bulletin board. The individual making the most errors has his name at the top of the list—those making fewer errors follow. Always be sure that workers are graded on errors as a percentage of total lines picked. The fastest order picker may appear to be the worst error maker, when this may not be the case. If you post an error-rate list, be sure also, however, to provide recognition and praise for those workers who made no errors.

Using Case Labels

Most warehouses use the same document for case picking as for broken case or repack picking. But using a pressure-sensitive label for case picking will reduce picking errors.

The case label contains this information:

1. Location of the product to be picked
2. Product name, description, size, and number of units per case
3. Customer's name and address
4. Matrix print number identifying the customer (if possible)
5. Delivery particulars (such as, "Do not deliver after 5:30 p.m.," etc.)

Case picking is usually done from a pallet in a rack, a pallet stacked on the floor, or from a multicase inventory. With a good location system, the order filler can go to the right location, verify that the item on the case label and the case on the pallet in that location are the same, and be reasonably sure that the proper item is being picked. The worker then peels the label and applies it to the case.

Because of the high dollar value of case picking as compared with most broken case picking, the cost of a quick check can be justified. If the case labels have been computer-printed, the only checking necessary is to verify that the product description on the label and the product description on the case are identical. If they don't match, the case must be held back until appropriate decisions are made on its disposal.

"Picking Rhythm"

"Picking rhythm" describes a method by which the order filler performs the job while simultaneously checking each step of the order-filling procedure.

Order-picking rhythm is somewhat like learning how to drive a car. When we learned to drive, we had to concentrate on each separate step. We had to start the car, learn to shift gears by depressing a clutch while applying pressure to the accelerator, and moving the gear shift lever. However, it wasn't long before we were able to drive automatically, with the individual steps performed almost as one.

This same approach applies to learning order-picking rhythm.

The order filler goes to the proper location and checks the order document against the locator sign on the aisle, section, and shelf. The next step is to compare the product on the shelf with the shelf label and the order document. If everything tallies, the order filler checks the order for the required number of pieces, selects them, and makes the final check of verifying that the number of pieces picked is the number originally required.

It will not take long before the order filler can perform each of these steps without thinking, just as he or she learned to drive a car.

Order errors can be reduced, if not to absolute zero, at least to a few hundredths of a percentage point. This is done, first, by eliminating the clerical errors that occur before the warehouse ever gets the order. Then warehouse order-picking errors are controlled by improving the stock locator system; clarifying item identification; avoiding confusion regarding quantity of items to be picked; and using a clear, legible picking document.

Improving several environmental factors also will cut errors. These include installing good lighting, providing convenient shelf locations, effective item placement, and restricting multiline order filling. Finally, it's up to management to hire competent order fillers, train them thoroughly, and recognize and reward error-free performance.

Part VIII

THE HANDLING OF MATERIALS

37

RECEIVING AT
THE WAREHOUSE

Receiving is a deceptively simple process in many warehouses.[28] The conventional wisdom says that order picking and shipping are the most labor-intensive activities in warehouse operations, and therefore they deserve and get far more attention than receiving. Yet a faulty receiving process can create as much trouble in a warehouse as poor order picking or shipping operations. This chapter could change your thinking about the importance of receiving. It will detail the process and the physical flow with potential variations in both. The importance of scheduling will be demonstrated, as well as pitfalls that can and do create trouble in receiving.

The Process: Eleven Steps

There are eleven sequential steps involved in effective receipt of merchandise at a warehouse. Not all are needed in every situation, but all of them should be considered in the planning of receiving.

Here are the eleven steps:

1. Inbound trucker phones warehouse to get a delivery appointment and provides information about the cargo.

2. Warehouse receiving person verifies the Advance Shipping Notice (ASN) and conforms it to information received by phone from the inbound trucker.

3. Trucker arrives and is assigned to a specific receiving door.

4. Vehicle is safely secured at the dock.

5. Seal is inspected and broken in the presence of the carrier representative.

6. Load is inspected and either accepted or refused.

7. Unitized merchandise is unloaded.

8. Floor loaded or loose merchandise is unloaded.

9. All unloaded material is staged for count and final inspection.

10. Proper disposal is made of carrier damage.

11. Merchandise is stored in assigned locations.

How does this process work in detail? In step 1, receiving should be allowed only on a scheduled basis, with every trucker making an appointment and then assigned an unloading time. Many warehouse operators believe that they are unable to operate a scheduled receiving dock. Yet most carriers appreciate the precision of a scheduled dock and will cooperate with any warehouse operator who insists upon establishing unloading appointments. While scheduling has obvious advantages for the warehouse operator, it also represents an obligation. When the receiver promises to handle an inbound load at 10:00 a.m., he can expect to be penalized in the form of detention if he fails to keep the appointment that he set. Obviously, running a scheduled truck dock requires the ability to measure and predict flow of work to allow the warehouse to hold to the set timetable.

With today's electronic capabilities, it is relatively easy to obtain a detailed manifest with every receipt of merchandise. The popular terminology is ASN, or Advance Shipping Notice. A growing number of warehouses have established a policy that no receipt will be unloaded without the ASN. This document represents the best way to be sure that the correct product has been delivered to the correct place. There is little excuse for surprises with today's capabilities for information transmission. The ASN should include seal numbers to verify that the load was not tampered with.

As the trucker arrives at the warehouse property, a receiving door is assigned for the vehicle. If the vehicle is a boxcar, instructions are given about correct dock placement.

Before unloading, it is essential to be sure that the vehicle is safely secured. With a truck, this means that wheels are chocked or an available dock locking device has been secured. This ensures that there is no danger that the trailer will move during unloading. In many warehouses, a supervisor is responsible for confirming that these safety checks are complete.

When a trailer's seal is inspected and broken, the truck driver should be present and should initial the inspection form to verify that the seal number conforms with the one listed on the ASN. Any exceptions should be noted at this time.

When the trailer or boxcar doors are opened, there is an initial inspection to determine whether the load will be accepted or refused. Sometimes there is a quality test made at this point to verify that the material on the inbound load meets the receiver's specification. For example, one hamburger chain runs a random test of the fat content of beef before unloading proceeds. If the product fails the test, the entire load is refused.

Many loads contain both unitized or palletized material and loose or floor-loaded material. Depending on how the stock is arranged in the vehicle, either the unitized material is unloaded first or the floor-loaded material is unloaded. In any case, both unitized and floor-loaded product must be counted and checked.

In most cases, the entire load is held in a staging area for final inspection. If the warehouse has a buying department on site, the buyer may wish to examine the staged load before it is put away. However, in most cases the bill is signed and the vehicle is released once a final count of the staged merchandise has been completed.

Before releasing the vehicle the next step is to account for any damage that was noted on the inbound load. Depending upon conditions negotiated between receiver and shipper, carrier damage may be immediately refused to a trucker, or it may be held for carrier inspection and recooperage. In either case, the number of damaged pieces on the load must be noted on the inbound bill of lading and acknowledged by the carrier representative (usually a truck driver).

The eleventh and last step is to stow the merchandise in an assigned location. This is a critical process to achieve storage efficiency. In many warehouses, the decision about where to store the merchandise is not made by management, but is left to the forklift operator who handles the inbound load. When this happens, the natural process is to find the quickest and easiest spot to place the merchandise. The result is a poor storage pattern that wastes space because no one made a management decision. It is common to find the same item stored in many different places within the warehouse because there was no storage planner responsible for choosing locations and assigning new ones.

Physical Flow

The typical movement of merchandise is from carrier to receiving dock to staging area to storage area. However, certain options in the process should be recognized. If there is more than one receiving dock door, which is the best one to use? In most cases, the best receiving dock door is the one that is closest to the area in which goods will eventually be stored. However, when the inbound load is to be cross-docked, the best location will be close to the shipping door where outbound flow will continue.

While most warehouse receivers insist upon staging before storage, a growing number of operators recognize that staging represents double handling and a waste of space and time. The option is to immediately assign a storage area and move the load directly from truck to storage area. If further inspection is needed, it can be performed in the storage area rather than in a staging area. Staging is necessary when material is to be cross-docked, and it may be highly desirable when a significant amount of merchandise is refused because of quality problems. When most receipts have no problems, a growing number of warehouse operators have recognized the economy of eliminating the inbound staging process.

Variations in the Receiving Process

There are several variations to the eleven-step process just described. One is blind receiving. When receipts are blind, the checker writes down the received quantity without reference to any paper that shows the expected quantity. Some believe that blind receiving fosters accuracy, since the person receiving the freight has no idea of what *should* be there and therefore no temptation to approve a written quantity without proper checking. When goods are unloaded on a blind receipt, the receiving tally is completed before it is compared to the ASN. If the tallies do not agree, a second check is made to determine whether there is an actual error or whether there was a mistake in counting the inbound load.

Another variation is the use of bar code technology for receipts. Each inbound package is scanned with a hand-held unit, and the result of the scanning is a tally that is compared to the ASN.

A third variation is to move material directly from vehicle to storage bay without inbound staging. This variation will save both time and staging space, and it is a reasonable step when experience has shown that there are few problems with inbound loads. In some warehouses, loads from certain vendors are always staged because of a history of problems with those vendors. Goods from other vendors who have proven to be trouble free are moved directly to assigned storage locations.

The fourth variation is cross-docking, which is described in Chapter 39.

Another variation is damage. Once damage is identified, there are three ways to dispose of this material. One is to return every damaged carton to the vendor. The second is to salvage goods that are repairable by repacking the good merchandise and isolating that which is not saleable. Finally, it may be necessary to destroy material that is beyond repair. When scrapping takes place, it is necessary to confirm the fact that the material was actually destroyed. Unfortunately, there are a few dishonest people who will promise to destroy material and subsequently offer it for resale.

The last receiving variation is the handling of customer returns. This is a very special kind of receipt, one that has many potential problems. Customers tend to be undisciplined, even to the point where some returns may be in reused boxes that are not even similar to the cartons that were used to ship the merchandise. Returns require immediate and detailed checking, since the returning party may be inexperienced in packaging goods for shipment. The first determination is whether the returned merchandise is still saleable. The second is to determine what additional steps, such as reconditioning, might be taken to make the merchandise saleable. Obviously returns require considerably more care than other receipts, so the planning process for receiving must include time and space to adequately process customer returns.

Equipment

The equipment used on the receiving dock typically receives harder use than most other warehouse equipment. The power equipment makes short runs with frequent starting, stopping, and maneuvering. Stationary equipment such as dock boards and dock shelters is subject to impact from highway vehicles and lift trucks. All of this equipment should be rugged in design and carefully maintained to stand continuing heavy use.

Here are three special factors to be considered when lift trucks are selected for receiving dock work:

1. Install the biggest, widest tires possible. Wider tires reduce damage to both warehouse and trailer floors.

2. Lowered height for lift truck masts should not exceed 83 inches. This height allows entry into the lowest of trailer doors. Specify a freelift upright for both unloading and loading of trailers. The freelift upright elevates the load without increasing the overall lowered height of the forklift mast, which allows you to remove double stack loads inside a trailer.

3. Specify a sideshift attachment that allows you to hydraulically move loads away from the wall of the trailer. This allows removal and storage of the load with less damage. It also allows more precise stacking.

The specifications for stationary dock equipment should include the following features:

1. Automatic dock levelers to provide adjustment for variances in truck heights. Dock levelers may include a hook to secure the leveler to the trailer and prevent movement.

2. Dock seals or shelters will shield inside dock area from outside weather, reduce heat loss, and prevent unauthorized entry into the warehouse.

3. Dock lighting should illuminate the inside of the trailer to help avoid accidents.

Locating the Receiving Function

Tradition says that receiving docks should be at one end of a building and shipping docks at the other where a production process is involved. The receiving area takes raw material that moves through the production process within the building until it reaches the end of the production line. There it emerges as finished goods which move out through the shipping docks at the opposite side of the building. This pattern of dock separation is readily justified by the fact that it conforms to the flow of materials through the plant as goods are converted by production from raw material to finished products.

However, for the warehouse that has no production function, a separation of receiving and shipping is not justified. When receiving and shipping docks are close together, the dock space can be used flexibly to allow additional doors to be assigned to either receiving or shipping, according to demand. Furthermore, when some goods are received for the purpose of cross-docking, the amount of space that must be covered in the cross-dock has a major influence on cost of the operation. When goods can be moved in one door and out an adjacent door, there is an obvious saving in materials handling. Figure 37-1 shows a building with truck docks that are side-by-side and jointly dedicated to receiving and shipping.

Another option is to use two sides of the building along one corner, thus minimizing the distance covered between one dock area and the other. Figure 37-2 shows this arrangement around a single corner of the warehouse.

In other situations, storage may be decentralized with separate functional departments within the same warehouse. In this situation, receiving and shipping docks must also be decentralized to keep

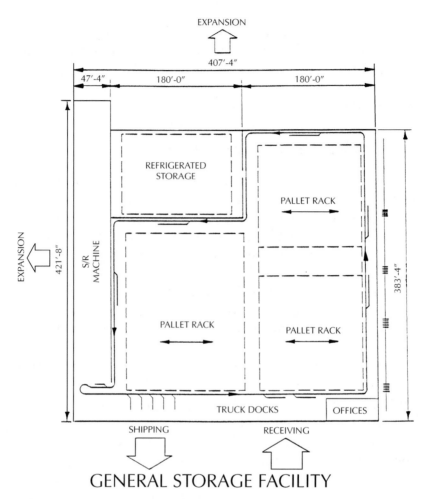

GENERAL STORAGE FACILITY

Figure 37-1

Figure 37-2

them as close as possible to the storage locations for the material involved. When docks are scattered, careful scheduling can be used to minimize the distance for moving materials from a truck dock to a storage area. Of course the situation will be even further aggravated if, through poor scheduling, an inbound truck at one end of the building contains material that is to be stored in the opposite end.

There are obviously a great many variations to dock locations. Frequently dock locations are absolutely dictated by the structural limitations of an existing building. In looking at the location challenge, the important thing to remember is that there really is no reason to separate receiving docks from shipping unless there is a manufacturing process involved. Traditionalists will provide numerous objections to combining shipping and receiving areas, but every one of them can be resolved with proper scheduling and warehouse discipline.

Scheduling of Warehouse Receiving

There are two critical points in scheduling warehouse receiving. The first is that it should be possible to handle nearly all receipts on an appointment basis. Common carriers are accustomed to making an appointment for an inbound load. Scheduled truck docks are not new to any experienced common carrier. When appointments are made and ASNs are provided, there should be no surprises in the receiving function.

Productivity gains may be available if scheduling is used to separate receiving and shipping. Assume that all warehouse receipts are handled on a first shift that operates from 6:00 a.m. until 2:30 p.m. Receipts that are programmed for immediate reshipment (cross-dock) are staged in the dock area, while other goods are assigned to storage locations. Shipping is done on the second shift, which begins at 2:30 and operates until 11:00 p.m. The second shift will reload merchandise that was cross-docked from the first shift, and it will handle all other shipments. Wherever possible, goods that were not put away on the first shift should be used for shipments on the second (unless strict FIFO rotation is mandatory). Separating

413

receiving and shipping by shift can overcome some of the objections of traditionalists who fear that there will be confusion if the same doors are used for both functions. By using time to separate the functions, the same doors are used, but not on the same shift. The warehouse operator gains the advantage of not hauling cross-dock receipts across the building, and the traditionalist gains the advantage of separating the receiving and shipping functions to prevent confusion. Where receiving and shipping functions are separated by time, it is preferable to have receiving on the first shift and shipping on the second, simply because some of the goods received will be immediately available to ship. Furthermore, improved customer service is offered when shipping operations can be handled late in the day. This provides time to include all of the available orders that came in during that day.

Pitfalls of Receiving

We observed earlier that receiving is deceptively simple. You should consider several situations where receiving can and does go wrong.

The first is acceptance of damage. When a warehouse operator receives and stores damaged merchandise, his failure to note that goods were damaged may cause him to assume liability for that damage. When goods are received in damaged condition, the damage may have been caused by the transportation company, or they may actually have been shipped in a damaged condition. The receiver may not know which of these two cases is true, and the claim will be resolved between carrier and shipper. It is most important that the receiver either refuse or segregate any goods that arrive at the warehouse in less than saleable condition. A Polaroid camera is an excellent way to document the existence of damaged merchandise upon receipt, and such a camera should be available at every warehouse receiving dock. Disposition of damage will vary according to negotiated conditions with shippers, receivers, and carriers. Ideally, inbound damaged merchandise should be refused to the carrier and immediately removed from the warehouse. Where this is not possible, the damage must then be held for inspection by a carrier representative or by the shipper.

If the receiver is asked to recondition damaged merchandise, the warehouse area where this reconditioning is performed should be located convenient to the dock where such goods are received. Poor control of inbound damage can cause significant damage expenses to a warehouse which should be avoided by effective inspection procedures.

A second receiving pitfall is the presence of in-transit theft. Unfortunately, goods are sometimes stolen from common carriers and the result is a shortage on the inbound shipment. Without careful counting, such a shortage might not be discovered. When theft is involved, there may be unitized loads with open spaces or voids that are found in the interior of an apparently full pallet. When this kind of theft occurs, it may be necessary to break down all or at least a sample group of inbound unit loads. Failure to detect in-transit theft will only encourage those who are responsible for it.

Certainly the most serious pitfall of warehouse receiving is collusion theft. Collusion theft is widely recognized in shipping. It exists when a dishonest truck driver and a dishonest shipping dock employee deliberately overload an outbound vehicle. Less recognized is the fact that exactly the same risks exist at the receiving dock. In this situation, goods are deliberately left on board the delivering vehicle, and a receipt for all of the merchandise is signed. It is just as easy, or even easier, to steal merchandise with collusion at the receiving dock as it is with collusion at the shipping dock. Yet many security programs ignore the theft exposure in the receiving function. One way to control this risk is through careful inspection of every inbound vehicle when the receipt has been completed.

Another pitfall in receiving is the need to recycle pallets or other load unitizing devices such as cages or other returnable dunnage used by some shippers. One example is the automotive industry, where widespread use is made of cages and other wood or metal platforms designed to unitize materials moving to an auto plant. These unitizing devices are expensive and must be returned to the shipper. When a receiving process fails to properly recognize and recycle returnable

415

containers or other damage, there is the potential for both financial loss and considerable dissatisfaction from the shipper. A good receiving procedure should include detailed instructions on the handling of any returnable materials.

A common pitfall is improper handling of cross-docking. When certain goods are designated for cross-docking and the designation is ignored, that merchandise may be moved into storage and immediately removed, with the result being a loss of time and a waste of labor.

The last pitfall is a failure to balance the workload between receiving and other activities. An old army expression, "Hurry up and wait," describes the pervasive frustration of military life. It exists in warehouses too, even in the receiving function. If the warehouse is staffed to meet peak conditions, there may be too many people to handle the work available during other hours of the day. By scheduling the receiving function as well as other functions in the warehouse, the operator can ensure that there is a steady flow of work throughout the day, and avoid "Hurry up and wait," with its obvious waste of people and equipment.

By avoiding these six pitfalls, you can improve the quality of receiving in your warehouse. When you look at reengineering the receiving function, consider the eleven steps of the process, and develop your own flow chart to stimulate receiving. Consider reasonable variations in the process that are necessary to fit the peculiarities of your own requirements. Be certain that you have the right equipment to get the job done. Consider the dock locations used and whether you have enough doors in the right places to handle receiving effectively. Since scheduling is an important part of the process, look at creative ways to schedule receiving so that it blends smoothly with the rest of the warehouse work. Finally, consider the potential pitfalls and look at ways to avoid these danger points.

38

SHIPPING

The ideal shipping routine is roughly the reverse of that just described for receiving. If an order destined for shipping is pulled from inventory by one employee and checked and loaded by another, the likelihood of discovering errors is greatly increased. Insisting on a shipping schedule will do a great deal to prevent unexpected peaks in workload or unpleasant surprises. A specified count of shipping labels or tags is a good way to create a double check of quantities in shipping. The shipping dock is also the last chance to discover mistakes made by order pickers.

Unit Loads

The grocery industry, through its trade associations, has continued a diligent search for an option to the standard grocery pallet. Ability to load palletized units is one of the best ways to reduce shipping costs.

Efforts to solve the unitizing problem in the grocery industry have centered on three options:

1. Develop a radically new design for a permanent warehouse pallet that can be used internally and readily exchanged.

2. Develop an economical one-way pallet that is designed to be shipped without exchange.

3. Develop a pool of pallets owned by a third party.

Perhaps the most significant design change for permanent pallets is a full four-way entry rather than the notched stringer used in the earlier standard pallet.

Doors Versus Time

One of the most frequent complaints heard from operations managers is that the warehouse does not have enough doors. Have you considered the possibility that the shortage of doors may really be the symptom of other problems?

If you do not schedule the arrival of trucks at your warehouse, many trucks will arrive at about the same time. The result is excessive waiting and needless pressure on your receivers and shippers. If your people say they don't have enough doors, is that true all day or just at peak periods? If scheduling were effective, wouldn't you then have enough doors?

Another major issue is the time it takes to load or unload each truck. The grocery industry achieves only 20% to 30% within one hour. Warehousing professionals know about the steps they can take to improve loading and unloading time. Unitized loads have been around for years, and we are very nearly out of excuses for not using them.

Is the problem that there are not enough doors, or have we failed to use the doors effectively? It is probably the latter, and you have the ability to correct that situation through improvements in scheduling of motor carriers and the unitization of cargo through pallet exchange or palletless handling systems.

Dealing with Shippers' Load and Count (SL&C)

As shipping and receiving agent for his customers (in the case of a public warehouse) or in his own freight (in the case of a private warehouse), the warehouse manager should know exactly what a shipper's load and count (SL&C) notation on the bill of lading (B/L) means. Such a notation can make it quite difficult to collect on in-transit loss and damage claims. On the other hand, a knowledgeable warehouse manager can take certain precautions which will either mitigate or remove entirely the effects of an SL&C notation.

An SL&C notation means that the freight was loaded and counted without the customary assistance of the carrier's truck driver. The notation is most commonly used when, for the shipper's convenience, the carrier drops or spots a vehicle at the shipper's dock and returns and picks it up (without inspecting or counting its contents) after it has been loaded and sealed.

There is a certain truck driver mythology surrounding SL&C. Drivers often scrawl "SL&C" on the bill of lading in the belief that this will have the magical effect of relieving the carrier of liability for in-transit loss and damage. Regrettably, this myth has wide acceptance among uninformed shippers and many untrained warehousemen.

The fact of the matter is that SL&C has absolutely no effect on carrier liability for in-transit loss and damage. Even if a load is traveling under a valid SL&C, the carrier's liability for in-transit loss and damage remains the same as it would have been without such a notation. What does change, however, is that the shipper or claimant's burden of proof is substantially increased.

The U.S. Supreme Court has ruled that although a carrier is not an absolute insurer, it is "liable if the shipper makes a prima facie case and the carrier does not meet its burden to show both its freedom from negligence and that the loss was due to one of the causes excepted by the common law rule." Exceptions have been expanded to include damage caused by (1) an act of God, (2) a public enemy, (3) inherent vice or nature of the product, (4) the public authority, and (5) the shipper's act of default. Making a prima facie case requires showing (1) the carrier received the goods in apparent good order (producing a carrier-executed bill of lading with a statement to that effect), (2) that the goods arrived in damaged condition, and (3) the amount of such damages.

Once a shipper has proved a prima facie case, the burden of proof shifts to and remains with the carrier. However, in an SL&C movement the shipper will obviously be unable to produce a carrier-executed B/L stating the goods were received in apparent good order. He will be unable to establish a prima facie case, and therefore be unable to shift the burden of proof onto the carrier. The point to remember, however, is that the only difference is the burden of proof. Should the shipper be able to establish, through a preponderance of evidence that the shipment was in good order and complete as described when loaded, the normal laws of in-transit loss and damage would then apply, that is, the claim would be much the same as in a non-SL&C shipment.

Here are some of the specific ways a warehouse manager can support shipper/claimant attempts to collect on claims for in-transit loss and damage in SL&C movements.

1. *Seals.* In the case of shortages in an SL&C shipment the carrier must show that the load was sealed at all times except while the shipper/consignee (or their agents) were supervising the (un)loading. Correct handling of a sealed shipment means that the receiver physically removes the seal. It doesn't mean that he stands on the dock and watches while the driver cuts the seal. The tally/ warehouse receipt should show "seal intact" only if the receiver has removed the seal himself; otherwise, the load should be considered as received with "seals not intact." If the seal is not intact upon arrival, the carrier cannot claim SL&C protection against shortage claims. If the seal is intact and removed by the warehouseman, it must be saved complete. Should a shortage be uncovered, the seal should be returned to the shipper for verification that it is indeed the same one applied at time of shipment.

2. *Improper loading.* If damages are caused by improper loading (in the case of an SL&C shipment), the carrier may not be held liable. The carrier must prove, however, that improper loading caused the damage. If, for example, an act of carrier negligence contributed to the damage, the shipper may still be able to recover. In addition, if the damage was caused by other than improper loading, the SL&C protection should be irrelevant. Therefore, when damage is discovered while unloading an SL&C shipment it is extremely important that the extent, nature, and circumstances be accurately and fully documented. Pictures should be taken and statements collected from the unloader and his foreman, and if possible, the driver. In other words, you ought not to assume that, since as it is an SL&C shipment, there is no point collecting documentation for a possible claim.

3. *Loading.* In those cases where, for the shipper's benefit, loading is done under SL&C there are certain precautions that may be taken to preclude a carrier's subsequent SL&C defense. Among them are:

- *Loading correctly.* The best way to avoid losing an in-transit loss and damage claim is not to have the claim in the first place. However, if a trailer must be loaded for later pickup by the carrier, that is, as an SL&C load, extra attention should be given to loading it correctly. In addition, that extra attention should be fully documented, for example, a loading diagram may be prepared showing what, where, and how the goods were loaded. This can be supplemented with pictures and a statement by the actual loaders and the foreman who inspects the load before it's sealed. In the event damage is discovered at unloading, the carrier must prove—not merely allege—improper loading. Documentary evidence such as noted here might outweigh the carrier's claim of improper loading and thus remove his ability to invoke SL&C protection.

- *Having the load inspected by the carrier.* If the load is sealed before being tendered to the carrier, the carrier has no obligation to break the seal and inspect the load—that is, unless the shipper (or his agent) specifically instructs the carrier to do so. Therefore, if possible, a notation should be placed on the bill of lading instructing the carrier, at pick-up time, to break the seal and inspect the trailer for proper loading. This should be sufficient to defeat the carrier's later claim that damage was due to shipper's improper loading.

- *Verifying SL&C notations.* When receiving, do not proceed as if the shipment were SL&C just because it is indicated on the bill of lading. There is always the possibility the driver at the other end added the notation after pick-up. If there is damage or shortage when the doors are opened, treat the task of documentation the same as if the load were not SL&C.

Who or what caused the damage is a matter that should be decided later on the basis of the evidence provided when the claim is filed. The warehouse manager's job is to ensure the claimant has the most complete and accurate documentation package possible.

4. *Not allowing the carrier to mark the bill of lading as SL&C.* If the carrier is present during loading but for whatever reason chooses not to observe the loading, do not allow him to later mark the bill of lading SL&C. DOT Safety Regulations require drivers to ensure, before moving a loaded vehicle, that the cargo is properly and safely stowed. The driver should be given an opportunity to check the load; if he refuses, both the refusal and the fact that an opportunity to inspect was available should be documented on the bill of lading. A note also should be made on warehouse records showing what the driver was doing while the trailer was being loaded.

5. *Reading the bill of lading correctly.* There are certain published commodity rates that apply in shipments where shipper loads and/or consignee unloads. These are not necessarily the same as SL&C. The carrier may not have been precluded from counting and inspecting the load just because the shipper did the actual loading. Notations on the bill that the shipper performed the loading should not, therefore, be automatically construed as SL&C.

It should be noted that all of the procedures described above apply primarily for shipments moving under common carrier rules. When your shipments are moving under a contract carrier arrangement, the rules might be quite different. Since deregulation, an increasing number of shipments are moving on something other than common carriers. When this happens, the shipper must carefully study the rules under which the freight was moved.

Shippers' load and count makes the claimant's task more difficult, but it does not make it impossible. Avoidance of abuses of SL&C and careful documentation of shipments made in this manner can still make claims possible.[29]

Controlling Damage Claims

Claims of damage due to improper loading are often difficult to defend. A Polaroid camera to record the condition of a loaded trailer can provide evidence to use when the consignee or customer claims the load was not properly stowed.

To be valid in a dispute, these photographs must be referenced. Therefore, it is wise to make a written notation on the reverse side of the photograph that shows the date the picture was taken and all details surrounding the shipment. If the carrier driver signs the photograph and acknowledges its authenticity, it's even stronger evidence.

Careful inspection of the trailer or box car presented for the shipment will eliminate the possibility of loading a defective vehicle.

Customer Pick-Ups

An integral part of many shipping operations is the processing of customer pick-ups. This kind of shipping represents a special challenge, because the order is typically loaded into a vehicle driven by employees of the customer's company. If shipping service is poor, the inconvenience is not filtered through a common carrier, but it is felt directly by your customer. Therefore effective handling of customer pick-ups can be critical to the success of any shipping program.

Certain elements are necessary for successful customer pick-ups. Ideally, no customer should come to the dock without an appointment. If the customer makes an appointment and provides advance notice of the goods to be delivered, the warehouse operator can and should pull and stage the complete order before the customer arrives.

One shipper promises to begin loading within fifteen minutes after the customer's arrival, provided the proper advance notice has been given.

Courtesy (or lack of it) as well as speed can make a lasting impression on most customers. Therefore, proper handling of this kind of shipping achieves an importance that goes far beyond the actual volume involved.

Perhaps the most important thing to remember about shipping is that it represents the last chance to avoid a warehouse error before goods reach the customer. In addition to watching for errors in picking, the warehouse manager should check on accidental shipment of damaged goods, errors in count, or shipping of goods not in first-class condition. One warehouse has this sign on the shipping dock: "The next inspector is our customer."

39

CROSS-DOCKING IN THE WAREHOUSE

Much has been written in on the topic of cross-docking. It has become increasingly clear that cross-docking has almost as many flavors as ice cream, and the flavor you choose has a great deal to do with the cost of the operation.

The equivalent of plain vanilla, and typically least costly for the cross-dock operator, is an inbound load that is marked and separated by outbound order. When received in this fashion, the operator simply maintains that separation as the goods are unloaded, then each order is staged for outbound shipment.

A more expensive flavor is an inbound load that is sorted and marked according to the stockkeeping units in the load, but not segregated according to outbound order. In this situation, the cross-dock operator must receive and account for the entire load, and then make up outbound orders according to a separate order manifest which should be provided with the load. This requires more work than the plain vanilla flavor.

A still more complex flavor is one in which the inbound material is labeled by outbound order, but not sorted or segregated. Here, the operator must unload the merchandise and perform the proper segregation by noting the destination labels and matching them with the shipping manifest.

Another flavor is one that requires that the inbound load be handled in combination with other merchandise that is already in storage at the cross-dock warehouse. The operator must blend merchandise in storage with new merchandise on the inbound trailer, a still more difficult operation than the plain vanilla.

The most complex flavor of all is one in which goods are separated neither by SKU (stock keeping unit) nor by outbound order, but are loaded at random on the inbound truck. Now the operator must perform two sorting operations, one to segregate SKUs and a second to fill customer orders. This flavor is probably the most expensive option for the cross-dock operator, though it may have the lowest cost for the shipper.

The flavor chosen has a great deal to do with the cost of the operation. Depending on how the freight is loaded, the shipper can either make cross-docking very easy or very difficult for the operator. Therefore, whether you are buying or selling cross-dock services, remember that you should be familiar with all five flavors and above all be certain that everyone involved know exactly which flavor will be tasted.

Success Factors

The most important success factor in cross-docking is the need to keep it simple. You can create a highly sophisticated and unprofitable cross-docking operation if you get carried away.

Since cross-docking is not storage, a third-party operator should charge an additional fee for any product that is in the facility for more than three days. Otherwise, the users would receive storage services without paying for them.

The three important ingredients for success in cross-docking are time, communications, and accuracy. With respect to time, it is essential that the cross-dock operator receives the detailed information on the load before the freight arrives. Failure to submit information in a timely manner will usually wipe out the cost advantages of cross-docking.

Quality of communications is as important as their timeliness. The cross-dock operator needs to know exactly what freight is scheduled to be received, when it is scheduled to arrive, and when the outbound load must be shipped. If the details are not accurate and complete, this will again destroy the advantages of the cross-dock operation.

Ample communications delivered on time are still useless if they are not accurate. Furthermore, the operator has an absolute obligation to provide equal accuracy in documenting the handling of both inbound and outbound moves.

Ultimately, your success in managing a cross-dock operation depends on three essential elements: receiving, developing quality warehouse people, and the office staff. We will consider each in detail.

Receiving

Goods in a cross-dock operation are received by several modes: truck, box car, or marine container.

The only efficient way to receive product for cross-docking is to make up the outbound orders as the inbound material is received. For example, let's say you receive products for nine different orders on one pallet, but as that pallet is handled, the goods are sorted onto nine outbound pallets, one for each order. When you are finished receiving, you should have all the product from that truck on individual outbound pallets. When you have received the last of the inbound trucks, its completion should allow you to fill the last of the outbound pallets. If there are irregularities in the form of over-shipment or orders canceled at the last minute, the excess stock is put into warehouse inventory.

In an efficient operation, one worker is taking the product off of the inbound truck and counting it. A second worker separates that product into its outbound orders, putting each case on the appropriate outbound pallet or marrying that product with material on other pallets that are going to the same destination. In an operation with considerable volume (over 500,000 pounds per cross-dock receipt) a conveyor system and racking may be desirable to improve the materials handling process.

Because cross-dock receiving takes a little longer than conventional warehouse receiving, it is important to widen the window for receiving operations, possibly handling receipts on multiple shifts in order to maintain customer service.

Warehouse Personnel

Cross-docking is a more intricate operation than some other kinds of warehousing. Your effectiveness in cross-docking is enhanced by minimizing the number of different times that product is handled. Accuracy is essential, and that depends upon people. Specific employees should be assigned to customers so that they can learn the details of each customer's needs. As the employee's knowledge of the product increases, both speed and accuracy will improve.

There are certain types of people who we find are most likely to be successful in handling this work. The successful operator, male or female, is usually between the ages of 18 and 28, married, and living with spouse and children, with everyone trying to get ahead. This worker may have had two or three previous jobs that were unsatisfactory, and now he or she is trying to be worth something and to earn a decent dollar. Many of these people have no previous warehouse experience, so they do not bring any bad habits. They can be trained to handle the cross-dock operation. They start off just one or two dollars over the minimum wage plus health and welfare benefits. Give them the incentive to learn, and they will to stay because the job is better than those they had before.

Taking the time to get to know your workers is very important. You may have a college degree and be skilled in your field, but you do not do the actual labor. That hourly worker who struggles from paycheck to paycheck is the one who will make your profit or get you in trouble. You must believe, and you must make them believe, that they are a vital part of your structure, and that their work is very important to you. It is their suggestions that will allow you to improve the way you handle the operation, but you will get those suggestions only if they truly care.

Try to know your people, to know names of spouses and children, their anniversary date, and their birthdates. Keep index cards that contain this information for quick reference. Each time you go into the warehouse, read one or two of these index cards and make it a point to look up those people and say hello and ask about their families.

You should expect your supervisors to listen as well as to give good directions. They are taught what the company expects them to do to accomplish our goals. Make sure they understand your company's philosophy. Be sure they will not mind being asked the same question one hundred times until a new employee feels comfortable. It is better to be asked questions than to let a new employee guess and do the wrong thing. Expect your supervisors to follow your example in taking steps to know the people in the warehouse.

Particularly in these days of corporate restructuring, this type of caring is more important now than it has ever been in the past. From time to time, a pizza bought by the company and shared with your people at lunch break can have a big influence on productivity. This is done easily in a small warehouse. In a larger one, you may accomplish it section by section over a period of time to cover all of the employees.

Office Staff

Timely and accurate performance in the office is as important to cross-docking as the work that is done on the warehouse floor. One office person should be assigned to each customer or category of the cross-docking operation (but this is not their only responsibility). When one office worker is dedicated to a particular group of cross-docks, you have the opportunity to form a close-knit working relationship between the office and the warehouse workers who handle that product. Aside from better teamwork, this relationship will enhance your staff's ability to solve new problems. Suggestions that may improve the cross-dock operation could come either from office or warehouse. Remember that your office people and your warehouse people probably know the customer's operation better than you do, and may well know it better than the customer's people do. There are certain details relating to cross-dock operations that do not change from week to week. Others do change, and this teamwork gives you the opportunity to refine operations and ensure that your people don't work any harder than they really have to. Always be receptive to the suggestions they give.

Pricing

For the third-party operator, cross-docking can yield revenues per square foot and revenues per hour that are two to three times greater than for a normal public warehouse operation. The value added in cross-docking is greater, and the customer is willing to pay for it. The third-party operator must charge prices that relate to the amount of work needed for each customer's operation.

It is dangerous to set prices across the board. Each customer must be priced according to the specific needs and volume associated with that operation. I suggest a ninety-day trial period with each new customer, with the ability to reevaluate both the system and the pricing at the end of the trial period.

While ninety days is an appropriate time to readjust prices and procedures, it actually takes about six weeks to move from startup to a smooth-running operation. It takes that long for you and your employees to learn the product and the individual characteristics of each client's needs. It also takes your client that long to get used to your operation. Try to have a client representative on hand in the warehouse during the first ten days of operation. You may negotiate a higher rate for the first six weeks with an agreed roll back for the second six.

Here are a few examples. Client A sends you between 150,000 and 200,000 pounds of product (7,500 to 10,000 cases) that is cross-docked over a 72-hour period. During that period, you need the full time of one office person, or 24 hours during the three days. You also need the time of two and one-half warehouse workers which will not exceed 50 hours in the warehouse. At the rates you charge, this total commitment of 74 hours will yield a profit between 18% and 30%. Your productivity is measured at between 2,027 and 2,703 pounds per man-hour, but remember that includes office hours as well as warehouse labor.

Client B has double that tonnage, or a range of 300,000 to 400,000 pounds (15,000 to 20,000 cases) on each cross-dock. In spite of the much greater volume, only 8 additional office hours are needed, for a total of 32. Instead of two and one-half warehouse employees, you now need four people for a maximum of 24 hours each. Total hours needed are 106, compared with 74 for Client A. With Client B, the profit ranges from 28% to 40%. The pounds per man-hour handled ranges from 2,830 at the low range of volume up to 3,774.

Volume has a significant influence on costs and pricing. It takes a certain minimum number of people to handle a typical cross-dock, but those same people are capable of handling more product in the same period of time.

Forms

There are six forms that can be used to document and support a cross-dock operation. The two illustrated are the pick ticket and the loading tally.

One *pick ticket* is prepared for each product code. It lists those customers receiving merchandise and the quantity going to each destination. The pick ticket is available as the inbound truck is received, and it is used to control the distribution of cases for outbound loads. A sample is shown in Figure 39-1.

A *loading tally* accompanies each bill of lading and is used to control the outbound shipments. It is filled out by outbound truck to show where each order is located within the trailer, starting with the nose of the trailer and ending at the rear. Figure 39-2 shows a sample.

The *pallet record* lists number and type of pallets received and shipped, and it warns that the customer will be invoiced at a specific price if comparable pallets are not returned. The purpose of this form is to police the pallet exchange function.

The *receiving tally* verifies that the shipment was properly received. It describes each product and provides space to tally the quantity of each product code.

Pick Ticket	Loading Tally
Product Code _____	Trailer No. _____ Loaded _____
	For Delivery _____
Customer Case Qty	**Driver:** This load must be checked by piece count and product code number, at time of unloading. Please report any discrepancies to the Dispatcher on OS&D form and return this tally to your office.
Chain A _____	
Chain B _____	

Pick Ticket:

Product Code _____

Customer   Case Qty

Chain A _____

Chain B _____

Chain C _____

Chain D _____

Chain E _____

Chain F _____

Old Inv. ____   Ttl Cs ____

Amt. Rcvd ____   New Inv ____

Loading Tally:

Trailer No. _____ Loaded _____

For Delivery _____

Driver: This load must be checked by piece count and product code number, at time of unloading. Please report any discrepancies to the Dispatcher on OS&D form and return this tally to your office.

B/L [| | |]
        Pallet
1. Store
B/L [| | |]
        Pallet
2. Store
B/L [| | |]
        Pallet
3. Store

Stop Offs
Nose ____
2. _____
3. _____
4. _____
5. _____
6. _____
7. _____
8. _____
9. _____
10. _____

Special Instructions:

Figures 39-1 and 39-2

A standard *bill of lading* controls each outbound shipment. Finally, an *invoice* is used to bill the customer and provides full references for inbound and outbound shipments and charges.

Achieving the Goal

An effective cross-dock operation provides timely and accurate redistribution of cargo without the need for storage. If everything is not done accurately and on time, the results will be disappointing. Information handling is as important as freight handling. In the last analysis, the whole operation depends upon the performance of well-motivated people. Everybody in warehousing will see more cross-docking in the next ten years than has ever been seen before. Ignoring cross-docking will not diminish its growth.[30]

40

SPECIALIZED WAREHOUSING

Most warehouses are designed to store bulk merchandise for volume shipments at ambient temperatures. But there are also some highly specialized warehouses. In this chapter we will look at four that are growing in importance and complexity.

The first is temperature-controlled warehousing, which has grown substantially in the past several decades with the growing popularity of frozen foods and fresh produce requiring temperature control. The use of temperature-controlled warehousing for nonfood products has grown with the increased distribution of chemicals that require refrigeration.

Second is warehousing for hazardous products, which may not have grown in popularity, but has certainly grown in complexity. As public awareness of the dangers of hazardous chemicals has increased, those warehouse operations that must safely store such products have experienced growing regulation and risk-management problems.

A new growth area in specialized warehousing is fulfillment, which is the handling of mail order and express shipments moving directly to the consumer.

Household goods storage has some distinctive differences from merchandise warehousing.

Let us examine these four special warehouses in detail.

Temperature-Controlled Warehousing: The Essential Differences

There are at least four kinds of cold warehouses. The standard freezer operates at 0 to −10°F; ice cream storage freezers operate at −20 to −25°F; blast freezers combine extreme cold with rapid air circulation to quickly freeze freshly packed products; and chilled warehouses hold product at 35 to 45°F. If you are designing a cold storage building, consider potential future uses and the cost of conversion. For example, if you are building a large chilled warehouse, design the building to allow for future conversion of part of the space to a freezer warehouse. The lower the temperature one wants to maintain, the higher the cost of the building.[31]

"Temperature-controlled storage" is usually cold storage, and its prime use throughout history has been to preserve foods. Before the development of refrigeration machinery in the 1890s, products were frozen by blowing air over salt and ice. Many of the cold storage plants were branches of ice plants. The quality of this cold storage was questionable, and some state laws required retailers to warn their customers of goods that had been in cold storage. Early cold storage warehouses were used primarily for dairy products, meat, and poultry, and they were filled on a seasonal basis. In the early decades of this century, most temperature-controlled products were chilled rather than frozen. Consumer-sized packages of frozen foods were introduced in 1929 and did not become popular until the 1940s. By 1970, freezer space comprised more than 75% of the total public refrigerated warehouse space in the United States. "Cooler" space is used primarily for fresh fruits and vegetables, dairy products, and eggs. There is increasing use of chilled storage for nonfood products such as plastics, film, seeds, and adhesives.[32]

One significant difference between temperature-controlled and dry warehousing is the cost of the facility. A freezer warehouse will typically cost two to three times as much as a dry storage warehouse of similar size. The cost difference in utilities is even greater, with electricity costing as much as five times the amount per square foot. So the successful temperature-controlled warehouse operator depends on excellent conservation and building maintenance to control these energy costs.

---

* From Operational Training Guide, International Association of Refrigerated Warehouses, Bethesda, MD.

The freezer room should be monitored by temperature gauges, one at eye level showing the temperature as you walk into the room, and others of the same type in different corners of the room. In addition to this, a temperature recorder with a weekly disk should be used as a permanent record. A 24-hour monitoring system that gives a warning of temperature fluctuations of three degrees or more should be installed. This cuts down exposure to liability for temperature fluctuations and fire. Putting plastic curtains inside the permanent doors helps in controlling energy costs.[33]

In a dry warehouse, walls may be thin steel panel, protected by a modest amount of insulation. In contrast, the walls of a freezer are an important part of the insulation system. A freezer may have six-inch foam insulation panels, clad by sheet steel both inside and out. The floor too is an important insulator in a frozen warehouse. A typical specification would be six inches of concrete poured on top of six inches of foam insulation. Below the insulation, to protect against heaving of the earth beneath the floor, heat is provided by piping warm ethylene glycol through a layer of sand. The insulation layer thus serves a dual purpose—keeping the cold in and the heat below out.

The roof of a dry warehouse may consist of nothing more than a thin steel deck with a small amount of insulation. Most temperature-controlled warehouse roofs will start with a steel deck, but the structure must be designed to hold refrigeration equipment which either is mounted on top of the roof or hangs below the ceiling of the warehouse. Above the deck is a layer of ¾-inch fiberboard. Above that is an additional 10 to 12 inches of foam insulation, and then another layer of fiberboard. The weather seal is a single-ply rubber roof protected by a layer of stone ballast.

The operation of lift trucks is affected by cold, especially the lower temperatures of a frozen-products warehouse. Because of tight insulation and recirculated air, internal combustion engines cannot be used. Electric trucks need modification to operate at zero degrees. Heaters are required for the electric contact points as well as heavier duty batteries to allow for a full shift use without recharge. The harsh conditions also produce additional wear and tear on the equipment. In this environment, thorough preventive maintenance is even more critical than in dry storage warehousing.

The batteries on electric equipment will last longer if they are charged every four to six hours rather than the eight hours recommended in normal temperatures. The warehouse that has both dry and cold space should have a plan to rotate fork trucks so that each truck spends only part of the time in the freezer. Stand-up lift trucks are somewhat easier than the sit-down types for people working at below-zero temperatures. Some trucks are specifically designed for this environment, with an enclosed cab to allow the workers some relief from the cold temperatures.

Workers should have a break-room where they can relax with coffee or soft drinks. A ten-minute break each hour will improve productivity for people working in the freezer. You should rotate workers from all areas of the warehouse so that each works for some of the time in the freezer.

Most warehouse operators provide protective clothing for employees. This includes insulated boots, gloves, and freezer suits. Working in a freezer results in a greater fatigue factor, since a significant portion of body energy is spent in keeping warm. In an ambient-temperature warehouse, a work crew is capable of handling an overtime or emergency assignment of well over 12 hours without significant loss of productivity or accuracy. Fatigue takes its toll in a much shorter time in a frozen warehouse. Most operators see a greater amount of sick leave among workers in a frozen environment.

The task of supervision is more difficult in a temperature-control than in a dry-storage warehouse. A supervisor in a dry warehouse can watch the loading dock and gain a good idea of what is going on throughout the warehouse. In a temperature-controlled warehouse, the dock is separated from storage areas by walls and doors to preserve the temperature in the storage rooms. A supervisor on the dock cannot see what is happening in the storage rooms. Effective supervision in the cold storage area requires additional foremen in the cold rooms. Supervisors have an even greater risk of health problems, because they may move in and out of cold rooms more frequently than the workers, and because they are less physically active.*

The nature of cold storage requires changes in the way work is scheduled. Because the cost of space is very high, staging of outbound orders may not be practical. Temperature-control warehouse operators tend to select most outbound orders as close as possible to the time of shipment in order to minimize space committed to staging. Product quality considerations may prevent staging outside the freezer area. When freight cannot be preselected, the operator cannot use staging as a scheduling buffer and when staging is difficult, performance of motor carriers is particularly critical. The cold storage operator must run a scheduled truck-dock.

Housekeeping is another function that is affected by the harsh environment in a frozen warehouse. Spills that are not cleaned up quickly may freeze, cause accidents, and stain the floor. The method for scrubbing a freezer floor is far more expensive than in dry warehouses, because it requires the use of a nonfreezing solution. The lowest-cost solutions are banned by FDA rules. Sometimes the only way to clean a floor in a frozen environment is to scrape it. The cleaning equipment, like lift trucks, is restricted to electric power. The cold causes similar maintenance problems and shorter battery life.

The temperature-controlled warehouse requires an extra measure of precision because the consequences of failure are serious. Failure to maintain temperature control of the product can have very costly consequences for the owner of the merchandise. Therefore the warehouse operator must not only protect the product, but also provide ample proof that such protection was always provided.

As the use of chilled and frozen products seems to be growing faster than the economy as a whole, it is likely that temperature-controlled warehousing will be a part of your future even if it has not been part of your past.

---

* From Jesse Westburgh, Citrus World, Lake Wales, FL.

Hazardous Materials Warehousing

Hazardous materials can be grouped generally into the following categories: flammables, explosives, corrosives, poisons, radioactive materials, and oxidizers. In addition, as the EPA's classification of materials continues to be expanded, those materials having "Reportable Quantities" are likely to increase.

For the warehouse operator who has never handled hazardous products, life suddenly becomes more complicated. The relationship changes between the warehouse operator and other concerned parties, forcing the operator to deal with agencies and individuals who would not be involved with other kinds of warehousing. Storage of hazardous materials creates increased responsibilities and liabilities.[34]

From an administrative standpoint, the distinguishing feature of hazardous material distribution is the regulation to which it is subject. There are several authorities who either have regulatory power over hazardous materials or with whom you should consult:

The Occupational Safety and Health Administration (OSHA) has regulatory power over warehousing hazardous materials with regard to protecting your employees from exposure to these products. OSHA's concerns are satisfied by taking the hazardous material information provided by the manufacturers on their container labels and Material Safety Data Sheets (MSDSs) and conveying that information to your employees through an approved communication program.*

The Environmental Protection Agency (EPA) regulates hazardous materials through its Superfund Amendments and Reauthorization Act (SARA) Title III Program dealing with Emergency Planning and Community Right to Know. Emergency planning is carried out through state and local Emergency Planning Councils, both of whom must be supplied with either the previously mentioned MSDSs, or an appropriate listing covering any hazardous materials you store. The local Emergency Planning Council could conceivably be repre-

---

* From an article by Lake Polan III, Allied Warehousing Services, Inc., in Warehousing Forum, Vol. 4, No. 4 © Ackerman Co.

sented by your local fire department. Any release of a hazardous substance must also be reported to both state and local councils.

The Department of Transportation (DOT) regulates the movement of all materials including hazardous materials. It has established 20 different classifications of hazardous materials, and it requires, among other things, that all materials be labeled properly. All trucks leaving your facility with hazardous materials must be appropriately placarded, and you must also properly annotate your bills of lading when hazardous materials are shipped. For the most part, warehouses accept the classifications provided by the manufacturer, but this is not always the case. For LTL (less than truckload) outbound shipments, special care must be taken. When combining products with different classifications, the quantities involved in the shipment may require a special "dangerous" placard to be placed on the truck. Overseas shipments come under regulation of several international agencies, some more stringent than DOT.

While many agencies regulate hazardous materials, they have little to say about how the warehouse operator actually handles or stores product in his own facility. Therefore, the most important source of procedural advice is the supplier, the product manufacturer, or the product owner of goods in a third-party warehouse. These are the individuals who should know the most about the product involved, and should provide the warehouse operator with comprehensive documented standards for safe storage and handling. The operator also has a responsibility to acquire a maximum amount of knowledge and experience with the hazardous product. The possibility arises that other product owners who share the use of the warehouse will be concerned about storage and handling practices that increase the risk to their products—storing plutonium near baby food, for example.

If you lease your space, your property owner will also take an interest in hazardous materials storage. While your lease will probably not prohibit this business, you will need to address such issues as indemnification, the adequacy of your sprinkler system, increased insurance costs, and your ability to return the building in usable condition at the end of the lease. "Reasonable wear and tear" does not contemplate a building left contaminated with methyl-ethyl-bad-stuff.

Most manufacturers, even if they self-insure, may use some or all the guidelines for storage developed by the National Fire Protection Association. NFPA membership represents both the manufacturing and insurance industries, with standards sensitive to the needs of the latter. Your insurance carrier conceivably uses NFPA standards as well, with the noted exception of the Factory Mutual Companies, and some others who have developed their own standards. You will need to establish a close working relationship with both your insurance agent and carrier(s) to help you evaluate the costs and risks of hazardous material storage, to determine that you can insure those risks and the price of the coverage.

State and Local Building Inspectors

If you plan new construction or are remodeling your facility to equip it for hazardous material storage, you may also need the approval of your local building inspectors to ensure that you conform to the requirements of a local or municipal building code. These codes may or may not conform to NFPA guidelines. Local codes can be more restrictive than state codes, but if they are less restrictive then state codes take precedence.

Because a number of different authorities are involved, it is inevitable that there will be inconsistency or even conflicting regulation or instruction. Therefore, a key to functioning with these different agencies is to determine the authority having jurisdiction.

Where public safety is primary, the "authority having jurisdiction" may be a Federal, state, local, or other regional department or individual such as a fire chief, fire marshall, chief of a fire prevention bureau, labor department, health department, building official, electrical inspector, or others having statutory authority. For insurance purposes, an insurance inspector, department, rating bureau, or other insurance company representative may be the "authority having jurisdiction." In many circumstances the property owner or his designated agent assumes the role of the "authority having jurisdiction."[35]

It is not unusual for the warehouse operator to find that the person responsible for regulation knows less about the product than the operator does. In such situations, it is wise to influence selection of the authority having jurisdiction, since this can have a dramatic effect on cost. Sometimes the warehouse operator becomes the authority by default, simply because no other authority is willing to make a decision.

Unfortunately, it is not practical to provide a simple "how to" that encompasses all the product variables and provides the storer with instructions on safe procedures. So where does the operator go to learn the best way?

If the storer is a third-party warehouseman, he will probably seek information from other third-party warehouse companies handling similar products. Storers of hazardous products typically indicate that they have this experience in their advertising or in warehouse and logistics directories. If you are a private warehouseman, you still might share information with third-party warehouse operators and with other operators of private warehouses. Even storers of competitive products are willing to share information about safe warehousing practices, simply because everyone is anxious to learn from others in this field.

The National Fire Protection Association (NFPA) is the leading authority on fire safety, and the cost of membership in this association is nominal.

Because regulation in the field is changing rapidly, there is a danger that the information you have is now out of date. To be sure that your warehouse is in compliance with current standards, it is wise to use the services of an agency specializing in interpretation of the Federal register.

Of all the information sources available, the best is the manufacturer of the product who should have the most information available about the safe keeping of that product.

Becoming a competent warehouser of hazardous materials requires a serious commitment in organization, manpower, and capital resources. The warehouse operator is exposed to increased risks, uncertain requirements, greater costs, and a potential regulatory nightmare. Successful hazardous materials warehousing requires an increased discipline throughout the organization, since the consequences of failure can be most severe.

For storers of hazardous materials, the first concern is safety, the second is service, and the third is cost. As the requirements become more stringent, only the most competent warehouse operators will remain in this line of business.

Fulfillment Warehousing

Six special features make product fulfillment warehousing different from public and private warehousing. These are the following:

1. The warehouse operator has direct interface with consumers.
2. Information requirements are instantaneous.
3. Order size is much smaller than typical warehouse orders.
4. The order-taking function at warehouse level is much more precise, particularly because it involves contact with the consumer.
5. Customer service requirements are different and typically more demanding.
6. The transportation function is more complex.[36]

Storage and materials handling functions are quite different. Fulfillment warehousing involves more than simple storage. Because of fast turns and low volumes, gravity-flow rack and high-security areas are almost always needed.

The handling function will also differ from that of the more conventional warehouse. While there may be some LTL shipments, there will be a much higher concentration of parcel service and mail movements and handling will include the metering of those shipments. The materials handling equipment investment will therefore have more emphasis on scales and meters to control outbound movements.

Paper flow for a fulfillment warehouse is more complex than for most other warehouses. A large number of orders are received by telephone. A significant amount of time is spent in handling customer returns. Because a fulfillment operator is dealing directly with individual consumers, the customer service function is particularly critical.

Some users want the fulfillment center to create invoices or even dunning notices. Accounts-receivable aging reports are another service that can be a byproduct of the handling of invoices. Nearly every fulfillment center must handle the major credit cards. Credit card authorization systems become a necessity. Many fulfillment centers handle banking for their customers, and a smooth relationship with the center's bank is a necessity to handle these accounts.

The best of fulfillment centers offer a 24-hour turnaround on orders. A few take two or three days to process and ship, but fast turnaround is becoming a standard. Because of this fast turnaround requirement, labor flexibility is needed to deal with seasonality and variance in work load. The best fulfillment centers maintain a pool of part-time workers who are available if a second shift must be added or if extra people are needed quickly.

A fulfillment center can be a major headache in terms of claims and theft. Product shipped can be misdirected by a dishonest person who is running a postal machine. Operating the postal machine is a vulnerable position, and care should be taken in selecting the individual who takes on this job. Many fulfillment centers negotiate the inventory variance to be allowed in advance, based on the customer's own experience with errors. Some users will allow a predetermined formula for shrinkage that is in line with their internal experience.

Typically, the fulfillment center must absorb the dollar consequences of all of its mistakes—the public warehouseman's usual limitations do not apply in this service. The fulfillment center normally needs more equipment for communications than for shipping. This includes the ability to handle credit cards and toll-free phone lines. Compared to the conventional warehouse, the order volume is extremely high. The ability to do a great many things with a computer is far more critical in the fulfillment center than in the conventional warehouse. Finally, the number of stock keeping units controlled is usually relatively high in a fulfillment center.

As companies constantly seek new ways to market their products, fulfillment is a warehousing service that is destined to increase in popularity.

Household Goods Storage

Although there are many similarities between the warehousing of household goods (HHG) and other merchandise, there are three essential differences that define the activities and the way they are managed. First, merchandise warehousing typically deals with *new* products. In contrast, the warehousing of household goods and personal effects typically involves *used* furniture. Second, HHG storage is almost always a part of the transportation contract, rather than a separately defined warehousing activity. Third, the ownership of the goods handled has passed to the end consumer, which personalizes the activity to a degree that does not exist in other types of warehousing.*

Because the HHG warehouseman is required to provide transportation, he always has a trucking capability; but this is not always the case in merchandise warehousing. The transportation contract often calls for movement over a long distance so the furniture warehouseman needs state and Federal authority. To make a profit on a long distance run, the HHG warehouseman also needs to find back haul loads for the return trip. To solve the back haul problem, HHG warehouse operators banded together in agency relationships, and many of these have evolved into national van lines. These associations provide the means of controlling equipment, and they also offer a division of revenue for services performed. HHG van lines have offered a closer bond and a more formalized sharing of information than any of the voluntary merchandise warehouse groups.

Modernization

Mechanization and unitization have changed household goods storage in ways similar to merchandise warehousing. With general

---

* From an article by Frederick S. Schorr, management consultant, Hilton Head, SC.

merchandise, materials handling evolved from case handling to the movement of palletized unit loads. With household goods, containers called vaults are used to eliminate the piece by piece handling of furniture and personal effects.

Because of the specialized storage requirements of household goods, the multistory buildings have retained their economic viability much longer. Some operators developed creative programs to produce revenue, including customer self-storage and record storage. There is a significant liability problem in handling used furniture, so self-storage is particularly attractive to the HHG firm.

Preparation of HHG for storage is typically labor intensive. Also because of the weight and configuration of many items, more than one person is required to perform the service. Thus costs for handling are a much larger part of the contract than they are for merchandise storage.

Household goods are more vulnerable to damage than most general merchandise. Furthermore, because the products are not new, the question of where the damage occurred is always a potential problem. The frequency and cost of claims is greater for HHG than for general merchandise. HHG warehousemen typically offer a cargo insurance which is available to the customer at extra cost. Many HHG operators provide a claim service to provide expeditious repair or adjustment for damage to the customer's goods. The best HHG operators have been leaders in the quality process. They seek customer feedback and work to offset negative perceptions of the HHG industry. Obviously, improvement of quality also translates into profit improvement.

Diversification

The large growth in types of services offered was driven by two forces. The first is the seasonality of the HHG business. Most moves take place during the summer months, which forces companies to scramble to find alternate sources of revenue the rest of the year. The annual fixed cost of buildings and equipment necessary to provide service during the peak months, as well as the desire to retain trained and experienced employees, has to be met. Second, since each HHG move is an individual transaction, requiring the services of a salesman or estimator, the HHG warehouseman was philosophically prepared to expand or contract his business on demand. Merchandise warehousemen typically are more concerned with occupancy in the large warehouses they have constructed.

Further diversification brings HHG warehousemen into receiving, storing, and delivery of products other than used household goods. These commodities fall into two categories:

1. *High value products:* items such as computers, office products, medical diagnostic equipment, and exhibits for trade shows
2. *New products:* furniture, fixtures, and appliances

Because HHG movers are skilled at providing a large variety of labor-intensive handling and storage services, it is natural that they diversified beyond the basic moving of furniture by expanding into new markets that require the same type of services. The movement of museums, libraries, offices, and plants often results in the storage of both used and new equipment. One of the fastest growing HHG segments is the temporary warehousing, consolidation, delivery, and installation of inbound shipments of furniture and furnishings for hotels, offices, and hospitals. Most of the major hotel chains contract directly with moving companies to receive and store their products, including everything from furnishings and carpeting to wall hangings, drapes, kitchen equipment, silverware, and china. Upon delivery these items are placed in the rooms and installed ready for use. This of course requires effective planning, coordination, and control between the builders, decorators, and warehousemen. The period between the completion of construction and the grand opening is usually "just in time."

Those HHG operators who diversified found it necessary to acquire expertise in a wide range of endeavors, greatly increasing the management skills required to get the job done.

However, no requirement places as much demand on management as the fact that in the movement and storage of HHG, you are dealing with consumers rather than managers. Although the arrangements may be made by a corporate manager, the goods are usually moved under the watchful eye of the owner. Both the beginning and end of the transaction take place away from the premises of the HHG storage company. Add to that the usual stressful nature of uprooting a family and it is easy to visualize the communications problems which can complicate the moving of household goods. Rather than a series of repetitious movements, each HHG transaction is unique.[37] Thus managing the exception often seems to be the rule!

Summary

The four specialized types of warehousing described here have grown in popularity and complexity. Continued growth of cold storage is promised by changes in our food consumption and cooking patterns. As public concern about dangerous chemicals grows, hazardous materials warehousing will continue to be in demand. The popularity of credit cards and marketing by catalogue and television will continue to create a growing need for fulfillment centers. Finally, household goods operators fill a unique niche which is gradually diversifying.

41

ORDER-PICKING

Accurate order-picking is typically the most important warehouse operating responsibility. Actual costs of an incorrectly picked order are estimated at thirty to seventy dollars per bad pick, not counting the customer dissatisfaction.

In many warehouses, order-picking is the largest single expense category in the operation. Good order-picking demands high levels of management in planning, supervising, checking, and dealing with personnel. The order-picking operation is not easily or economically automated. Even with all of our automation advances, order-picking often remains a manual operation because brain, eye, and hand coordination have not yet been equaled by any machine.

Because of its high labor content, order-picking presents the greatest opportunity for error reduction. A good order-picking document is the first step for accurate and efficient picking.

Seldom do warehouses receive goods in the same quantities or packaging required for shipping. Shipments from the warehouse must be orders assembled from stock, since economic order quantities to the warehouse are seldom the same quantities that customers actually purchase.

Four Methods of Picking Orders

The job of selecting orders can be divided into at least four categories: single-order-picking, batch picking, zone picking, and wave picking.

Single-order-picking is the most common means of selecting an order. One order picker takes a single order and fills it from start to finish.

In batch picking the order picker takes a group of orders, perhaps a dozen. A batch list is prepared that contains the total quantity of each stock-keeping unit (SKU) found in the whole group. The order picker then collects the batch and takes it to a staging area where it is separated into single orders.

Zone picking is the assignment of each order selector to a given zone of the warehouse. Under a zone picking plan, one order picker selects all parts of the order that are found, for example, in aisle 12, and the order is then passed to another picker who selects all of the items in aisle 13. Under this system, the order is always handled by more than one individual.

Wave picking is the division of shipments by a given characteristic, such as common carrier. For example, all of the orders for UPS might be grouped together into a single wave. A second wave would pull all of the orders destined for parcel post, and still other waves would select shipments routed by other carriers.

Quality in Order-Picking

Quality means performance to standards and nothing more. Warehouse management must design, implement (through training), and then insist upon standards of order-picking that are error-free. Those managers who *know* there are going to be errors because there always *have been* errors will *always* have errors.

On the other hand, error-free order-picking is done by those who believe in zero error and plan for it to happen. Any order-picking error could be a customer lost, never to be regained.

Order-Picking Forms

A good order-picking form is one that has only essential information on it, that is, customer identification, order number or date, location of items to be picked, item description, and a specific quantity to be picked.

Newer order-picking systems based on computer assistance should have the built-in elements of a good manual system. When you implement automation, choose a system that will accomplish your objectives. Consider these principles of automation: define your needs in detail, get the software that meets those needs, and then look for hardware.

The use of color coding in the order-picking operation will reduce errors. Color coding should be used in any operation where it can be applied.

To reduce errors, be sure that the same terminology is used for the same items throughout the system in the warehouse.

Systems for Order-Picking

When setting up your order-picking system, plan on generating data that will allow you to measure performance once the system is operating. Some common performance ratios used for order picking are the following:

$$Orders\ per\ hour = \frac{Total\ number\ of\ orders\ picked}{Total\ labor\ hours\ used}$$

1. If orders are uniform, this is a valid measure, but if the number of items or quantities varies greatly from order to order, the following ratio will be more indicative of performance:

$$Lines\ per\ hour = \frac{Total\ number\ of\ lines\ picked}{Total\ labor\ hours\ used}$$

2. This ratio is a more accurate measure of the work performed by the picker as each line represents a task. Lines picked per hour is probably the most commonly used measurement ratio.

The order-picking system consists of pick slots (locations) where the product is available for selection in the quantity called for on the picking document. The location of the pick slots depends on the system, but they must always provide the necessary picking identification and be physically conducive to low fatigue and error-free picking. In other words, put the most popular items between waist and eye level.

Order-picking is seldom done by only one method. Variances in package size and configuration, picking quantities, stocking quantities, and inventory requirements often necessitate more than one system. Most order-picking operations are hybrids of three order-picking methods, listed here in order of system complexity:

1. *Unit-load* picking is done when a pallet load of product is pulled from stock. An example of unit-load picking is a major appliance warehouse.

2. *Case-lot* picking is the selection of full cases of a product. However, the order is less than a full-pallet unit-load. Case-lot picking is best done by staging a unit load in a pick line and pulling case quantities until the unit load is depleted.

3. *Broken-case* picking is done when less than full cases are called for by the customer's order. This kind of picking may be done from shelving or flow rack, depending on the size and volume of orders.

Fixed versus Floating Slots

Where do you locate the stock? The shortest travel route in order-assembly sequence will yield the lowest picking cost.

Fixed locations are assigned to each product and the product is always located in the same place. Simplicity and elimination of errors are the greatest advantages of this system. The disadvantage is the waste of space that occurs when a slot is reserved for unstored product. A fixed-slot system can be combined with a preprinted order form to produce a simple order-picking system that uses the stock numbers as locations. This system is ideal where the order quantities are small.

Floating slot systems use the next available location in the warehouse for storage instead of reserving an assigned area. This random location system requires that a precise locator system be instantly available for fast moving items.

Floating slots are likely to increase picking travel and thereby erode order-picking efficiency. Travel will be reduced if random storage is arranged by zones of activity. Keeping fast movers together in short-travel locations will help control travel.

Order-Picking Methods

Order-picking can be manual, power-assisted, automatic, or it may be a combination of these methods.

A manual system uses two- or four-wheel hand trucks or carts pushed through the pick line and hand-loaded.

A powered system uses unguided or guided vehicles to transport and/or elevate the warehouse worker through the pick line. Pallets, carts, or other containers are manually loaded by the order picker.

An automatic system uses the computer to guide the picker to the pick location, elevate him to the proper pick height, instruct him as to the pick location, and indicate the proper pick quantity. Automated picking may be any combination of these, accomplished through computer control.

Single-Order versus Batch Picking

Single-order picking requires the picker to assemble the total order before moving on to another one—in other words a complete pass through the order-picking area for each order to be picked. Single-order picking has these advantages:

- Maintains single-order integrity.
- Simplifies the picker's job.
- Avoids rehandling or repacking.
- Provides fast customer service.
- Allows for direct error checking and establishes direct error responsibility.
- Is highly efficient when the number of SKUs per order is small.

Single-order picking has these disadvantages.

- Requires full order-picking route travel for all orders.
- Doesn't allow for speed-picking of large quantities of an individual item.
- Requires the highest number of picking personnel for a given number of orders.

Batch picking is selecting of the total quantity of each item for a group of orders. In a breakout area, batches are resorted into the quantities for each order. Batch picking has these advantages:

- Reduces travel to pick the total quantities of a group of orders. Picking travel time can be reduced as much as 50%.

- Minimizes picking time for quantities of an item.

- Permits volume picking from large quantity or bulk storage, thus reducing the need for constant restocking of the pick lines.

- Provides a second check of the quantity picked by comparing the batch picked against the individual quantities in each order.

- Improves supervision by concentrating the final order-assembly in a smaller area.

Batch picking has these disadvantages:

- A second pick, or a distribution of the picked quantity, is required to fill individual order requirements.

- Space is required for the distribution and order-assembly operation. Additional equipment may be required depending on the size of the batch-pick area.

- Individual orders are open until the entire batch of orders is complete.

- Counting is done twice and differences in count will require reconciliation time.

A variation of batch picking is to have the picker first pull the total item quantity and then place the proper quantity in separate slots or tote bins for each order. The order picker is now doing single-order-picking in a batch-picking mode. When combining these two methods of picking, higher skill is needed since the potential for error is increased. A prepack check may be desirable if a pattern of errors develops. This picking method should be done only with well-trained and experienced personnel.

Zone System of Picking

Zone picking is accomplished by arranging pick lines in zones that handle similar types of items. For example, a "family" series such as carburetor parts may be located in one zone.

The arrangement of the zones can provide for an order-assembling system in which each zone is used to build each order. Three zone arrangements may be used:

1. *Serial* zones are arranged in sequential order. The order picker must always go from zone A to zone B, then to zone C, etc.

2. *Parallel* zones are an arrangement of independent pick lines, perhaps with one on each side of the aisle. The picker need not take them in any particular sequence.

3. *Serial/parallel* zones are those where a number of serial zones are arranged in a parallel configuration.

Zones have the advantage of bringing together similar products in family groups. Where an order may be pulled from one family, this system arrangement expedites the picking process. Zones are very adaptable to use of flow-rack systems.

Designing Your Order-Picking System

In designing your system, consider how restocking will be accomplished. A minimum of one day's picking requirements is usually kept in the pick area. When restocking volume begins to equal the picking volume, it may be time to pick from the bulk supply area.

Scheduling of order-picking depends on the necessary lead time to produce the picking documents. Order-picking systems work best when documentation is done on the day preceding the scheduled pick date.

Consider "reverse engineering" your system. Take the finished orders that are to be shipped and start walking from the shipping point back through the flow process. As you move "upstream," see if the step just completed or the one before it could be eliminated.

Equipment selection is easier to determine when approached from the output need rather than the input possibilities.

The following outline will help you develop your order picking operation:

1. Define your objectives by listing the customer's expectations and requirements.
2. Define your objectives in terms of:
    A. Cubic displacement of loads received, stored, picked, and shipped.
    B. Volume of packages stored, picked, and shipped.
    C. Flow rate of receipts, stored and picked, and shipped material in each classification.
3. Find out the state-of-the-art by requesting literature, specifications, and installed project reports from equipment suppliers.
4. Establish a set of operating specifications that say what you want to accomplish, not how you want to accomplish it. Operating specifications should encourage ideas and alternatives.
5. Ask two or three suppliers to review and bid on your objectives, giving them access to all of your information.
6. Evaluate the system's total economics for its life cycle. Be sure to place a fair value on flexibility and resale value, since you can seldom be sure that your product or needs will not change.
7. Purchase at the value price, not at the best price. All equipment needs service and supplier support. If you deal with the lowest bidder, it is wise to add something for the risk you run.
8. Implementation programming for a new order-picking system should be a part of the entire process and should start with the beginning of the project. Planning implementation with the entire warehouse staff involved will ensure a successful operating system.

Hardware

Effective order-picking involves a combination of both storage and materials handling equipment. The order-pick function essentially has two options: either move the picker to the stock or move the stock to the picker. The most common means of doing the latter is through a carousel or a computer-controlled stacker crane. To some extent, a gravity-flow rack also moves stock to the picker, since it allows merchandise to slide from the back to the face of the rack.

When the picker must move to the stock, the storage layout should be arranged so that this movement is minimized. In a high-cube warehouse this is something done by establishing pick mezzanines that allow some order pickers to stand on an elevated platform and pick from racks at a higher level.

Ergonomics

Defining ergonomics is a little like defining an accordion.[38] If you ask somebody what an accordion is, they will make some hand signs to describe it without telling you how it really works. We often do the same thing with ergonomics. Consider three definitions. Ergonomics is the study of job demands from the perspective of what tasks workers can safely perform. It is also the science of designing machines, tools, furniture, and work methods for maximum human comfort and efficiency. Third, good ergonomics is the business of helping people work smarter and not harder, and arranging the work so that people will minimize the possibility of excess fatigue and/or personal injury.

Unfortunately, many managers today look at ergonomics as a way to stay out of trouble with the Federal government. In our opinion, a more important motivator should be the improvement of your company's accident and safety record and a reduction in health benefit costs. On-the-job injuries are expensive, and one trade group reports that the average medical cost for a back injury is $30,000. If good ergonomics can significantly reduce your injury record, that should be a stronger motivation than fear of Federal regulators.

Perhaps the most important motivator is a gain in productivity. When people work smarter, they are more productive. When management arranges the workplace to reduce fatigue, your people can do the same amount of work with less effort and therefore move more pounds of freight every day. Good ergonomics can and should have a payback in improved productivity.

How to Improve Ergonomics

The most obvious step to improve ergonomics in the warehouse is to avoid conditions that cause an awkward or strained situation in manual handling of merchandise. Perhaps the worst strain can be caused by twisting. (Referred to by NIOSH as the asymmetric multiplier.) Avoid any situation that requires that the worker engage in a twisting motion, particularly if it is done while lifting or handling cartons. Avoid or at least minimize the situations that require the worker to stoop to the floor or to reach overhead, particularly when lifting heavy cases. Avoid walking by placing merchandise in a position where the number of steps taken in manual handling is at least minimized. Reduce repetitive motions that can cause injury, commonly referred to as cumulative trauma disorder (CTD). Even in the office, certain motions repeated often enough can cause an injury such as carpal tunnel syndrome. What can you do to avoid these awkward motions? The most important step is arrangement of stock in the warehouse. Those items that move the fastest or that are the most difficult to handle should always be placed in the "golden zone" which is between belt height and shoulder height of the average individual. This allows the stock picker to grab and move the case without stretching or bending. It is obviously impossible to place all merchandise within the golden zone, but at least that which moves fastest should be in this area.

There are several practical steps that can improve ergonomic conditions. Some grocery warehouse people place two extra pallets on the floor beneath the first loaded pallet. This puts the lowest case on the floor eight to twelve inches higher than it would be if the two extra pallets were not used, and it therefore reduces the risk of strain from leaning down to the floor to pick up the lowest case. Use a case hook when an order selector must pick cases from the back of second level slots. The hook allows the selector to move cases to a safe position without standing on equipment or product.

Common sense can reduce walking. If order-picking is done in a "Z" pattern, the picker selects from one side of the aisle and then immediately selects merchandise directly across the aisle. When a pallet jack is used for order selection, position the pallet so that it is two to three feet away from the merchandise to be picked. If it is farther away, unnecessary steps are needed. If it is closer, the picker will have to twist while moving merchandise from the storage pallet to the selection pallet.

Rotating jobs every few hours has several advantages. Obviously job rotation will allow workers to do different tasks and minimize the possibility that a repetitive lifting situation could cause injury. It also has other advantages such as the ability to cross-train and develop new skills, and to avoid job boredom.

In some warehouse situations, injury is closely related to fatigue. Uncontrolled overtime can therefore contribute to injuries and accidents. Improved scheduling and judicious use of part-time workers can enable management to reduce the amount of overtime.

Jobs should be designed for micro-breaks, and workers should be trained to use these breaks to avoid undue fatigue and to plan ahead. Workers should be encouraged to stop moving periodically and to study the remaining work and consider the best way to get the job done. In other words, they should be taught to work smarter and not harder.

Training and job review are absolutely essential. Have all of your order pickers been taught the best way to select orders without undue fatigue or wasted motion? Are your supervisors properly trained on ergonomics so that they know how to teach workers the best way to get the job done?

Documentation is equally important. One company designed a methods checklist for order-picking. This checklist is not only a good training tool, but also an excellent piece of defense evidence if that company should face inspection by Federal regulators.

Proper slotting is an essential key to improved ergonomics. How do you ensure that the right merchandise remains in the golden zone? Unless management controls the putaway function and keeps the right materials in the golden zone, there can be no assurance that an ergonomically proper stock arrangement will remain that way. Control of inbound putaway is an essential aspect of good ergonomics.

Exercise can be as important as training to keep your people in shape to avoid injuries. Some warehouses start the day with aerobic exercises done as a group by everyone in the workforce. In one situation, management as well as workers go to the exercise session each morning to set the example. While the exercise program is not mandatory, most users find 100% participation.

Summary

Order-picking efficiency is dependent on planned storage of material to be picked. This planning minimizes the distance traveled and the pick time involved. Labor content for order-picking is usually the highest for this job compared to all the others in the warehouse. Therefore, picking offers the greatest opportunity for cost reduction through improved layout, better methods, and faster equipment. Nothing is more important than improving picking accuracy. The importance that accurate order-picking has on customer relations makes it a concern of senior management.

Accurate order-picking should be recognized as a major cost but also as a valuable sales asset.

42

STORAGE EQUIPMENT

A warehouse is more than a storage building. To handle cargo, it needs equipment. And it is the selection and use of that equipment that may spell the difference between profit and loss.

The operator may choose from a vast array of machinery and hardware designed to improve the efficiency of both handling and storage. The equipment choice is usually governed by the following criteria:

- Degree of flexibility desired for different uses
- Nature of the warehouse building
- Nature of the handling job—bulk, unit load, individual package, or broken package distribution
- Volume to be handled by the warehouse
- Reliability
- Total system cost

Defining the Job

It is easy to overlook the lowest-cost space in any warehouse—that which is close to the roof. Most contemporary buildings are high enough to allow at least a 20-foot stack height, and some allow substantially more. Fire regulations typically require that high stacks be at least 18 inches below sprinkler heads, but with this knowledge you can and should calculate the highest feasible stack height in the building. Then determine whether or not it is being used.

Saving space usually also saves time, since storage in a more compact area allows picking travel to be reduced. The prime justification for storage equipment is to increase cube utilization. Such equipment ranges from the simplest pallet rack or shelving to a rack-supported building designed for a stacker crane.

"Live storage" equipment not only increases the use of cube but also moves material when a movement is needed. One example is gravity flow rack, which both provides storage for merchandise and allows it to move from the rear of the rack to the picking face.

Improving Storage with Racks

Because the storage rack is relatively simple, it is easy to overlook ways in which storage capacity can be greatly increased by using a rack. The most common storage rack found in warehouses is the three-high rack system. Installed throughout a warehouse of 100,800 square feet, this rack will allow up to 6,930 pallet positions as shown in Figure 42-1. Yet, in many warehouses, there is enough cube to permit this system to go higher, so the rack can be replaced with a higher one or rack extensions fitted as shown in Figure 42-2.

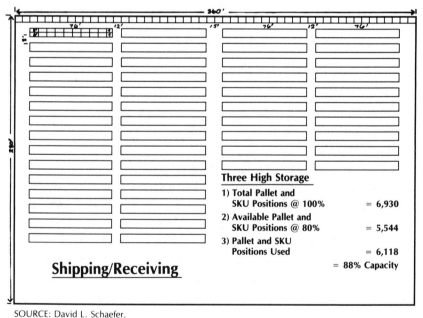

Three High Storage

1) Total Pallet and
   SKU Positions @ 100%   = 6,930
2) Available Pallet and
   SKU Positions @ 80%   = 5,544
3) Pallet and SKU
   Positions Used   = 6,118
                               = 88% Capacity

Shipping/Receiving

SOURCE: David L. Schaefer.

Figure 42-1

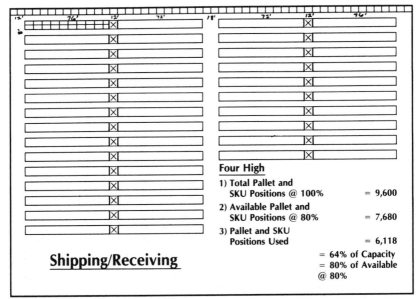

SOURCE: David L. Schaefer.

Figure 42-2

Once the rack is extended, it also is possible to bridge it over with cross aisles and further increase storage capacity. If this is done, the storage capacity shown in Figure 42-2 is increased 39%, from 6,930 pallet positions to 9,600.

Reducing Number of Aisles

One way to reduce the number of aisles is a rack system designed for double deep storage. The only disadvantage is the possibility that overcrowding and insufficient volume may cause one item to be blocked behind another. As shown in Figure 42-3, this will increase the total pallet positions to 12,560, nearly double the amount shown in Figure 42-1. However, because the double-deep rack denies access to the inside pallets, the total number of stock-keeping unit (SKU) facings available is less than the plan shown in Figure 42-2. Therefore, the double-deep system would be used only when storage capacities must be increased without increasing the number of SKU facings.[39]

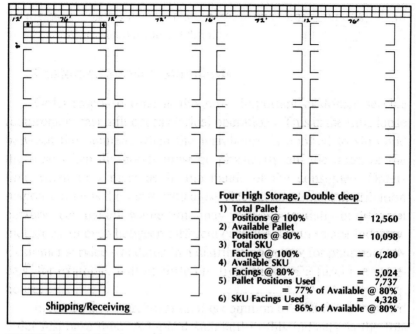

SOURCE: David L. Schaefer

Figure 42-3

Another means of improving space use is movable storage racks. There are roller-mounted racks which can be shifted sideways to create an aisle whenever needed. This equipment is costly, but it is justified by the space it saves. It may be impractical for a very fast-moving warehouse, because access to some facings must be blocked while the rack is rolled to create an aisle on the opposite facing. Mobile racks can be moved to close and open different aisles. A warehouse that is busy enough to require all aisles to be open at once should not use the mobile rack.

Other Types of Pallet Rack

Greater storage density can be achieved by using a "drive-in" rack in which each load is supported by a flange that grips the edge of the pallet. While the drive-in rack achieves maximum density, it may do so at a sacrifice in handling efficiency as the driver guides his lift truck through the narrow alley between rows.

464

Tier rack is a self-supporting framework that covers the unit load and permits freestanding high stacks supported by the rack structure rather than the merchandise. Tier rack is frequently used for stacking auto tires or other products having no packaging or structural strength.

Some storage racks use gravity to move stored loads from the back of the row to the face of the rack, permitting retrieval from the aisle. This is known as live storage.

Stacker-crane installations have been designed to heights of over 100 feet. A stacker crane can be adapted to either bulk or small-lot storage. A computer-controlled crane is particularly useful where random-lot storage or order-picking is required. In such a system the computer memory stores each item's location and the device is programmed to pick merchandise with minimum travel.

Live Storage—Gravity Flow Rack

Gravity flow racks, unlike conventional static shelving, slope from the back (feed-in side) toward the front (picking side). The flow rack will vary in depth (front-to-back) from as little as four or five feet to as much as 10, 12, or even 20 feet.[40]

Gravity flow racks are made in several different styles for ease of flow from back to front. The original flow racks were made from fairly conventional shelving with sufficient tilt so that the products fed into the back would slide through to the front. Guide rods were sometimes used.

Subsequently, gravity flow racks were made with small nylon wheels, then with wheel tracks. The use of wheels enhances the "flow" characteristics of the rack and thus reduces the required angle of declination. Reduced decline is desirable because it allows the elevation of the feed-in side of the rack to be lower.

You can achieve higher order-filling rates using gravity flow racks because of two factors: (1) If the backup or reserve cases can be located behind the pick location instead of to the side or on top of it, then the actual pick surface can be greatly reduced; (2) flow rack allows you to consolidate a greater number of pick facings within a limited number of running feet of aisle. Thus, an order filler standing in front of the rack can select a greater number of items without walking further down the aisle, so travel between picks is minimized.

Here is an example of how gravity flow racks work. Assume a particular product has a case dimension of 12 inches wide by 24 inches deep by 12 inches high. The broken-case (less than full-case) demand for this product ranges from three-and-a-half to four cases. If four cases are put on conventional shelving (no more than 24 inches deep) two lineal feet of shelving and at least 27 inches of shelf height (allowing for working clearance) are needed to accommodate the four cases.

With gravity flow racks, however, a single facing, 12 inches by 15 inches high, provides access to the first case. The additional three cases are lined up behind the front case, thus saving about 75% of the rack's facing area. If the "layback" type of gravity flow rack is used, reaching into the open case is simplified and less overhead clearance is required.

When to Use Flow Racks

Gravity flow racks are best for fast-moving products, and when the order-filling requirements allow the order filler to pull from a number of pick positions with little walking. Properly designed, with intelligent item placement and careful item selection, flow racks should increase order-picking speed considerably.

Since order-picking speed depends not only on the number of lines ("picks" or "hits"), but also the space between lines, the speed of order-filling is improved by reducing the facing of each line. Thus, all facings should be made with the short side of the carton forward rather than the wide side. It is important to use the left-to-right span of each rack shelf to the utmost, and this requires careful initial arrangement of cartons.

Vertical space between shelves also is important. To waste as little space as possible, cartons of similar height should occupy the same shelf. For ease of picking, it's best to put tall cartons (packages) on the lower two shelves, and shorter packages on the upper shelves.

Order size also is a primary factor. Small orders that require a long walk between lines can reduce the effectiveness of a gravity flow rack.

What Flow Racks Will Not Do for You

Flow racks are not the answer to all problems. Sometimes flow-rack installations do nothing to increase order-filling speed or save labor cost. On the contrary, they cost the warehouse operator more money and take up more space than conventional static shelving.

What's the difference between an effective flow-rack installation and one that wastes money and space? Typically, flow racks will provide little—if any—benefit unless the orders to be picked have enough lines in them to make the flow-rack technique work. For example, you frequently find ineffective flow racks in security areas where items in the small, confined area simply aren't fast-moving enough to warrant flow racks. Generally speaking, items picked less than 20 times per month should not be in flow racks.

Two or three sections of flow rack with one hit every four or five feet—three or four hits for the whole flow-rack system—will be picked at about the same speed as if the merchandise were placed in static shelving. While flow-rack picking *can* enhance order-picking speed, the flow-rack benefits depend on minimizing the space between hits, and having enough lines so the order filler can establish a fast pick rate.

You must also consider an additional, though small, cost—the difference in shelf-loading time (and labor). Flow racks take more time to load than shelves.

Any operator who installs a system to hold 100% of the items in the warehouse will have a cosmetically beautiful system with marginal utility. The best order-picking system is a hybrid, using flow racks for those items having high activity, and shelving or other systems for those with less activity or uncommon sizes.

The Carousel—It's More Than a Merry-Go-Round

There are two options for selecting merchandise in a warehouse. The most popular is for the worker to move to the merchandise.

The second option is offered by the carousel: the worker stands or sits and the product moves to that worker.[41]

Carousels are simply mobile storage units. They will not magically provide you with operational improvements. What is required is a thorough analysis of your application needs versus expected carousel solutions. While carousels can yield significant productivity improvements, they are not ideal for everybody.

If your application involves the storage and retrieval of *a product*, then you should evaluate carousels. Consider the nature of the product, its size and weight, how and in what quantity it is to be picked, pick cycle versus replenishment needs, available pick time, storage space available versus storage space needed, security requirements, and hazardous material requirements. Some common carousel applications are:

- Distribution order picking
- Manufacturing in process storage
- Progressive assembly
- Pallet pick

Distribution order-picking is the most popular use of such equipment. It is the selection of items to fill customer orders.

Some factories will use the carousel as a space savings means of holding parts that are in process. *Manufacturing in process storage* is then retrieved from the carousel as needed for the final assembly process. Computers and other electronic equipment must sometimes be tested before being approved for shipment. The carousel provides a test storage space for each unit without requiring aisles to access each tested item.

Some *progressive assembly* processes require minor variations for each assembled unit. Because the carousel allows individual items to be called in a varied pattern, it can permit greater work flexibility and higher throughput in less floor space.

A larger carousel designed to hold entire pallets will facilitate *pallet picking* and or pallet load building.

Three Kinds of Carousel

Horizontal. A horizontal carousel is a series of storage bins linked together in a closed loop. These bins commonly are constructed of wire with adjustable shelf levels. The bins can be of almost any size but the most common are 24 inches or 36 inches in width, 18 inches or 20 inches in depth, and 7 feet to 10 feet in height with carrying capacities of 800 to 1,500 pounds per bin. Greater capacities are available if needed. Normal rotation speed design is in the range of 60 to 80 feet per minute.

Horizontal carousels normally range in lengths from 20 to 80 feet. The desired length is most often determined by the available space and the pick-rate requirements.

Controls range from manual to fully computer-controlled systems that can be integrated into the complete manufacturing or distribution control system.

Horizontal Rotary Racks. These are simply one-level horizontal carousels stacked on top of each other. Each level can rotate independent of any other level to potentially produce extremely high pick and replenishment rates. Very few applications need or can justify the cost of this type of carousel. These units can match standard horizontal carousels in the range of product sizes and weights that can be handled. Normally only upper level integrated computer systems are used to control this type equipment.

Vertical. Think of a standard horizontal carousel turned so that it rotates perpendicular to the floor. Enclose this in a sheet metal housing and you have the concept of a vertical carousel. These units normally are approximately 8.5 feet to 10 feet in width, 4 feet to 6 feet in depth, and 8 feet to 35 feet in height. The selected size is determined by space and pick-rate requirements. Capacities vary but are available up to 1,200 pounds per level and 25,000 pounds per unit.

Benefits of Carousels

The improvements you can expect from a carousel system compared to a shelf, rack, and/or floor storage system are:

- *Better space utilization*: No aisles are needed with carousels, since the merchandise moves to the staging area. This elimination of aisles can create a dramatic improvement in the amount of space needed for each item that is stored.

- *Improved pick accuracy*: Newer carousels have computer control to greatly reduce the likelihood that the wrong item will be selected.

- *Increased productivity*: Maximum productivity can be achieved by assigning two or three carousels to a single operator. While one carousel is spinning, the operator selects material from another.

- *Improved inventory control and security*: Picking and restocking accuracy increases because the carousel presents either the correct item to be picked or the correct bin for restocking. Since the carousel system records where product is picked and who picked it, shrinkage tends to decrease. Also, the system will have access only by authorized personnel.

- *Better management*: All workers have specific work stations and job functions are readily understood and measured.

- *Improved morale and image*: Operators take pride in having a system to manage. This carousel can also be a marketing tool to show customers the commitment the company has made to accurate product shipment.

Conveyor Systems

Conveyor systems are employed in manufacturing and are frequently used to handle the interface between a production plant and a plant warehouse. There are seven factors that should be taken into account when selecting a conveyor:

1. Product or material to be handled
2. Its outside measurements and physical characteristics
3. How many and what flow rates are involved?
4. What is the conveyor to accomplish specifically?
5. How large an expenditure is justified?

6. Is the product fragile, or is noise a problem?

7. Are there any other restrictive factors involved, such as space available or atmospheric conditions?

Types of Conveyors

The hand-pushed monorail consists of a single overhead rail that may be a standard I-beam or a rail incorporating a contoured, hardened lower flange on which a trolley runs. The load is carried on one or more trolleys. A similar unit with power can be adapted with dispatching controls, block systems, and other automatic features.

The power-and-free conveyor consists of a trolley conveyor from which is hung a second set of rails to carry the free trolleys and the loads. A chain engages and pushes the free trolleys below. Several different types of mechanisms are available that permit the power chain and free trolleys to be coupled together. These allow stop–start control of one load independently of all other trolleys and loads on the conveyor. This important feature makes it possible to change horizontal and vertical direction easily so the free trolleys may be switched into alternate paths, and also allows the use of drop or lift sections. Power-and-free conveyors can be used to sort and provide live and in-process storage.

The dragline conveyor, or tow conveyor, is a power-and-free system turned upside down. The power chain under the floor is engaged by dropping pins mounted on the front of four-wheeled carts through the chain slot. Track switches allow alternate routes or destination spurs.

Skate-wheel conveyors are gravity conveyors of two or more rails with cross-shafts on which skate wheels are mounted. Skate-wheel conveyors can be set up on portable stands with power boosters (belt conveyors) at intervals.

Gravity roller conveyors have tube-type rollers mounted on the fixed-axle shafts instead of skate wheels. Though the roller conveyor is more expensive, it also is more flexible for handling many types of packages. Some are powered and are then called live rollers.

The difference between the belt conveyor and the belt-driven roller conveyor is that the belt runs on top of the rollers. Belt conveyors are widely used in order-assembly operations. Because they provide a continuous moving, flat surface, they are more suitable for order-assembly when small or odd-shaped packages are to be handled.

Chain-on-edge conveyors are used for handling four-wheel dollies. This type of conveyor is usually found transporting dollies and loads through manufacturing operations. The two-strand chain conveyor is vertically flexible and often is used for dipping loads, going up into ovens, or other process operations.

The slat conveyor has steel or wooden slats fixed between the strands generally at every pitch. These are used mainly to handle heavier loads, or in cases where the unit load could cause damage to a belt due to projections on the bottom of the load.

By putting tilting slats on a single or double-strand chain conveyor, it is possible to sort at fast speeds. The tilting slats tilt on command, dumping the load into specific slides, chutes, runout conveyors, bins, or wherever it is required.

The Automatic Guided Vehicle Systems (AGVS) Alternative

As the price and availability of automatic guided vehicle systems improve, such systems will replace many existing conveyors. AGVS, described in Chapter 43, combine the automation of a conveyor with the flexibility of a mobile vehicle. While the route the vehicle follows is a fixed path, it is easier to change that path with an AGVS than with a conveyor.

Summary

The equipment just described is capable of both improving cube utilization for storage and in some systems moving that stored material to a more convenient spot. Much of it is justified by savings in both space and labor.

43

MOBILE EQUIPMENT

Widespread use of the forklift truck had revolutionized warehousing practices before the middle of the 20th century, enabling warehouse operators to justify erecting sprawling one-floor buildings to replace the more compact multistory structures used in the past. Many kinds of mobile equipment are used in warehouses today, but most are variants of the common forklift truck.

Choosing Lift Trucks

A few years ago a national hamburger chain advertised that its sandwiches could be prepared in 256 different ways. Lift trucks are far more complex than sandwiches. If you calculate all the options of power source, operator location, lift attachments, vehicle characteristics, and brand options, thousands of different choices are available. Within this confusing scene, how can you reach an orderly decision in selecting equipment? By examining each of the options in order, you can arrive at a format for selecting the lift truck that will work best in your warehouse.

The most readily available materials handling cost source of power source is the human body. Some low-lift pallet trucks provide lift from a hand operated hydraulic pump with the push and pull provided by the operator.

However, external power is required for most materials handling machines, and the two major options are internal combustion engines and electric motors.

Figure 43-1 illustrates the advantages and disadvantages of each power source.

There are four fuel sources for internal combustion engines:

	Advantages	Disadvantages
ELECTRIC	*Better indoor air quality *Less maintenance *Lower power source cost *More frequently the power source for narrow aisle trucks	*Lower load/lift capability *More complicated refueling *Lower horsepower *Poor training or negligence can ruin batteries *Higher initial equipment cost
INTERNAL COMBUSTION	*Lower initial equipment cost *Better suited to multi-shift operations *Easier to refuel *Better for long runs, high speed, ramps and rough terrain *Higher horsepower *Higher load/lift capability	*Requires good ventilation due to CO2 emmisions *High maintenance *Potential high fuel costs

Figure 43-1

propane, gasoline, diesel fuel, and compressed natural gas. Propane has constantly grown in popularity because it generates very little exhaust and therefore is safer and cleaner to operate within a warehouse. Gasoline and diesel engines emit significantly greater amounts of exhaust and odor, and they are therefore preferred for outdoor operation. Compressed natural gas is cleaner, safer and cheaper than propane, and is available through the same gas lines used to heat your warehouses. Fuel availability is likely to influence the choice between the four alternate internal combustion engines.

Operator Location

There are three options in operator location: walkie (also referred to as operator-walking), rider, or man-up.

Walkie trucks are the most economical lift trucks to buy or to operate. These vehicles are often referred to as "pallet jacks." They are designed to facilitate horizontal travel of pallets or unitized loads of product. They are also used for order-picking from floor locations. These trucks elevate the load only a few inches above the floor. Since fork spread is fixed, the pallet size must be standardized in order to avoid pallet and product damage. Electric walkie trucks reduce fatigue and are frequently used where loads are heavy or distances to be covered are longer. One type of walkie truck offers higher lift, permitting the operator to raise the forks up to 13 feet for stacking or placement in storage racks. Some models have extended forks to handle two pallets instead of one. A few will handle four pallets, two deep and two high.

The rider truck is the most commonly used lift truck. This vehicle allows the operator to ride along with the vehicle. While more costly than the walkie truck, the rider truck provides greater speed and comfort. Within the rider category, there is the option of a seated vehicle or a stand-up vehicle. Mounting and dismounting is typically easier in a stand-up vehicle, but of course there is greater fatigue since there is no seat.

The man-up truck has operating controls on a platform adjacent to the fork carriage, allowing the operator to move up and down with the forks. Most man-up trucks have a guidance system so the operator controls only the up-and-down movement, not the steering. Because the operator can pull a few cases from a pallet without removing the entire pallet from a storage rack, the man-up vehicle will save significant time in selecting small orders. Further, the order picker has significantly better visibility because he or she can move up close to the item being selected. Some models have two sets of controls which allow the order picker to operate the truck either from a loading platform or from the truck chassis.

Types of Lift Attachments

Lift attachments represent the widest range of choices in lift-truck specifications. Forks remain the fastest and simplest means of moving product.

Most of the options to forks are designed to eliminate the use of pallets. The most common of these is a device to handle slipsheets, sometimes referred to as a push–pull attachment. This attachment is designed to grasp the extended lip of a slipsheet and use it to pull the load onto a set of wide platens or forks that support the slipsheet in transit. A push device then reverses the process to remove the slipsheet from the truck carriage.

Palletless handling is also achieved with carton or roll clamps. These are vertical paddles designed to grasp the sides of cartons or paper rolls and lift them without any pallet or loading platform.

Some highly specialized attachments grab and lift cargo with a vacuum cup, a magnet, a top-lift device, a boom, or a revolving carriage. Each of these alternates is more costly and more difficult to use than the fork attachment, but for many products the specialized attachment may be faster, safer, or more versatile. Most importantly, these devices allow handling of unitized loads without a pallet.

In selecting between the options, the truck buyer should learn whether or not the attachment has been used successfully by other warehouse operators handling the same or similar products. There have been cases where significant damage was done through misuse of specialized attachments.

Conventional or Narrow Aisle Trucks

The common lift truck with a 3,000 pound capacity is the most popular vehicle found in warehouses, and it typically requires a 12-foot aisle. A few very narrow aisle trucks will operate in an aisle only a few inches wider than the truck itself. These trucks are designed to eliminate the need to turn the entire vehicle to place merchandise in stacks. Figure 43-2 outlines the advantages and disadvantages of selecting narrow aisle lift trucks.

Brand Selection

Industrial vehicles are available from a wide range of manufacturers and the proliferation of brands can cause confusion for the buyer. It is useful to review the annual equipment directories published by two magazines, *Modern Materials Handling* and *Materials Handling Engineering*. Each of these directories provides a guide to alternate brands for each type of truck and attachment.

In making a brand selection decision for your warehouse, at least five priorities should be considered.

Priority one is the quality of the local dealer. This quality is measured in both quantitative and subjective terms. On the quantita-

	Advantages	Aisle Width Capacity Disadvantages
Conventional Aisle Truck	Industry standard truck. Easily transferrable to various tasks. Greater load stability. Greater load capability	Requires 12-foot aisle. Decreases available storage space.
Narrow Aisle Truck	Reduces need for counterbalance trucks. Decreases required aisle width. Increases available storage space.	Less flexible for various tasks. Lower load capabilities. Less load stability. May require "super flat" floor.

Figure 43-2

tive side, consider the financial stability, parts inventory, and service record of the dealer being considered. Even the best of equipment will malfunction occasionally, and your dealer's ability to correct the malfunction quickly is of prime importance. Don't hesitate to talk to that dealer's other customers to be sure that the service record is as good as advertised.

There is a qualitative measure involved in attitude. Are you convinced that key people in the dealership are truly dedicated to providing quality service for the equipment they sell? Attitude can often be detected without making a customer survey, and the buyer should visit the service department of the dealership and form careful judgments about the performance of the people who work there.

A second priority is standardization and the ability to substitute vehicles. If there are three lift truck brands in your warehouse, there are three different dealers to be called. Furthermore, each truck has operating characteristics slightly different from those of a competitive brand, and an operator who is accustomed to one brand may be more accident prone when using a truck with different controls. Therefore it usually makes sense to standardize on one brand of truck.

Reliability is the third priority. In most operations, the cost of down time is far more significant than the price differences between competitive equipment. A cheaper lift truck that is unreliable is always a bad bargain. There are two measures of reliability—your own past experience with a brand and the testimony of other companies using the same kind of truck.

The fourth priority is ease of training the operator as well as the operator's satisfaction. This does not mean that a warehouse manager should let the workers make the brand decision for forklift trucks. However, it does mean that if every worker in the warehouse tries to avoid using one particular machine, there may be some problems in connection with operator satisfaction that simply cannot be ignored. Nearly all materials handling dealers offer training courses, but some are better than others. Training quality should also be examined when making a dealer decision.

The last priority is cost. The initial cost of the lift truck is always less than the labor and maintenance costs connected with the vehicle during its useful life. Therefore, choosing on the basis of initial price is probably the worst possible way to make the brand decision.

The marketing of handling equipment is fiercely competitive, and trade magazines must be influenced by advertising budgets. Few magazine writers could point out weaknesses of a lift truck featured in ads. While lift truck technology has changed slowly, any article written five years ago about selecting equipment would have significant differences from one written today. Changes in technology and engineering will create some new options and some new points of emphasis in the selection decision. Fundamentally, the buyer is purchasing a tool, and the design and value of that tool must be related to the warehousing job to be done.

Justifying Narrow-Aisle Equipment

As operators have sought ways to save space in warehouses, lift-truck manufacturers have produced variations on the truck that allow it to operate in narrower aisles. The oldest and still most commonly used mobile lift truck found in warehouses is the "sit-down" counterbalanced forklift truck, referred to here as the common lift truck. Two variations are stand-up trucks which gain stability by using two outriggers extending in front of the driver to provide balance in handling a load. Three other variations have a mechanism that allows the load to be turned without turning the entire vehicle.

Each of the variations costs more than the common lift truck. How do you justify paying a premium for equipment? Here are the five types of mobile lifts to consider:

1. Common lift truck
2. Single-reach truck
3. Double-reach truck
4. Turret truck
5. The Drexel truck

1. A 3,000-pound capacity *common lift truck* is the most popular type found in warehouses. This truck, which costs about $20,000 with battery and charger, requires a 12-foot aisle to allow a right-angle turn to stack merchandise in rows facing the aisle. It is very versatile. In addition to stacking goods in storage, it is used for unloading from vehicles, hauling loads from staging areas to storage bays, picking orders, and loading vehicles. Many operators use this truck for virtually all the work done in the warehouse. It is relatively easy to learn to operate, and because it is so commonly used, it can be readily bought or rented from a wide variety of lift-truck dealers.

2. A first step in saving space is the *single-reach truck*, which can be operated in an 8-foot aisle, or 4 feet less than the aisle required for the counterbalanced type. The single-reach truck is priced in the neighborhood of $25,000.00, or about $5,000.00 more than the common lift truck. The truck uses a scissors-reach mechanism which moves the fork carriage forward into the storage pile, and somewhat greater skill is needed for the operator. Because the single-reach truck is used in many grocery distribution warehouses, it is usually easy to find replacements in dealers' rental fleets. However, the truck is less versatile than a common lift truck, because its smaller outrigger front wheels are not designed for crossing dock plates. The user must therefore have other equipment for loading and unloading trucks and rail cars. Unlike the common lift truck, the single-reach truck is available only with electric power.

3. The *double-reach truck* has a scissors-reach mechanism which extends twice as far as that of the single-reach truck. This allows a warehouse operator to store two pallets deep. The two-deep rack configuration allows twice as much product to be retrieved from the same aisle, but it is effective only where nearly every item in storage has a volume of at least two pallets. If fewer than two pallets are stored then the operator must either waste space or time by blocking one pallet with another pallet containing a different item. Eventually, the rear pallet must be unburied for shipping. The double-reach truck requires an aisle of 8.5 feet or about 6 inches more than the single-reach truck. It costs about $5,000.00 more than the single-reach or about one and one-half times the price of the common lift truck.

4. The *turret truck* has a fork carriage that rotates and eliminates the need to turn the entire truck. Turret trucks are available either with a conventional cab or with a "man-aboard" cab which allows the operator to ride up with the load. Turret trucks usually require a guidance system for safe operation in narrow aisles, and with the guidance system, the equipment cost is $65,000 to $80,000 per unit, depending on lift height. The design of the turret prevents the use of a tilt mechanism for the mast. Therefore in high stacking (over 22 feet) the truck cannot be effectively operated unless the warehouse has a "super-flat" floor with variances of not more than $\frac{1}{16}$ inch. It is difficult, expensive, and sometimes impossible to change conventional floors to "super-flat" floors. The turret operates in an aisle of about 5.5 feet, or 6.5 feet less than the aisle in which the common lift truck operates.

5. One competitive alternative to the turret truck is produced by *Drexel*. The Drexel swivels the entire mast rather than the fork carriage, and it swivels only to the right. Therefore the operator must turn the truck around to select from the opposite side of the aisle. Unlike the turret truck, the Drexel has a tilt mechanism to compensate for imperfections in the floor. This truck can be acquired for about $60,000. It can be used without a guidance system, though it is more productive with the guidance system because aisle widths can be minimized and the driver can operate without the need to steer.

With a price range of $20,000 to over $80,000, a buyer must justify paying a premium for narrow-aisle trucks. If you operate a 200,000-square-foot warehouse with 40,000 square feet devoted to work aisles, the use of the very narrow aisle equipment might reduce the aisle space by half, or save an additional 20,000 feet. Assuming that you need at least two narrow aisle trucks, this means that you would be spending up to $60,000 extra per truck (the cost of the guided turret truck versus the common lift truck) or an extra $120,000 to save 20,000 square feet of space. This equals $6.00 per square foot, which is certainly less than the construction cost of additional warehouse space today. On the other hand, consider the fact that the turret truck is usually not used for loading and unloading trucks, which means that you need extra trucks for this task. Also, the turret may not be as fast to operate as the common lift truck. If the turret is slower, the space saved in the warehouse must be balanced against the extra time used to operate the truck.

Also consider the possibility that a super-flat floor may not be available. If stack heights are over 22 feet, conventional floors may prevent the use of the turret truck, regardless of cost. The Drexel, because it operates in a similar aisle width without the need for a super-flat floor, appears to be most easily justified. Unlike the turret, these trucks can be used for the full variety of warehouse tasks.

Automatic Guided Vehicle Systems

Automatic guided vehicle systems (AGVS) have been used in manufacturing plants for decades, but their use in warehouses is much more recent. AGVS have moved into the warehouse thanks to their improved technology and relatively lower cost. Further, when the cost of new real estate is far higher than the cost of some "used" real estate, AGVS are justified by their capability of making an obsolescent building practical.[42]

There are two kinds of automatic guidance systems: a vertical system which automatically places stock selectors at the right spot to move into high-rack positions; and the more common AGVS, which is a horizontal system with path selection following a wire buried in the warehouse floor. It is the horizontal system that will be described here.

A signal transmitted by the wire buried in the floor activates the electronics in the AGV.

Centralized computer systems can program the AGV to follow a specific path by deactivating certain wires and activating others. Sometimes the same vehicle can be steered automatically—following the wire path—or manually—off the path. One guided vehicle may carry one unit load, or a whole train of trailers may be attached to an automatically guided tractor. By using the tractor-trailer system, up to 50,000 pounds of cargo can be horizontally transported behind the guided vehicle.

There are four main types of AGVs:

1. A towing vehicle designed to haul a string of pallet trucks or trailers

2. A unit-load transporter designed to carry one individual load of up to 12,000 pounds. These vehicles are designed to accept loads delivered from a guided fork truck or from a powered or nonpowered conveyor or load stand.

3. An automatic guided pallet truck, similar in appearance to the conventional jack used for order selection in grocery chains. These pallet trucks can handle up to four pallet loads or a total of 6,000 pounds.

4. A light-load transporter, a smaller vehicle designed to handle lighter loads such as parts or mail

The heart of the AGVS is an onboard microcomputer that controls and monitors vehicle functions, giving each vehicle the ability to travel independently and automatically to a programmed destination.

Typically, the guide path is a wire loop that is embedded in a groove cut into the concrete floor. When the system is installed, the groove, ⅛ inch wide by ⅜ inch deep, is cut with a diamond concrete saw; the wire is placed in this small trench, the groove is filled with epoxy resin which dries to the same color and hardness as the original concrete. In one facility, a 1,500 to 2,000-foot path was installed in less than a week. Installation can be done either by the vendor or the user, and does not need to affect the day-to-day operations of the warehouse.

If layout changes occur in the warehouse, the paths can be altered. When a path is changed, the wire from the old path does not have to be removed.

In operating the system, a warehouse worker takes a load from the home area, places it on the guided vehicle, moves the vehicle onto the guide path, keys in the destination address, and pushes a start button. Once this is done, the computer takes over. The vehicle moves to the designated drop location, deposits the load at a specific address, and the returns to the home area for a new assignment.

The chief advantage of an AGVS to a warehouse is that the guidance system eliminates travel time by warehouse workers. A value can be put on this travel time through the use of generally accepted warehouse standards. In a warehouse standard developed originally by the U.S. Department of Agriculture, a time of 7.7 minutes is needed for each 1,000 feet of travel. If one assumes that the fully burdened cost of a warehouse worker is $25.00 per hour or $.42 per minute, this means that each 1,000-foot trip costs $3.23. If the $35,000 microprocessor is to be paid back in three years, it will have to eliminate 684 miles of travel per year in order to be justified. Larger warehouse operations will find it possible to eliminate at least this amount of travel. For example, if there is a 300-foot travel from storage rack to loading dock, the elimination of 27 round trips per day will cut somewhat more than 684 miles of travel per year.

Another advantage of AGVs is their flexible use in different warehouse layouts. There are buildings available today at bargain prices because their layout makes them of marginal value for conventional warehousing. In some cases, these are older buildings with docks that are not convenient to storage areas. Others may be multi-story buildings that cannot be operated economically with conventional equipment. Through the use of AGVs, possibly in combination with vertical conveyors, inbound loads can be automatically dispatched to a storage address without intervention by people. In such cases, the use of AGVs will allow the economical use of a building that would be quite uneconomical with conventional handling equipment.

In addition, a benefit that is frequently overlooked is the ability of AGVs to save space. Because tracking of the vehicle is so precise, lanes between loads can have as little as 4-inch clearance between passing vehicles.

What to Look for in Mobile Equipment

The critical element in making a purchase decision for mobile equipment is economics, but a cost study is not limited to the purchase price.

One forklift truck advertisement shows how the cost of fuel, maintenance, and operator's salary in six years (10,000 hours) will exceed the purchase price of the truck by more than 1,000%. It is false economy to purchase materials handling equipment on the basis of lowest initial price. Reliability must be measured both in terms of maintenance cost and risk of down-time. The conventional lift truck's greatest "reliability" is that it can be easily replaced by another lift truck. Some conveyors and cranes cannot. What will you do when the equipment breaks down? The time to consider this question is before the equipment is ordered—not after the operation is running at capacity.

In choosing mobile equipment, you must also anticipate future requirements and whether the equipment can be adapted or profitably resold when market conditions change your product or the way in which you handle it.

44

APPROACHING WAREHOUSE AUTOMATION

Making a profit from investment involves taking risks. Experience enables these risks to be evaluated and the benefits assessed. The main problem with investment in warehouse automation is that it is frequently a *once-in-a-lifetime experience* and it is easy to make a mistake. The secret of success is meticulous planning and the use of an appropriate level of technology, taking into account the difficulty of predicting what the various operations within the warehouse may be required to do.

One mistake is not considering the possible changes in throughput and thus failing to justify the cost. High technology carries a greater initial cost and lower variable cost than low technology. This means that although the unit throughput cost may be favorable to advanced technology at the planned volume, that cost rises rapidly if the throughput is reduced. Confidence that planned volume levels will be maintained is a necessary prerequisite for the use of advanced technology.

Another frequent mistake is to compare the cost of a high-technology operation with one using existing methods, without considering ways the methods could be improved by simpler and more flexible techniques. A project advocate can play an important role in generating awareness and enthusiasm, but may overzealously pursue automation without considering the alternatives.

To make radical changes in methods as a response to an immedi-

---

This chapter was written by John Williams, a materials handling consultant from Hampshire, England.

ate problem is a recipe for disaster. Technology changes should be included in the corporate strategy plan and play a part in achieving your objectives. Properly planned and implemented, and with methods chosen to match future strategy, technology will enhance modern warehouse operation. With an increasing tendency toward inventory consolidation in larger and higher throughput warehouses, there is increasing opportunity to use automated techniques.

The Relevance of Technology to Warehouse Operations

Two basic warehouse operations are storage and sorting. When the storage operations are virtually nonexistent, we have a cross-dock warehouse in which sortation systems are almost mandatory, particularly for small packages.

For now, let's consider the operation in a more conventional warehouse that receives and stores goods on pallets, selects orders in carton or broken-carton quantities, and ships to multiple locations.

Receiving

Vehicle unloading is almost certainly by lift or pallet truck. The main uncertainties are fluctuating throughput volumes and the degree of control that can be achieved over inbound deliveries. The extent of standardization of pallet type, unit load size, extent of overhang, and degree of pallet damage will all be factors in determining whether unit loads can enter an automatic system without repalletization.

Depending on the distance to be moved and throughput, it may be feasible to convey unit loads from the unloading dock to storage zones. The more flexible automated guided vehicle system (AGVS) is likely to be a better solution in many applications.

Common to all methods will be pallet identification by bar-coded label. The greatest flexibility is achieved by using bar-coded labels in conjunction with onboard radio frequency terminals and scanners, which communicate with a warehouse management computer. This proven method is the benchmark against which the benefits and costs of more advanced technology should be measured.

Reserve Storage

Reserve storage is comparatively easy to automate with reliability because it is normally a pallet-in/pallet-out operation. The advantages are very narrow aisle (VNA) operation combined with the cheaper space obtained by use of height. Because 80% of a building's cost is in the floor slab and roof, higher buildings reduce the cost per cubic foot in general.

But beyond certain heights and depending on building codes and wind loads, costs increase. Rack-supported buildings are an attractive alternative to conventional structures, but rack-supported warehouses must be built with some care because the stress normally carried by the building structure must be carried by the racks. Rack distortion may be sufficient to make a stacker crane system inoperable.

A reduced number of storage locations is one of the benefits of computer-controlled reserve storage. For example, quarantine items may be secured and their removal barred to those who haven't the necessary password to operate the computer procedure. In a similar manner, reserved merchandise may be accumulated and held until released.

With complete automation (high racks and stacker cranes) allowing stacking heights up to 120 feet, there is a safety problem if order picking operators work at the same time in the same area. One solution is to handle manned order-picking on the day shift and allow crane operation only at night. High-stacking cranes have problems of aisle transfer. Below 40 feet the flexibility of the VNA lift truck and its complementary order-picking truck must be considered.

Order Selection

For a long time to come human order pickers are unlikely to be replaced by more automated methods. However, the increased emphasis on preventing back injuries will stimulate redesign of some warehousing systems, leading to reduced-weight cartons or to cartons being presented to the operator in a manner that eliminates excessive bending, stretching, or body twisting. This may involve mechanized carton lifting devices or increased use of *goods-to-picker* systems rather than conventional selection from racks.

There are two basic types of order selection systems using the human operator. The first and most common is the *picker-to-goods* system. The operator on a *man-up* lift truck or crane can be automatically moved to the next picking location. In a *goods-to-picker* system, the operator remains at a workstation and pallets or cases are moved to and away from the workstation under computer control.

Completely automatic picking systems are used for very high throughput picking. In these systems, cartons are loaded into gravity lanes and fed on to a conveyor by computer-controlled gate mechanisms.

Staging for Shipping

There are variations in the ways cartons are moved from the order-picking area to be sorted into outbound loads. In the simplest systems, the order picker moves the accumulated order in a bin or on a pallet to be deposited at a staging area. At a higher level of technology, tow conveyors, automatic guided vehicles, or sorting conveyors can be used. In some applications, order pickers accumulate a group of orders by walking through one warehouse zone, which stores a portion of the stock-keeping units (SKUs) in the warehouse. Simultaneously, pickers in zones are selecting merchandise from the same group of orders, all identified by bar-code labels that specify the staging location. Some systems use a carton conveyor that is built into the order selection crane.

It is not difficult to see some of the problems that can arise in balancing workload between zone pickers and the effect that these imbalances can have on shipping operations. Matching the throughput volume with product mix, order size variations, and picking zones is the toughest part of the balancing act. You may require computer simulation to get the design right.

The Benefits of Automation

If you expect to justify warehouse automation solely by the reduction in the cost of handling and storage, you are likely to be disappointed. The justification for advanced technology often lies elsewhere.

One of the most frequent justifications for the use of automatic storage and retrieval systems (ASRS) in high-rise buildings is the reduced warehouse footprint, which permits higher storage volume on a restricted but strategically located site. It is not difficult to see the investment justification when the costs of site change or increased transport costs are taken into account. Another benefit of automation can be better resource use. The machines work at their own tempo, without fatigue. Multishift operations become more feasible, and this can be an important element in the cost justification.

Another often-mentioned justification is freedom from industrial relations problems. The lower skill levels required for conventional warehousing contribute to high personnel turnover or militant unions with restrictive practices. With fewer hourly workers, and with the increased use of engineering and process control skills, higher wages can be paid and recruitment policy can be more selective. Selection and training costs may be higher per person, but the number of people is reduced.

The use of human operators becomes subject to the law of diminishing returns with increasing warehouse size. Put quite simply, workers get in each other's way; yet restricting workers' zones produces the problem of work balance and load scheduling. Many warehouses see dramatic savings in the number of operators and lift trucks as a consequence of introducing automation. Compare these savings with a higher hourly wage and higher maintenance cost for automation equipment. It may also be necessary to pay separation costs for those who lose their jobs as a result of automation.

People cause errors and one of the benefits of automation is improved data capture which results in improved inventory control, greater customer satisfaction, and reduced inventories. However, the same savings can be obtained at lower capital cost by the use of radio frequency terminals in conjunction with bar-code readers.

People are often the cause of fires. The fewer people working in an area, the lower the risk of fire. However, fire insurance underwriters are concerned about the *chimney* effect of high stacks and narrow aisles. Storage racks often require a higher level of fire protection. Energy costs are increased to power additional equipment; but to offset this, heating and lighting costs can be minimal.

Some argue that the use of advanced technology improves a company's image. The creation of a *high-tech* image may be attractive to investors or to the manager who wants to work in a progressive company.

The Risks of Automation

Getting It Completely Wrong

The biggest risk is that the system will not do what you want it to do. Why should this be? It is extremely unlikely to be bad technology because, for the most part, equipment is reliable and well proven. It is most likely to be because you have failed to predict what you require the system to do.

Warehousing is subject to many variables. Throughput can fluctuate seasonally, daily, or hourly. Order size can vary, as can the number of lines per order. The number of SKUs and size of packages can change. In selecting equipment, omission of a factor from a performance specification is likely to cause failure. High-technology systems are inflexible and you can't adjust for your errors or omissions by increasing personnel or adding a few more lengths of conveyor.

It is important to get the specifications right. You can get help from outside— consultants and the equipment suppliers will be only too pleased to advise. However, it is important that your operations managers should be responsible for planning. After all, no one understands your business better, and they have the best possible incentive to ensure that the systems work.

There can be no hiding behind the excuse, "Don't blame us because it doesn't work; we had nothing to do with it." If they are to be given the responsibility, operations managers should learn from the experience of others who have undertaken the same task by visiting similar warehouses.

They also need time. Experience shows that nearly everyone regrets that they did not allocate more time to the planning stage. System suppliers complain that they are pushed into contracts with earlier implementation dates than they consider desirable, only to find that the schedule cannot be met because the customer changes the specification. This is a clear case of *haste makes waste*.

Be sure that a task force structure considers the operational requirements and views of all the departments. Line managers seldom have the ability to carry out this work, even if they have the time. Inevitably it will be carried out by a group working in a staff relationship. A task force with representatives from each department and reporting to a committee chaired by the chief executive can ensure that the essential integration of views takes place. It also demonstrates the on-going commitment of the chief executive and heads of departments to the project.

Failing to Contract Successfully

Once you know what you want, you are ready to choose your contractor. During the earlier stages of the investigation, you will evaluate various equipment suppliers, making visits to installations and discussing their performance. A number of specialist subcontractors provide the building, the racking, lift trucks or stacker cranes, conveyors, computers, and controls. Resist the temptation to coordinate these yourself. Instead, choose a prime contractor with a clear legal responsibility to fulfill all your specifications.

Check all references and investigate the prospective contractor's financial status to be sure the company can complete the project. Contractors have been known to go bankrupt as the project sinks into difficulties.

Be sure the contract stipulates the maximum down-time that you are prepared to accept over a period of time. System reliability must be a matter of legal definition, including a rated throughput per hour, per shift, and per week.

Because the project must be completed and handed over on time, a time schedule should be written into the contract, including penalty payments in the event of failure.

It is at this stage that the additional time spent on the specification and preplanning stage pays dividends. Preplanning avoids last-minute alterations that can lead to contractual amendments and disputes. There is considerable opportunity for misunderstanding in the complicated exchange of information in major automation projects and some projects do end with litigation. To prevent misunderstanding, all communication links should be made formal and detailed records carefully maintained.

Overestimating Your Organization's Ability to Cope with Change

This applies not only to the acquisition of new skills, but also to enabling managers to understand the issues involved and to make decisions.

Skills that may not exist in the organization are certain to be required. These include architectural specifications, mechanical system design, project management, computer systems, and process control. In addition, experience will be necessary in dealing with human relations problems resulting from change. These skills can be obtained by training or recruitment. Consultants may provide specialist advice and knowledge based on their wider experience, but they should never be used to impose a solution on the organization. These consultants should be members of the project team, and in no way should their presence usurp the ultimate authority of the operational managers.

Many operators are apprehensive of changes and some will be unable to make the transition to meet the challenge of using new techniques. This at least necessitates a planned communication and training program, and it might even suggest including a human relations specialist on the project team. Some organizations have completely rotated their personnel or organized retirement programs for older workers. A system can only be as effective as the skills of people allow it to be.

In Conclusion

There is a large range of available technology for warehouse automation. The availability of technology is not an issue nor, for the most part, is the ability to design reliable systems. The problem is that business requirements and people are not completely predictable. Move with caution if you want to reduce the risk of failure.

45

PALLETS AND UNIT LOADS

Some attempts to unitize marine freight go back to very ancient history. Unitized handling as we know it today really had its impetus with the technologies developed to supply the military during World War II. At that time, the industrial truck was still in its earliest development stages. A wood platform or pallet was recognized as the best way to allow the industrial forklift truck to transport a number of small pieces of freight. The first such platforms were skids, consisting of a deck of wide boards nailed across two or more runner boards which raised the platform enough to allow the forks to move underneath and pick it up. Because of the need for stability in high stacking, the skid evolved into today's warehouse pallet when additional boards were nailed across the bottom of the runner boards to provide stability. Pallets were and are primarily made of wood, simply because wood is the lowest cost construction commodity available in most countries.

Standardization

Skids and pallets came in a wide variety of sizes, with the most common ones measuring as little as 3 by 3 feet for low-volume grocery products up to 4 by 8 feet for stevedore pallets used in marine transportation. Early in the 1960s, one warehouse operator made a survey of grocery chains in his community to explore the possibility of exchanging pallets. At that time, each food chain had a storage rack system based upon a different size pallet, and standardization seemed impossible.

The 1960s saw the acceptance of a standard pallet specification in the United States. Overseas, such standardization had come much earlier. The United States military had primarily used a standard pallet, and huge quantities of these pallets were war surplus in 1945. The Australian government had a particularly large supply of materials handling equipment left over after the war. The Australian government formed a Commonwealth Handling and Equipment Pool, later abbreviated as CHEP. When CHEP adopted a standard size pallet, Australia became the first place in the world to achieve a standardization of pallet sizes as a legacy of its military surplus.

Within the United States, General Foods deserves prime credit for the pressure it created in the early 1960s for a standard pallet. This standard (Figure 45-1) specified the size and spacing of the boards on the top and bottom deck as well as the size of the runner boards or stringers. The General Foods pallet was to be made of hard wood with a specified weight per unit of approximately 80 pounds. Notches on the stringer boards allowed the pallet to be

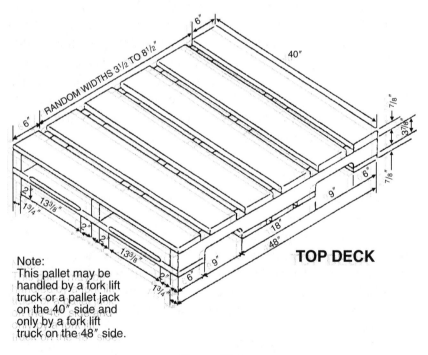

Note:
This pallet may be handled by a fork lift truck or a pallet jack on the 40" side and only by a fork lift truck on the 48" side.

Figure 45-1

entered from all four sides, though it is handled most easily from the 40-inch face with its larger openings.

Because of its influence on both suppliers and its customers, General Foods had achieved growing acceptance of a standard pallet when the program was transferred to Grocery Manufacturers of America (GMA). Through the cooperation of its members, GMA finished the job of establishing the General Foods specification as an industry standard. Wholesalers and chains adapted their materials handling and storage rack systems to the new standard pallet. Subsequently, the program was transferred to a Grocery Pallet Council, and then efforts to maintain standards disintegrated. The hardwoods specified in the original General Foods design became scarce and expensive, because up to half of America's hardwood lumber was being consumed by the pallet industry. To cut costs, many pallet users found cheaper types of wood. Others made changes in thickness and spacing of deckboards. Eventually the Grocery Pallet Council disbanded, and today the only pallet standard that is preserved is the length and width dimensions. The 48-inch by 40-inch pallet remains the most common size used, and the predominant size used in the grocery products industry. Most storage rack systems are designed to accommodate the 48-inch by 40-inch pallet.

The Problem with Pallets

Wooden pallets have always been the primary method of unitizing freight for mechanical handling. Thanks to the standardization move initiated three decades ago by General Foods, about half the pallets used in the United States conform to the 48-inch by 40-inch (1.2 by 1 m) dimensions.

Unfortunately, there is little other good news with respect to pallets. Efforts to enforce a standard specification for pallet design disintegrated years ago, and pallets used today are seldom uniform in construction quality.

While users are creative in developing pallet patterns, unit load footprints do not always conform to the size of the pallet. Significant overhang or underhang is likely to be a source of damage.

Many pallet users prefer to use lowlift trucks (pallet jacks) for order selection, but such equipment is capable of damaging pallets if the bottom deckboards are not aligned with the wheels of the truck. Product overhang may be the cause of such misalignment.

There are six significant problems with the wood pallet used today:

1. It is too heavy to be handled safely. OSHA regulations state that one person should not lift over 50 pounds, but most wood pallets weight 65 to 80 pounds.

2. Pallets create a space problem. When space is critical in a trailer or container, the 9-inch thickness of conventional pallets causes capacity reduction.

3. Pallet storage is a safety problem. High piled stacks of empty pallets will create an intense fire which will defeat some sprinkler systems.

4. Pallets are a sanitation problem. If stored outdoors, they are likely to be infested by insects, birds, and rodents. Wood pallets cannot be steam cleaned.

5. Quality control of pallets is difficult. Only an expert can determine by visual inspection whether pallet lumber is properly dry and of specified type and quality.

6. Pallets are typically not in good repair. If repairs are not made as damage occurs, broken pallets will remain in use and cause both personal injury and product damage.

New Kinds of Pallets

In recent years, recycling has become a growing concern in industry. A few pallet manufacturers are experimenting with materials considered to be superior to wood in their recyclability.

Presswood pallets are recyclable and are produced from low-grade timber stands which are normally not economical for other commercial purposes. They use nearly 100% of the tree's biomass, compared with less than 30% biomass usage in producing conventional wood pallets. Plastic molders use injection molding technology to manufacture pallets that are recyclable and can be made using recycled plastic.

Presswood and plastic pallets meet these Grocery Manufacturers of America's design requirements: true four-way entry capacity, 48-inch by 40-inch dimensions, weight of less than 50 pounds, sufficient deck coverage to prevent damage, constructed of reusable or recyclable materials, and sanitation and drainage capabilities to handle meat and produce.

One plastic pallet is similar to a conventional pallet (top and bottom decks with rails between), but the pallet has several advantages. The pallet has a nonslip surface, molded banding slots which eliminate the need for banding clips, and a tongue and groove system for easier stretch film wrapping. Plastic pallets are especially cost effective in closed-loop systems because they are more durable and are also safer, having no protruding nails or splinters. Because they are plastic, these pallets are weather and moisture resistant. The pallets interlock for ease of stacking and storage.

Presswood pallets are molded as one continuous unit from a mixture of wood fiber and synthetic resins. The design is unlike conventional wood pallets in that it has no bottom deck boards. Instead, it has a single solid deck from which legs protrude. This design makes the pallet nestable, requiring one-fourth the storage space of conventional wood pallets. The nestable feature does more than save storage space—it decreases fire risk because it eliminates the air space between conventional pallets. Similarly, the design improves handling efficiency because one forklift can hold up to 60 nested pallets (three times the number of typical wood pallets that can be carried per load). The solid deck has drainage grooves and the legs have drain hole plugs to prevent product contamination. These pallets meet quarantine regulations for overseas shipping. The solid deck with its rounded edges eliminates product sag. It minimizes the likelihood of worker injuries because there are no nails or splinters. Presswood pallets have been compression tested to 25,000 pounds and are available in a range of load-bearing capacities. Those pallets that have sufficient strength to support themselves in pallet rack or drive-through rack are in the higher price range.

Presswood and the plastic pallets are just two examples of the innovative pallets being produced today. For warehouse managers seeking a new pallet system, the choices are numerous and varied. But with some effort, the search can yield durable, easy-to-use pallets both business managers and environmentalists will find acceptable.

Alternatives to Pallets

While the wood pallet is the oldest and easiest method of handling unitized loads, it is not the only method available. The first developments of substitutes for pallets began as early as the 1940s, when materials handling manufacturers cooperated with paper companies in developing a thin and disposable shipping platform known as a slipsheet. Constructed of high-tensile laminated paper, the sheet is much thinner than a wood pallet. However, it cannot be lifted with conventional forks without damage. Handling equipment manufacturers developed a special device known as a "push–pull" attachment. The device is designed to grip a protruding tab of the slipsheet and pull the entire sheet with its load onto flat metal plates that support the load while it is being transported. At destination, a plate pushes the loaded slipsheet onto the floor of a trailer or onto the floor of the warehouse. Figure 45-2 shows the slipsheet and push–pull attachment in use.

Another special device which has been in use for over several decades is the carton clamp or grab truck. Figure 45-3 shows a clamp truck with a load. This truck has paddles that squeeze sides of the load enough to allow it to be lifted. This device was originally designed for cotton bales and rolls of paper, and some clamps are designed to rotate the load so that it can be inverted or placed on its side. The clamp truck works best with relatively large and sturdy packages. Because it requires side pressure to lift the unit, improper pressure adjustment or packaging that is not designed for clamp use can cause the clamp to damage the product. In the early days of clamp truck use, extensive damage to home laundry appliances occurred when a warehouse operator used improperly adjusted clamps that dented each unit as it was lifted. The device must be

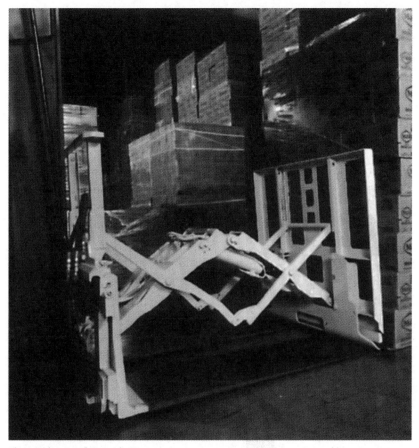

Photo courtesy of Cascade Corporation

Figure 45-2

carefully maintained, since it can also be damaged by operator abuse. Some warehousemen have used clamp attachments to pick up everything from bicycles to bailing wire, causing damage to both the clamp attachment and the merchandise handled. Because of the way it functions, the clamp truck does not require any loading platform. It can pick up a load and transfer it from one loading platform to another. There is some space loss in both storage and freight vehicles because a few inches of side void are left to allow access by the arms of the clamp attachment.

501

Photo courtesy of Cascade Corporation

Figure 45-3

Tradeoffs in Unit Handling

As you consider the various options in unitized handling, it is well to consider the advantages and disadvantages of each.

The wood pallet is the fastest and simplest way to store and transfer unitized loads. It is easier to operate an industrial truck equipped with forks than with any other kind of loading attachment. The generous sized openings in standard pallets allow them to be safely stowed and retrieved even when they are in a high stack. However, if the pallet is to achieve full savings in materials handling, it should be transported with the load from origin to destination.

If the material is hand stacked from a warehouse pallet to the floor of the truck, or if the process is reversed, the loading or unloading process will take more than six times as long as unitized loading. One time study of the process showed that manual loading of 22-pound cases will be done at a rate of eight pallets or 480 cases per man-hour. The loading of full pallets can be accomplished at the rate of 50 pallets per man-hour or 3,000 cases per hour.

If full pallets are loaded, the shipper has the option of donating them or taking other pallets in exchange. Pallets are costly, and most users of them are reluctant to donate a platform that costs between $6.00 and $10.00. However, if they exchange pallets with the trucker, vendor, or supplier, they run the risk of trading a high-quality pallet for a cheap pallet. Very few people can recognize the difference between the durable and expensive species of wood and the cheap ones, or well cured versus green wood. Few operators have the time or the ability to be sure that the thickness and spacing of deck boards and bottom boards conform to a standard specification. As a result, there is no practical way to enforce the design standard on warehouse shipping and receiving docks. Industry frustration with "junk" pallets increased after efforts to police a standard system collapsed.

Fire protection is another significant problem. Underwriters consider high stacks of wood pallets to be one of the worst hazards in a warehouse, since the spaces between boards allow combustion to create a "flue effect" which can result in a very dangerous fire. Therefore, preferred risk underwriters typically restrict the height of stacks of empty pallets in a warehouse. To avoid wasting valuable storage space, warehouse operators may move empty pallets outdoors. But when they are outdoors, pallets are subject to contamination and deterioration by birds and rodents. They are exposed to the elements and are further deteriorated by rain, snow, or ice. Finally, wood does not lend itself to steam cleaning or other effective cleaning methods.

Slipsheets or are more compact and less costly than wood pallets. Their compactness allows them to be stored in a much smaller space, and because of their construction they do not present the fire hazard of a wood pallet. However there are not standard specifications for these devices either. Slipsheets can be made of everything from paper to plastic. One major brewing company uses plastic slipsheets and recycles and remanufactures the sheets at its own facility. Slipsheets require special equipment and special handling. They cannot be handled by the conventional lowlift pallet truck which is designed specifically for wood pallets. Therefore when slipsheets loads are received at a company that uses conventional forklift and lowlift equipment, they must be transferred to a wooden pallet.

Carton clamps don't work well on every kind of load. Unitized loads of very small boxes are not practical for clamp loading, since the smaller cartons are more likely to drop out of the center of the unitized load.

New Ways to Use Pallets

Pallet leasing has made slow but steady progress in the past several years. One leasing company has introduced a block pallet to replace the original design. This pallet is more easily handled from all four sides. Figure 45-4 shows the difference between the two designs.

The leasing company assumes the responsibility for repair and standardization, thus relieving the user of two major headaches. Unfortunately, some users simply dislike the concept of leasing equipment, so pallet leasing is unlikely to ever achieve full acceptance.

A different approach is offered by one company which has a pallet management service to handle inspection, sorting, repairing, cleaning, and inventory control for large users of pallets. The service company maintains the pallets without owning them.

As the options of pallet leasing or disposable shipping platforms are considered, the user must also balance the advantages and disadvantages. Empty wood pallets, whether leased or owned, are still a fire risk when stored in quantity indoors. If outdoor storage is used, the contamination and deterioration problems remain. The lessor may assume much of this burden by keeping idle pallets at its own depots.

A leased pallet made of some material other than wood could be an attractive alternative, particularly if this would allow indoor storage and steam cleaning. However, this alternative has not surfaced yet.

Most of the national publicity about the unit handling problem

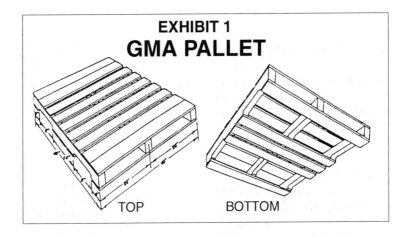

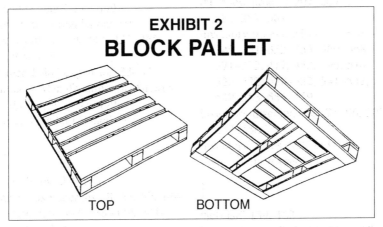

Figure 45-4

has focused on the grocery industry. Yet many other industries use unitized loads and have similar problems and interests. It seems clear that the challenge of unitized loads will be resolved in the next few years, simply because there is so much attention being given to it. The eventual result is likely to significantly change many warehousing operations.

46

DEALING WITH DAMAGE

Losses from accidental damage, casualty, or pilferage are not only costly to industry, but can also easily wipe out the productivity gains made elsewhere in the company. Loss and damage in the distribution system are a major source of waste in American industry today. And there is no indication that this waste is declining.

Causes of Damage in Physical Distribution

Product damage in distribution usually results from three sources: failure of packaging, mishandling in transportation, or damage done in the warehouse. Even in the best of warehouses there will be some damage.

The packaging engineer usually develops a compromise between the lowest cost carton and one that is so strong as to be completely damage-proof. If the compromise moves too far toward low price, the resulting package will not protect against the normal warehouse. The warehouse operator has a unique opportunity to report on the results of the packaging engineer's work.

The warehouse operator ultimately has two concerns: controlling the damage caused within the facility, and identifying and controlling that which occurs outside. Warehouse workers must understand that external damage not identified at the receiving dock will subsequently be charged to the operators of the warehouse. In some warehouse operations, all damage found on receipt is rejected to the delivering carrier, but in most the warehouse operator takes damaged goods in for repacking or repair.

Since most external damage occurs in transportation, we list six typical causes of transportation damage:

1. Faulty loading practices
2. Abrasion from shifting within the vehicle
3. Shock damage from bouncing or sudden starts and stops in transit
4. Leaks in the cargo van or boxcar
5. Transfer handling damage
6. Improper blocking and bracing in a rail car

Disposal of Carrier Damage

When goods are damaged by common carrier, the warehouse operator's damage-control procedure might include these seven steps.

1. Segregate all products in which damage is evident or suspected, including both damaged merchandise and damaged packaging. On those inbound loads where a carrier representative is not present (rail cars or unattended trucks), take photographs of the damage while the freight is still in place on the carrier vehicle.

2. The carrier's bill of lading should be noted with the quantity and nature of suspected damage. If a carrier representative is available, he initials this notation as his acknowledgment of it.

3. Warehouse receiving tallies should contain similar notations.

4. In a public warehouse, the owner of the goods should be notified.

5. After it is tagged for identification, each piece of damaged merchandise should be moved to a segregated area.

6. After inspection, merchandise that has suffered damage to packaging only should be recartoned and returned to stock. In some cases minor repairs to merchandise can be made at the warehouse; more severely damaged merchandise should be shipped to a repair center or scrapped.

7. Finally, the total cost of dealing with the damage should be calculated and reported. This cost includes handling, storing, and processing damaged merchandise.

Concealed Damage

Certainly the most vexing problem in controlling damage in transportation and distribution is concealed damage. No one learns that the product is damaged until it gets to destination and the packaging is removed. Frequently, this is after the product has moved through all the distribution channels and is in the hands of the ultimate consumer.

The damage is concealed because the outside of the container offered no hint that its contents had been damaged. When this happens, the party discovering the concealed damage returns it for credit, and a claim to recover this loss goes back through the system. The delivering carrier, after determining that damage did not take place in the consignee's own handling operation, may attempt to involve prior carriers or prior warehouse operators in a sharing of the claim. Sometimes, a compromise is adopted to spread the claim cost among the several parties who might have caused the damage.

Three kinds of evidence are helpful in dealing with a concealed damage claim: the delivering carrier's bill of lading, photographic evidence of the damage, and an invoice indicating the value of the item or the cost of its repair.

Warehouse Damage

Warehouse damage is inevitable. Any warehouseman who claims to be running a "damage-free" facility is either suffering from delusion or is untruthful. The forklift driver who never damages anything is probably working below optimum speed. Normal handling and storage functions will cause a certain amount of damage in any operation.

U.S. Department of Agriculture officials studied warehouse damage and found 66 causes. Of these, six accounted for 50% of the damage. These six are listed below, with the damage rate per 100,000 cases shipped:*

---

* Total 82.1 cases per 100,000. *Distribution Warehouse Cost Digest,* Vol. 15, No. 1. © Alexander Research and Communications Inc, New York.

Dropping cases in aisles	16.1 cases
Protruding nails in pallets	15.8 cases
Damage by forks of lift trucks	14.4 cases
Damaged during storage	14.4 cases
Damaged in filling racks	12.8 cases
Damaged in removing from second level rack slot (order picking)	8.6 cases

Warehouse damage is most prevalent at corners of stacks, in narrow aisles, or at the face of a load (through a collision in stacking). Faulty palletizing or dropped packages may also cause warehouse damage. If merchandise is piled above its packaging limitations, the stack will collapse. Shifting may cause the corners of boxes in a unit load to be misaligned; this in turn reduces the stack strength and may cause the pile to collapse. High humidity also is a frequent cause of package failure, since most corrugated materials will be weakened by moisture. Roof or sprinkler leakage may soak packages and cause them to fail. Occasionally, cross-contamination of stored goods can be caused either by odor transfer or by chemical reaction.

By definition, warehouse damage is the result of human failure. Such damage is minimized through training, proper maintenance of handling equipment, communication of correct storage and handling practices, and knowledge of the sources of chemical reaction or cross contamination.

Reducing Warehouse Damage

Following these seven steps will reduce damage in most warehouse operations:

1. Use storage racks for bagged materials to eliminate damage caused by leaning stacks.
2. Prevent nail damage by covering pallet faces with light material such as a sheet of plywood.
3. Buy pallets with high nail retention.
4. Don't use forks that are too long and protrude on other side of a pallet.

5. Maintain warehouse training programs.
6. Remove damage as soon as it is discovered.
7. Maintain proper aisle widths.

A great deal of the damage found in most warehouses is ultimately caused by carelessness. Greater concern can be developed through keeping score on the results of damage control and communicating those results to warehouse employees. Highlighting your damage ratio is a vital part of developing this concern. Results should be publicized at employee meetings, or used as the basis for contests, with prizes awarded for the best record in controlling damage.

Recooperage and Repair

When cases of packaged goods are damaged, it is frequently possible to salvage much of the damaged shipment by removing the inner cartons that are broken and replacing them. This process is frequently referred to as recooperage and must be done meticulously. Damaged food products are likely to attract insects or rodents. Products that are stained, but otherwise undamaged, are still considered unsalable. Fumes or odors from damaged products could cause contamination to good merchandise stored near them. Liquids can spill and stain other containers. For this reason, any repackaging operation must be designed to handle the product cleanly and separate good stock from bad.

Some product damage in a warehouse can be repaired so perfectly that it is undetectable. Major appliances, such as refrigerators or washing machines, sometimes suffer minor dents or scratches that can be repaired by using processes similar to automotive body work. Metal office furniture can be repaired with equal ease. Again, quality control in such repair is absolutely essential, since the consumer who receives an apparently new product that is actually damaged will be an extremely unhappy customer.

At times it is more economical to sell damaged goods to a salvage firm than to engage in either recooperage or damage repair. However, damage disposal must conform to your marketing policy.

If slightly damaged product is allowed to be sold at a discount to salvage firms, the ability to market that same product at list price might be impaired. For this reason, many manufacturers strictly control the salvage and disposal of all damage, even products judged to be a total loss.

Protecting Damaged Merchandise

One of the toughest problems in any warehouse storing damage-prone products is to be sure that powders or liquids do not leak from damaged containers and stain or spoil other merchandise in the warehouse. With many products, the consequences of leakage can cause a great deal of additional damage.

Some warehouse operators have been very innovative in developing systems to control secondary damage. When damage of sensitive materials is discovered, it is important that the damaged cases immediately be removed and segregated from good stock. At times, this removal is not easy, because it could involve depalletizing for which there is no labor available. One answer is the development of a drip pan, a piece of sheet metal a little larger than a pallet with corners turned up and sealed to hold powder or fluid. In this way, leakage from the damaged pallet is held in the drip pan and not allowed to migrate through the warehouse. Another answer is the use of fiber or metal drums as a temporary repository for damaged cases or bags. Like the pan, the drum keeps the spillage from causing further damage.

One showcase warehouse handles nothing but carbon black, one of the most sensitive products from a standpoint of secondary damage. Carbon black is the same material used as a toner in photocopy machines, and bags of this material, when broken, can cause substantial discoloration of goods stored nearby. In this warehouse, every lift truck is equipped with a portable vacuum as well as broom and dust pan. When a leaker is discovered, cleaning and controlling the leakage has higher priority than completing the shipment. By placing the emphasis on damage control, the carbon black warehouse maintains superior housekeeping with a very dirty product.

Storage of Damaged Products

Damaged products stored in a warehouse are an embarrassment, so most warehouse operators keep their damage storage where it's not seen by visitors. Unfortunately, however, management doesn't see it either, and the accumulation of damaged goods is overlooked.

The best way to control the cost of storing damage is to keep down the quantity in the warehouse. And the best way to control the quantity is to store it in a conspicuous spot.

Loss and Damage Guidelines

The primary method of measuring loss-control performance is by conducting a warehouse operations audit. But before making an audit, be sure that the following conditions exist:

1. Aisles are clear, well-marked, and sufficiently wide to allow equipment to maneuver.

2. Special guards are used to protect racks and columns, and guided aisles have guide rail entries.

3. Block stacking is not leading to load crushing or leaning stacks, and loads do not overhang the pallet.

4. Floor-loading, rack-loading, and structure load limits have not been exceeded by recent changes in material or equipment.

5. Equipment is not being overloaded or operated at excessive speeds.

6. Front-to-back members are installed in selective pallet racks to prevent loads from falling through openings.

7. Masts on industrial trucks fit easily through doorways and passageways.

8. Safety screens are located beneath overhead conveyors.

9. Hazardous and flammable materials are well marked and are stored, handled, and transported according to regulations.

10. No sagging load beams and no bent upright trusses are present in the storage racks.

11. Materials handlers have undergone formal training in the use of handling equipment.

12. Preventive maintenance programs are used for all handling equipment.

Damage Control Ratios

Using damage-control ratios is a useful damage consciousness-raising exercise. Here are some examples:

$$Damaged\ loads\ ratio = \frac{Number\ of\ damaged\ loads}{Number\ of\ loads}$$

$$Inventory\ shrinkage\ ratio = \frac{Inventory\ investment\ verified}{Inventory\ investment\ expected}$$

Other statistics that can be used are "Percentage of Accident-Free Operators," and "Number of Accidents per Industrial Truck-Operating Hours."

Damaged Loads Ratio

The damaged loads ratio provides a measure of the loss due to poor handling. The ratio should be computed at each stage of flow throughout the system, beginning with receiving and ending with shipping. A load is defined as damaged if there are *any* damages to it.

It is not likely that individual equipment operators will maintain accurate damage reports, even if they are not at fault. The time required to prepare the reports and a lack of interest in the task combine to create an atmosphere that is not conducive to reliable reporting of damaged loads.

Random samples should be made periodically to ascertain the damage percentage on loads being staged in receiving, inbound inspection, packing, and shipping. In addition, when perpetual inventory audits are performed, a damage report should be repaired.

For example, in a receiving department pallet loads are staged awaiting movement to storage. An audit of the loads is performed with the following results:

Number of loads	64
Number of damaged loads	4
Number of cases	670
Number of damaged cases	12

Preventing Warehouse Damage: A Questionnaire

As you work to reduce damage in your warehouse, the checklist below may provide some loss prevention ideas:

1. Is some of the "warehouse damage" actually unreported carrier damage? If so, what steps can be taken to stop the acceptance of damaged freight?

2. Has some damage been caused by splintered or damaged pallets? What steps are being taken to control pallet quality?

3. Is damage concentrated in certain items? If so, is there a packaging problem with those items which might be corrected?

4. Have maintenance problems in lift trucks caused damage, such as protruding forks, improperly adjusted clamp attachments, or defective slipsheet attachments? If so, what can be done to improve lift truck maintenance?

5. Are warehouse workers providing prompt reporting and feedback about the existence and cause of warehouse damage?

6. Has damage been caused by narrow aisles and overcrowding?

7. Has the percentage of product suffering damage moved up or down since last year?

8. What other changes in warehousing practices might reduce warehouse damage?

Because damage is unpleasant, management as well as warehouse workers sometimes may try to sweep it under the rug. Yet, dealing with damage is part of good warehouse discipline—and remember that the warehouse operator now has a golden opportunity to submit his diagnosis of its cause. Because damage always represents waste, improved control represents a primary way to improve overall warehouse productivity.

47

REVERSE LOGISTICS
IN THE WAREHOUSE

The traditional management of logistics systems, including warehouses, has emphasized the movement of materials from sources of supply to points of consumption. Warehouses are part of the flow process, but we usually assume that the flow goes only from source to consumer.

Until the past few years, concerns about the flow of materials from consumer back to source were almost unheard of. However, recent years have brought new challenges, including the need for reversing the traditional logistics flow. Hence we here spotlight the challenge of returns.

Three Reasons for Returns

The first of these new challenges was improvement of consumer protection through quick and accurate product recalls. A production error, a design failure, or even a case of deliberate sabotage may cause the need to isolate and return all units of a given production lot to ensure the safety of the consumer.

Perhaps the best-known example was the case of deliberate poisoning of a popular pain killer several years ago. Similar incidents have taken place with contaminated food items, faulty automobiles, and chemicals that have a production defect. When the fault or defect is discovered, the manufacturer usually isolates the problem to given production lots and limits the recall to those lots instead of recalling the entire production item. Some food manufacturers make test recalls to verify the capabilities of warehousing and trucking supplies to handle a recall correctly. The need for speed and precision in the process is obvious, particularly when the defect could threaten the lives of consumers. In this situation, the prime stimulus for product recall is consumer safety.

A second major motivator for returns is avoiding the degradation of the environment. This seems particularly popular for widely used chemical products, such as plastic packaging. There is concern about the depletion of landfill sites and the consequent rising cost of solid-waste disposal.

A proven way of reducing the volume of waste is to recycle and reuse certain packaging materials. This may involve the sorting of recyclable material by product type: plastic, ferrous metals, other metals, and paper. Once sorted, the recyclable material moves via various routes until it is returned to a manufacturing point where it can be remanufactured into new packaging or other similar products. Precision may be less critical in this kind of return, since consumer safety is not involved. However, the concerns of environmentalists have created almost as much public pressure for recycling of packaging as we have always had for the recall of defective consumer goods. In some cities today, laws require that a vendor of a major appliance return the appliance carton to relieve the consumer of the substantial disposal problem of the shipping container. The problem of course is not eliminated, but only transferred from consumer to vendor.

The third motive for returns is conservation of assets. For years before we had environmental concerns, pallets were exchanged rather than donated to the buyer. The exchange reflected the fact that the vendor simply couldn't afford to give away an expensive pallet with each unitized load. Similar motives caused the development of refillable beverage bottles decades before anyone was worried about landfills. The bottles were expensive and, in fact, there was a time when a popular soft drink came in a bottle that had a higher value than the fluid inside. The consumer was not forced to return the bottle, but was simply penalized for failure to do so.

The auto industry has reusable packing techniques, such as steel racks to hold engines, metal bins and cages to hold a wide variety of small auto parts, and several devices to unitize tires. In each case, the warehouse, the trucker, or any other party in the logistics process has an obligation to return this permanent packaging material so that it can be used again. The auto industry has been notably successful in imposing the discipline needed to protect reusable packaging so that it can be used many times before being discarded.

The Role of the Warehouse Operator in Returns

Warehouse managers, both third-party and private, are key players in the return process.

Consider first the safety recall. The speed and accuracy of the warehouse manager in identifying, isolating, and returning the lots that are potentially dangerous is of critical importance. The warehouse manager who fails in the process is likely to be severely disciplined, even if the failure is only in a test recall. Manufacturers who face this kind of safety risk have little tolerance for lack of accuracy in a product recall. When faced with a recall, the warehouse manager must approach it with all the seriousness and concentration of a fire or disaster drill. Done wrong, the consequences could be just as grave.

The warehouse operator can be particularly creative in the case of environmental returns, since some warehouses are able to provide value-added services such as compacting, bailing, shredding, grinding, or other processes. These will lower the cost of storing and transporting the material that is to be recycled. The economy of performing such services as close as possible to the origin of the material flow is obvious. Therefore, we can anticipate that a growing number of warehouses will have equipment designed to compact material being returned for recycling.

The most important thing the warehouse can do in handling conservation returns is to perform the function at the lowest possible cost. Since the motive for the return is conservation, the manufacturer seeks to minimize costs in all phases of the return process. Sometimes the material could be stored outdoors at a lower cost, but outdoor storage is an eyesore, and such storage also exposes the materials to contamination. The best way to reduce costs may be not to store the material at all, but to be sure that it is promptly moved in an otherwise empty vehicle that is returning to the recycle point.

Reconfiguring the Warehouse for Returns

New layouts and new equipment are likely to appear in warehouses that are involved in handling returns.

In some cases, returns will involve disposal of waste materials by burning. Then the warehouse operator will need to acquire and operate incinerators. Layout planning must include finding the best location for the incinerator and developing a pattern of flow to that machine.

Other types of equipment may be needed to process or to temporarily store recyclables and waste materials. These include bailers, shredders, and compactors. The arrangement of this equipment within the warehouse or on the warehouse grounds must be carefully considered to avoid problems of odor contamination, spillage, or other housekeeping problems. The layout should also be designed to provide optimal layout and flow of materials to and from the machines. This flow should include adequate areas for segregation and sorting of waste materials.

Dissimilar materials must be isolated for recycling purposes, and the facility should be designed to facilitate the sorting process. In some cases, warehouse facilities will be redesigned or reconfigured to handle recycling and waste management in the receiving areas.

Security Issues

Sometimes a product return is required for stock balancing or to allow minor repairs. One small-appliance manufacturer allows dealers to return unsold merchandise to a central warehouse on a periodic basis. This good-will gesture is done to remove some of the inventory risk and encourage dealers to adequately stock the product line. In other cases, slightly damaged merchandise is moved to a central collection point for refurbishing or repair. Stock balancing and repair returns may involve costly merchandise, and sometimes the merchandise is not returned in original containers. When this happens, it is not easy to obtain an accurate count and prevent pilferage.

Therefore, a warehouse handling such returns must be configured to provide maximum security for the returned products. Goods in makeshift packages should be reconditioned immediately so proper count of the merchandise can be maintained. The handling of valuable returns represents a significant security risk, but it can be controlled through careful storage practices.

Reconditioning and Repackaging

One of the consequences of handling returns is the frequent necessity to repair, clean, recondition, or repackage the returned merchandise. Accomplishing this may require the establishment of a small assembly operation within the warehouse. Packaging, cleaning, or repair may be accomplished by moving the product down an assembly line that contains the appropriate materials. This may require significant reconfiguration of warehouse design.

Running a Warehouse in Reverse

One third-party warehousing company has developed an unusual and effective specialty—the handling of returns. This unusual process can best be described as an order-picking distribution operation running backwards.

One operation is dedicated to a major retail chain. Destination shipments move to 2,100 different vendors, though usually only 850 of these are active at any one time.

The process starts when a consumer takes a purchase back to the retail store and claims that it is defective. The retail store ships pallet loads of returned goods to the return center. For security reasons, these pallet loads are unitized and sealed with a specially printed stretchwrap that is designed to provide evidence of pilferage if it exists. Before shipping, the retail store attaches a bar-coded "license plate" to identify each item. As the loaded pallets are received at the return center, they move down a conveyor lane where operators examine each item, scan the license tag, and attach a special sort label that provides additional information in alphanumeric codes.

Returned goods are separated into three types. Type 1 merchandise is that which fits in normal warehouse plastic tote bins. Type 2 is boxes that are too big for totes and weigh up to 40 pounds. Type 3 is product over 40 pounds and requiring a "team lift." Once put in categories, products then move to a "home slot" or staging area until a full pallet load has been accumulated. Products that are improperly packaged are repackaged at the return center to be sure that the product will be protected from handling or transportation damage. Irregularities fall into three categories. Red tag items are those that should be destroyed. Green tag items are those that should have been salvaged at the store. Research items are those that require further checking and verification in a special research department. Once a full pallet is accumulated, it is placed in storage rack to wait for a "vendor cut," which is permission to ship product back to the original source. The vendor cut depends on a minimum quantity by weight, dollars, or time which is specified by the vendor. Pallets awaiting vendor cut have a pallet identification tag which is color coded to show the month. A precise locator system allows the operator to know exactly where every item is staged.

Outbound movements to the vendors are in full pallets, again sealed with the special security stretchwrap. The carrier is required to sign only for the total number of pallets, with the presumption that a sealed pallet has an accurate count.

To ensure accuracy, an audit group does a recheck of ten outbound orders out of a total of 90 to 200 orders that move each day. The operation experiences an error rate of less than 1%, and a significant portion of this is errors that originate at the retail store.

The inventory in this warehouse turns more than 50 times per year. An on-site representative of the retailer is available primarily to handle problems of communicating with the stores that originate the movements to the distribution center.

In another city, the client is a manufacturer. Their returns center accepts merchandise from all retailers handling cosmetic products. The center is tied in with central information systems, one controlled by the client and the other by the warehouse. One role of the return center is to prevent abuses of the return privilege by retailers. Another is to inspect and check the quality of returned goods. When saleable stock is isolated, an automatic count device puts the prescribed number of units in a tote bin which is subsequently repacked into a box. Some merchandise is consigned to a secondary retailer who specializes in selling off-price items. This product still contains price labels from other stores, and it is packed in generic boxes for shipment to the secondary retailer. Other merchandise may be designated first quality, and it is repacked in original factory boxes for shipment to primary retailers.

Before the center was established, the client destroyed every product that was returned. Today, the return center puts value back in the product, either by restoring some to the primary market at regular prices, or by consigning other products to a secondary market at reduced prices. A portion of the product must be destroyed, and the return center contains a shredding machine that pulverizes the product before it is hauled to a landfill.

Another task handled by the center is the correction of promotional errors. This typically involves repackaging when the carton, literature, or product contains inaccurate or unsuccessful ad messages. The product is in good condition, but the existing packaging is not usable.

Product returns have always been considered as a warehouseman's nightmare. Yet a specialist can turn the process into an orderly activity that restores value to products that previously could only be destroyed.

Keeping It Simple

A valuable series of scholarly works deals with the concept of reverse logistics. Some of this writing makes the process appear more complicated than it is. Logistics people have handled certain kinds of returns, such as beverage bottles, for many decades, so there need be no mystery about the process.

In essence, returns are primarily common sense. Worries about the environment and consumer safety have created a new degree of urgency about the process, often because of the involvement of government and consumer protection agencies. Despite this, the essential components of the process have changed remarkably little. The next time you see some ponderous writing about reverse logistics, just think about the return of soft-drink and milk bottles.

Part IX

HANDLING OF INFORMATION

48

CLERICAL PROCEDURES

The clerical section of a distribution center is frequently the most undervalued part of the system. It often receives an inordinately small amount of attention from management. While entire magazines, books, and trade shows inform management about stacker cranes, storage racks, and other hardware associated with the storage and handling functions, relatively little attention is given to the most critical function of all—the clerical operation.

Most warehouse operations that are in trouble can trace the difficulty to malfunctions in the clerical operation. This is especially important because of the "middle-man" role of warehousing in the typical distribution system. Prompt and accurate communication for both shippers and consignees using the warehouse is critical to the success of any warehouse operation.

While major improvements in warehouse materials handling methods have occurred during the past several decades, the technological change in office procedures has been even more pronounced. The prime cause was the introduction of electronic data processing equipment, followed by its miniaturization and substantial cost reduction.

The emphasis on computer-oriented information systems has placed greater responsibilities on the clerical personnel at the warehouse. The interface between the warehouse and the user's computer has become increasingly important. Since clerks now routinely feed information directly into warehouse users' computers, the potential for disruption through clerical error is greatly increased.

Any warehouse clerical system should be designed around a

profile of the user's needs. The operator must know what kinds of information are needed, and when.

Customer Service Standards

Order-response time is the most important customer service requirement that influences clerical operations. This is the time lapse between the moment when the warehouse is notified to ship and the time when the goods must be physically out the door, or the time when they must be in the hands of the consignee. Order-response time is the most critical service factor at most distribution centers. Yet, the allowable time may vary substantially in different industries, or even between different customers in the same industry. Customer service standards in a distribution center for pharmaceuticals, for example, will be different from those of a furniture warehouse.

In third-party warehousing, it is common to establish a standard order-response time. A typical standard in this industry is the following:

1. All orders received by noon are shipped on the same day.
2. All orders received after noon are shipped on the following day.

Location of Clerical Center

The clerical function need not be handled at the warehouse, since today's data communications equipment provides almost unlimited location options. Clerical work can be performed at a remote location, with bills of lading sent to the shipping dock or via facsimile transmission. Receiving information can be reported via the same means of transmission and perpetual inventories can be maintained at a location outside the warehouse.

With today's capabilities in processing and transmission of information, a warehouse can be controlled from a clerical center that is miles away from the storage building. Yet, most managers want the clerical function under the same roof, primarily since clerical personnel frequently participate in inventory checks and other activities that involve the cooperation of both the office and materials-handling employees.

Office Environment

The atmosphere and appearance of an office play a vital role in its effectiveness. When the office is isolated from the warehouse, an insular attitude may develop among the clerical personnel. Since effective customer service requires complete cooperation between the clerical and materials handling personnel, poor office location can deter teamwork.

Security for the office area is frequently overlooked. Today's office contains electronic equipment small enough to be easily carried out. Few warehouse offices handle cash or negotiable securities, but they do handle documents and records that are difficult or impossible to replace. Backup systems stored separately should be available to replace records that may be lost or destroyed.

Since the warehouse office is usually the first place customers are received, it should be a showcase for the entire distribution operation. A pleasant work environment also will pay off in improving office morale and efficiency. Ample lighting, effective use of colors, and control of temperature and noise levels must all be considered in developing an effective work environment.

Computer printers create noise problems unknown in offices a few years ago. These machines should be isolated or surrounded with sound-dampening enclosures. A carefully planned office layout is the best means of controlling noise levels. As more computers use the quiet laser printers, noise promises to become less of a problem in the future.

Organizing the Clerical Function

There are two methods of organizing work in a warehouse office—either by clerical function or by customer account.

In the clerically oriented system certain individuals perform specific tasks for every user of the warehouse. For example, one individual might be in charge of preparing receiving reports. Another prepares all the bills of lading, and still another has duties limited to inventory record-keeping. The clerically oriented system permits a person with limited skills to be employed in repetitive work. In effect, it is a variation of a manufacturing assembly line. Just as one assembly line worker may do nothing but install wheels, the receiving clerk will do nothing but prepare receiving reports. As in an assembly line, such repetition can be tedious, but the boredom can be relieved by rotating responsibilities and cross-training personnel on different jobs.

In public warehousing individual users or customers are sometimes referred to as "accounts." Most public warehouses will assign certain clerks to handle one or a group of accounts. That clerk will perform all of the clerical functions in connection with specific accounts. These may include receiving reports, bills of lading, and supervision of inventory records. When the customer calls, there is one person in the office who maintains a hands-on familiarity with all the clerical functions of that account. The same practice can be used in a private warehouse, though in this case the "account" is the outbound shipment consignee. The account-oriented system requires greater versatility and skill on the part of each clerk in the system. On the other hand, it affords a more personalized approach to a customer who wants to talk with someone who has total familiarity with that customer's transactions.

Interaction with Marketing

One activity of the clerical department in most warehouses is interacting with the users' sales staff. Most warehouse users have recognized the desirability of separating warehouse workers from sales personnel in order to maintain maximum efficiency of both. Yet, the demands of sales personnel must be balanced with the shipping capacity of the warehouse to develop a customer service system that is both predictable and realistic.

For example, if warehouse shipping capacity is 20 truckloads per day and shipment requests for 25 truckloads are received for a given day, some compromises must be made. So the clerical coordinator's job is to merge the demands of salesmen with the capabilities of the warehouse. When releases exceed capacity, the coordinator will develop a schedule approved by the user to designate which orders will be shipped and which will be deferred. This avoids friction and the time lost if sales personnel communicate directly with the warehouse personnel.

Manual and Electronic Data Processing

In the pre-computer era warehouse inventory control systems consisted of hand-posted ledgers and manually prepared shipping and receiving documents. Visible record card files were developed to make it fast and easy to use inventory records. These well-designed systems allowed the user to see low stock or low activity at a glance.

With today's array of computers, a wide degree of automation is available for inventory control, word processing, or communication. One important advantage of computers is their ability to speed order-response time. Nonetheless, there have been faulty computer installations in which order-response time has been slowed rather than expedited.

Batch processing is a means of increasing data processing efficiency by first accumulating a group of orders and then processing all of them at one time. Batch processing uses the computer more efficiently, but it will do so at a sacrifice in order-response time unless the batches are processed frequently.

JIT (just-in-time) systems and faster response times have substantially reduced the popularity of batch processing.

The need to increase clerical capacity is a frequent motive for changing to an automated system. Some automated systems can handle far more line items and orders per day than manual systems.

A fringe benefit of an automated system is the requirement for rigid standardization of clerical procedures. Enforcing the discipline necessary to operate computers has side benefits in most warehousing operations, if an adequate level of standardization can be maintained.

Control of Automation

Automated systems are never as automated as they first appear. Correcting the system when it develops trouble is one drawback. When incorrect information is produced, someone must be able to quickly diagnose and correct the problem. This is not easy to do in a complex system, and failure to do so can have disastrous results in a warehouse operation. Keeping backup data is a necessity for reconstructing the pieces when the system goes awry.

Loss of inventory control is a disaster in any warehouse clerical system. This could occur if inventory records are destroyed and no backup records are available. Under such circumstances, the only recourse would be a physical count of the inventory. Such failures might be caused by deliberate tampering with data processing systems. Effective computer security is therefore very important.

Measuring Clerical Costs

The best standard for comparison between various clerical systems is a cost per unit. This may be expressed as a cost per bill of lading, a cost per billing line, or a clerical cost per unit shipped. By developing a clerical unit cost for both existing and proposed systems, it is easy to determine whether a new system is justified.

Though more sophisticated systems may be installed for non-cost reasons, such as improved response time or accuracy, a true cost of these should still be determined. In this way management can establish tradeoffs between a proposed new system and an existing one.

Management Information

A bonus benefit of computerized clerical operations in the warehouse is the development of management information that can aid in controlling warehouse operations. The operations manager can gain information on productivity and labor expended for each warehouse account. Payroll reports help to control overtime and measure materials handling productivity. In a public warehouse, cash flow is controlled through reporting on aging of accounts receivable.

Clerical systems have changed as much as, or more than, any other aspects of warehousing. Regardless of the proliferation of equipment, ultimately the system depends on well-motivated personnel.

49

COMPUTERS AND WAREHOUSE MANAGEMENT

The use of computers in warehouse management is taken for granted so often that we fail to recognize the countless instances where no system exists, or the one that does exist does not function well.

Some Definitions

Hardware is the computer equipment itself, the central processing unit, and its peripheral devices.

Software is the general term to describe all written instructions used to control the hardware.

Application software is a set of programs that enable a computer to carry out a specific function for an end user. Examples of application software are inventory systems, order processing systems, and locator systems.

Operating system software is a set of software programs to control the computer and allocate computer resources as required. The operating system controls and directs input from the user and output from the application software. Commonly used operating systems include MS-DOS/Windows®, Windows® 95 (and its successors), and UNIX®.

Program development software is used by programmers to translate the instructions (sometimes called *source codes*) into the *object code* required by the computer. Examples are Progress, BASIC, PASCAL, FORTRAN, and COBOL. Programs executed by computers are written in binary code, a series of ones and zeros.

Firmware describes software that is integrated with the hardware and is always present whenever the computer is used. Some firmware is implemented into the electronic circuitry of the computer. Other firmware is located in memory chips that can be changed to update the instructions. One example of firmware is the program that activates when a computer is first turned on and manages the system startup.

The warehouse management system is a computer-based program that allows goods to be physically controlled, identified, and monitored from the time they arrive at the receiving dock until they are loaded onto a delivery truck. One of the most important features of such a system is its ability to locate and trace every item.

Hardware versus Software—Then and Now

In the early days of the computer industry, there were a few major vendors of hardware. All sold proprietary software systems along with the hardware. Not only the hardware, but also the operating systems and even the programming languages and development tools, were unique to each manufacturer. Application software written for one computer system typically would not run on another brand of computer without an expensive conversion. There were often incompatibilities among computer models, even within a single manufacturer's product line. A programmer who had developed skills on a particular system could not cope with another system without considerable retraining.

The cost of even a modest computer system was vastly greater than it is today. Hardware systems alone ran into six figures and frequently into a seven-figure purchase price. Indeed, monthly rental costs for most computers were typically higher than the purchase price of many business computers today. Most managers felt that the cost of a wrong decision in the choice of hardware could spell financial disaster.

Therefore, in the first two decades of computer use in business, hardware received primary emphasis. Many companies bought computers and then had to find software that was compatible with the hardware. This meant that choices were very limited after the hardware was selected. The amount of packaged software was very small, and finding a software program specifically designed for warehousing was usually impossible. For a user who did not have application software and could not find a suitable package, the choice of computer systems was based upon which hardware vendor offered the best product and the best software development tools and services. The salesperson was often poorly informed or otherwise motivated to sell a particular application package that was less than ideal for warehousing. The choice frequently depended upon other software, operating systems, program development software, and database management systems.

In 1980, the distinction between micros, minis, and mainframes was clear. Micros were small desktop units limited to a single user. Mainframes were large, expensive units capable of supporting dozens to hundreds of users. Minis were mid-range systems, capable of supporting from a few to a few dozen users.

Today, the micro is steadily pushing the mini out of the marketplace. Most people are familiar with the single-user micro, the desktop "personal computer," an indispensable business tool. But today's higher end micros are capable of supporting dozens, or even a hundred users with the proper operating system. Most warehouses today do not need a mainframe or even a mini, but can satisfy their needs with a micro. The opportunity for cost savings and increased productivity is staggering.

Which Comes First—Hardware or Software?

While hardware may have been emphasized in the early days of computers, without any question today the software comes first. However, while ten years ago the focus was on application software, today the emphasis is on operating systems and program development software.

There are occasional exceptions. A company with a substantial investment in application software may upgrade its hardware by choosing a computer that is compatible with its existing software.

In absence of strong reasons to the contrary, the typical user today chooses an industry standard operating system, along with program development software compatible with that system. The hardware can be any brand that will function with the software system.

Choosing Warehousing Software

If your warehouse does not have acceptable software and if you do not wish to spend the time and money to develop a proprietary system, you have many choices of software systems. The problem is to find everything that might be applied and then to evaluate the alternatives.

Sadly, even today some managers rely on a hardware salesperson to tell them what application packages are available. There are many information resources, but the most complete is an annual guide to logistics software published by Arthur Andersen and Company.

The first step in evaluating application software is to define the functions you need performed. Then evaluate available packages based on their ability to deliver the functions you need. This process should reduce the number of software packages that you consider.

When an apparently suitable application package is selected, the next step is to check references. The references are to current users, and at least some of these should have an operation similar to yours. This user check should include on-site visits to some of the companies given as references. Observe the package in a real world environment, meet the people who use it and learn first-hand what experiences others have had in applying the new system.

In checking references, a key point is the quality and availability of systems support. No matter how well the system works on the day it is installed, changing conditions in your warehouse will require appropriate changes in the information system. How you will be able to adapt to those changes depends on how much help you can get from the software company that sold you the system.

Consider where your company will be if the software supplier should go out of business. Do you have access to the source code to protect yourself in the event that the software supplier is not available at some future date? A number of successful systems have met disaster because of lack of support after the system was installed.

Barriers to Implementation

Although state-of-the-art for both hardware and software would permit one or a network of personal computers and packaged software to be used to support a fairly sophisticated warehouse management system, a distressing number of warehouses do not have such a system because other departments have demanded that the warehouse management system be integrated into information systems covering procurement, inventory control, production, sales, or other general business functions. Therefore, because the MIS (management information services) department has not had time to plug the warehouse into the larger control system, many warehouses have no management system or one which is primitive compared to today's state-of-the-art.[43]

Choosing a Warehouse Management System

Before you investigate further use of computers, it is important that you make certain your operation is functioning well now. Why? Because if you try to add computer-assisted processing to a sloppy or disorganized warehouse, the end result will be much less than you can achieve otherwise.

It might take only a few days to review your current operation. This can be done by your own team or in conjunction with a logistics consultant who has experience in doing these reviews. Work practices should be standardized and work flow streamlined as much as possible within current capabilities. Storage practices should also be looked at and "honeycombing" (unused empty space) reduced to the minimum. A thorough review will improve your operation today by weeding out inefficient practices. The review will ensure that your new system is based on using best practices in operating your warehouse. It helps you avoid the trap of changing your business to accommodate the method of operation described in a particular software package.

Be sure you allow ample time to select, modify, and install a new system—it usually takes 9 to 18 months. But start to upgrade warehouse practices immediately, so that you can gain productivity while you are going through the process of installing a new computer system.

Selecting a New System

When you have completed the diagnostic operations review and upgraded your current warehouse practice, you will be ready to start the selection process. The intent of your search should be to find a system that enables you to operate your warehouse the way you want to conduct your business, but in a more efficient manner with higher service levels and lower costs.

An effective approach to the selection process is to establish a task force that is responsible for selecting the system of choice. The team should be made up of at least one person from operations, a representative from your MIS or Data Processing area, and a financial person. If you want a larger group, select another member or two from other areas of the company.

To ensure that your team stays on track, you should identify and describe your current customer service practices. For example, most stated customer service policies include criteria for *product quality, order fill rate, accuracy,* and *order cycle times.*

A statement of requirements is the central reference point for your software search. Writing this statement can be a very slow process for a group whose members are working less than full time at the effort. Allow three to four months to complete this statement. The time spent will be well worth the effort if the end result covers all of your requirements and details the specifics of each area listed below. The statement must provide a clear description of the system you expect to eventually install, and it should specify efficiency level, accuracy level, and the customer service standards you must have to remain competitive.

The components of the Statement of Requirements should include:

- *Interface requirements:* This should detail the data that will need to be exchanged between other existing systems in your company as well as systems that you are planning to add in the future.

- *Transaction volume:* You should identify in detail the type and number of transactions that you expect the new system to process on a daily basis. Include the range of volume from low to high and the number of warehouses involved if more than one. A projection of volume for the next five years will give you room for growth.

- *Hardware criteria:* If your company prefers or already owns a hardware platform and peripheral equipment, this should be stated. If not, acceptable alternatives should be identified.

- *General requirements:* This should be a summary of the breadth of the system expected to be implemented and should describe an outline of the final system. Such areas as geographic coverage, functions included, material handling equipment to be utilized, and reporting requirements should be stated in this general section. Detailed requirements can be specified in the following sections.

- *Inventory location and management requirements:* In this section there should be a description of the system requirements for inventory practice, physical inventories, cycle counting, storage locations, tracking, reporting, and other company needs relating to inventory.

- *Receiving requirements:* All needs of the receiving function should be detailed here and should describe the products, advanced notice of delivery, types of receipts, exception receiving, use of bar-code labels, and other specific needs for your company.

- *Put-a-way requirements:* The project team will need to decide what requirements are appropriate for this section and could include direct put-a-way, exception handling, inventory updating, cross-docking capability, and confirmation of put-a-way.

- *Order management requirements:* Depending upon the make up of your customer orders, you will want to specify the type of order groupings, order statistics, reporting, and tracking needs for your ultimate system.

- *Replenishment requirements:* Most warehouses will have some type of replenishment needs, and this section should describe such requirements as replenishment based on FIFO (first in, first out), or on current demand as well as confirmation of replenishment.

- *Picking requirements:* Your particular type of business, product, and order mix will determine your picking method needs. This section should include requirements for such methods as sequential, wave, and/or batch picking. It could also include provisions for segregating orders, staging by order type or loading sequences. Confirmation of pick can be identified as a need.

- *Labor and work flow management:* With the data captured, there will be opportunities to track and report on labor requirements as well as work flow through the warehouse. This is one of the most important capabilities of the computer-assisted systems, since it can enable management to control the resources of the facility on a timely basis and adjust quickly as conditions change.

- *Shipping requirements:* This section should identify your needs for such techniques as sequential loading, order staging, exception reporting, trailer load closing, and interfaces with document preparation.

How Do You Locate the Right Software Company or Package?

Once you have completed the Statement of Requirements and have approval to proceed, you must identify software companies that provide a system that meets your specific needs. Many companies provide "package" systems. There are also companies that provide "custom" systems. Other software houses supply "semicustom" systems, which are modular and can be modified to meet your particular needs. Generally, the package systems are least expensive and the custom systems are the most expensive. Unless your system needs are very unusual, one of these systems should meet your requirements at an affordable cost.

By reviewing the advertisements found in most distribution or computer literature and by talking to your counterparts in other warehouses and companies you will be able to identify software vendors that are capable of meeting your particular requirements. If you seek a logistics consultant to assist you in this process, sources can be identified through professional associations such as the Council of Logistics Management or the Warehousing Education and Research Council.

When the potential suppliers have been identified, you should interview several of them, using your Statement of Requirements as the basis for discussion. Usually four to six interviews will give you enough information to prepare a project proposal. Try to find a system that can meet your requirements with 15% to 20% modification or less.

How Long Does Implementation Take?

When final approval has been given, the design, modification, and installation process can take nine or more months depending upon the extent of the changes to be made and the number of facilities involved.

If you follow this process, you will have installed a computer-assisted warehouse management system that will provide a more efficient, more cost-effective warehouse with high customer service levels. Many companies experience productivity improvement as high as 30% to 40% above current levels along with comparable cost reduction levels.

The model shown on Figure 49-1 provides the topics that should be included in a typical Statement of Requirements.[44]

Figure 49-1

THE XYZ COMPANY

A Model Warehouse Management System—System Requirements Outline

1.0	General Requirements	4.7	Purchase Order Reconciliation
1.1	Support Multiple Geographic Locations, Warehouses And Sections Within Each Warehouse	4.8	Receipt Inspections
		5.0	Put-A-Way Requirements
		5.1	Direct Put-A-Way
1.2	Support Item Definitions As Well As Single Item & Mixed Item Quantity Locations	5.2	Exception Handling
		5.3	Inventory Updating
		5.4	Confirmation of Put-A-Way
1.3	Support Definitions of Equipment, Storage And Human Resources	5.5	Cross Docking
		5.6	Combined Processing
1.4	Support Work Measurement Criteria And Track Performance Against Those Criteria	6.0	Order Management Requirements
		6.1	Order Groupings
		6.2	Order Statistics
1.5	Support Random Location of Inventory, FIFO Processing and Multiple Items Per Location	6.3	Reporting
		6.4	Tracking
		7.0	Replenishment Requirements
1.6	Provide Management Reporting Capability	7.1	Replenishment on FIFO Basis
		7.2	Replenishment Based on Current Demand
1.7	Provide Ability to Track Orders by Status	7.3	Replenishment Confirmation
1.8	Provide Inquiry Capability	8.0	Picking Requirements
1.9	Ability to Handle Exceptions and Unusual Situations	8.1	Sequential Picking
		8.2	Other Than Wave Picking
1.10	Support Interfaces to Material Handling Equipment And Radio Frequency Devices	8.3	Segregation of Picks
		8.4	Confirmation of Picking
		8.5	Use of RF Equipment
1.11	Support Use of Bar Code Labels	8.6	Plan Warehouseman Work
2.0	Inventory Location/Management Requirements	8.7	Order Integrity
		8.8	Exceptions
2.1	Item Identification and Tracking	8.9	Efficient Picking
2.2	Physical Inventories	9.0	Labor Management Requirements
2.3	Support Cycle Counting	9.1	Identify Resource Requirements
2.4	Inventory Accuracy	9.2	Work Flow Monitoring
2.5	Item History	9.3	Operating Levels
2.6	Product Profile Analysis (ABC Analysis)	9.4	Management Reports
2.7	Allocated and Available Inventory	10.0	Shipping Requirements
2.8	Quarantine or "On Hold" Inventory	10.1	Fluid Loading
2.9	Storage Locations	10.2	Order Staging
2.10	Storage Location Definitions	10.3	Confirm Complete Orders/Trailers
2.11	Storage Location Control	10.4	Identify Exceptions/Shortages
2.12	Storage Fill Rate	10.5	Print Manifests and Other Required Documents For A Trailer Load
2.13	FIFO Control		
2.14	Support Container Sizes and Equipment	10.6	Interface With Inventory And Invoice Systems
2.15	Reporting Capability	10.7	EDI To Intended Designations
3.0	Inventory Location/Management Requirements Assumptions	10.8	Transfer of Loads
		10.9	De-assign Product or Orders
4.0	Receiving Requirements	10.10	Picking Status
4.1	Advanced Notice of Delivery	10.11	Verification of Orders/Items
4.2	Different Types of Receipts	11.0	Work Flow Management
4.3	Availability of Items	11.1	Daily/Shift Workload
4.4	Exception Status, Handling And Reporting	11.2	Exceptions
		11.3	Work-In-Progress
4.5	Bar Code Labels	11.4	Interfaces
4.6	Multiple Receipts	11.5	Future Workload

Summary

Warehousing has enjoyed some remarkable success through effective use of information systems, but also from shocking failures. In March, 1996, *Information Week* described a warehouse disaster in an article titled "Melt Down." Fortunately, the successes have outnumbered the failures. The careful warehouse operator should be able to improve the odds that a system will be successful.

50

ELECTRONIC IDENTIFICATION

In the entire field of warehousing, perhaps no technical development has shown greater potential than electronic identification. There is more than one way to "machine-read" information from a package, but today the most common method is bar coding. The concept of bar coding has been around for decades, but it is just in recent years that its full potential in warehousing has been realized.

Electronically reading and copying a symbol from bar coding is faster and more accurate than any manual processing. A government study showed that scanning bar-code systems was 75% faster than entering the information through typing, and 33% faster than entering the information on a ten-key data board. Keypunchers had an accuracy of 98.2%, but in a study of over 1.25 million lines of bar code symbol data an accuracy level of 99.9997% was achieved.

Bar Codes—What They Are and How They Work

Bar codes are a grouping of lines, bars, and spaces in a special pattern. This pattern can be read by a machine, which communicates with people or other machines. The bar code itself can be applied when the carton or package is manufactured and printed. The cost of applying or printing the code at that stage is virtually nil, since package lettering has to be printed anyway. The code can also be applied on-site, perhaps at the receiving dock of the warehouse, by a small printing machine.

Once applied, the code can later be read or "captured" by a bar-code reader or scanner. Scanners of various sizes are available. Used in the warehouse, this method of electronic identification offers many advantages, and there seems little doubt that new applications will be developed in the years to come. At the receiving dock, a code can be read and the identification used accurately to update the inventory. Since many inventory errors occur because of mistakes in receiving, this application has great potential for improving accuracy. As the product is identified, a determination can be made as to its best storage location, and immediate instructions relayed back to the receiver to indicate the best storage address for the inbound item. When a physical inventory or a location check is made, electronic identification will ensure that items are identified correctly as they are counted or located. When it is time to ship, automatic identification drastically reduces the possibility of shipping error through misidentification of goods. Further, if every item is scanned as it moves past a loading conveyor, count can be verified. In those operations where broken case order-picking is involved, bar-code identification can be particularly valuable in verifying that the right number and assortment of inner packages are included in each master carton. Finally, electronic identification can be used as input for automated sorting or picking equipment. As the identification of each package is determined, instructions can be given to the sorting equipment.

The most common bar codes in use are: Code 3 of 9, Interleaved 2 of 5, Code 128, Code 93, Codabar, UPC (The Universal Product Code), and the Plessy Code. (See Figure 50-1 for an analysis of Code 3 of 9.)

Scanners

Machines that read or "capture" the data contained in a bar code are called bar-code readers or scanners. They are classified as follows:

- Portable scanners
- Laser scanning guns

- Fixed-beam stationary scanners
- Moving-beam stationary scanners

Scanners are mobile and can be carried to a receiving area, a racking operation, or a bulk storage area. They can be used on a lift truck anywhere in the facility. A clerk can communicate with someone who is out of the office through a visual display terminal (VDT).

The bar codes are read by shining a beam of light on the code. A hand-held terminal can have a pen-type wand for contact reading (to read, the wand must touch the bar code), or can be equipped with a laser gun for distance reading. Scanners use either a moving or fixed beam. Portable scanners are lightweight (two to three pounds), are programmable, and have visual display capability. Their principal use is for capturing data on packages not moving on a conveyor. Portable scanners can be used for order selection, inventory counting, cycle counting, location product verification, and stock replenishment.

Stationary scanners can be positioned on the side of conveyors to scan bar codes of passing cartons and transmit the information to a computer. Stationary scanners can also be positioned above conveyors. Whatever their fixed position, these scanners use grid patterns of light to pick up bar code signals and transmit the data to any point for decoding.

Stationary scanners can accurately read bar codes on objects traveling at high speeds. For variable carton heights, the overhead scanner is equipped with a lens that adjusts to the various carton heights.

One advantage of stationary scanners is that they can read bar codes without someone having to feed the material through a given point. This is also done through the flip-lens photo eye, which consists of two overhead scanners. One is a single-line laser; the other sends a multidirectional pattern of laser light in the form of "Xs" that scan the carton. No specific placement of the carton is required, because with the cross-weave pattern the bar-code symbol can be read in any position.

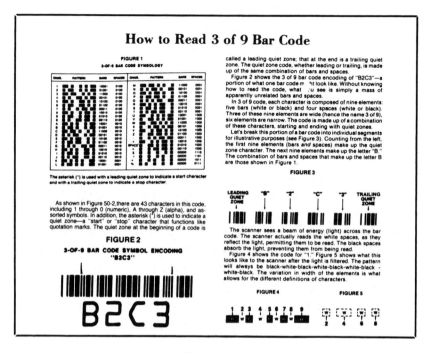

Figure 50-1

Figure 50-2

Endnotes

1. From Volume 11, #8, *Warehousing Forum,* © Ackerman Company, Columbus, OH.
2. From an article by Penny Weber, Genco Distribution System. Published in Volume 8, Number 7, *Warehousing Forum,* © Ackerman Company, Columbus, OH.
3. From an article by Frederick S. Schorr, published in *Warehousing Forum,* Volume 4, Number 11, © Ackerman Company, Columbus, OH.
4. From an article by Lee P. Thomas, published in Volume 19, Number 2 of *Warehousing & Physical Distribution Productivity Report,* © Alexander Communications, Inc., New York.
5. From an article by John T. Menzies for *Warehousing Forum,* Volume 3, Number 8, © Ackerman Company, Columbus, OH.
6. Ibid.
7. The flow charts and subchapter are taken from Volume 7, Numbers 18 and 19 of *Distribution/Warehouse Cost Digest,* © Alexander Research & Communications Inc., New York.
8. By Thomas L. Freese, from an article in Volume 11, Number 6, *Warehousing Forum,* © Ackerman Company, Columbus, OH.
9. From an article by Daniel C. Bolger, P.E. in Volume 11, Number 5 of *Warehousing Forum,* © Ackerman Company, Columbus, OH.
10. From *Injuries to Warehouse Workers,* by the U.S. Bureau of Labor Statistics.
11. From an article by Joel Sutherland, Con Agra-Monfort in Volume 9, Number 3, *Warehousing Forum,* © Ackerman Company, Columbus, OH.

12. By Howard Gochberg, from *Warehousing Forum,* Volume 8, Number 11, © Ackerman Company, Columbus, OH.
13. From an article by Eugene J. Gagnon, Gagnon & Associates, from Volume 7, Number 1, *Warehousing Forum,* © Ackerman Company, Columbus, OH.
14. This section is from an article by Theodore J. Tierney Esq, Vedder Price Kaufman & Kammholtz, Chicago, *Warehousing Forum,* Volume 3, Number 4, © Ackerman Company, Columbus, OH.
15. From an article by Barry Shamis, *Warehousing Forum,* Volume 7, Number 4, © Ackerman Company, Columbus, OH.
16. By Dr. Bernard J. LaLonde of The Ohio State University, from Volume 7, Number 9 of *Warehousing Forum,* © Ackerman Company, Columbus, OH.
17. Adapted from research from Professors Tom Speh and James Blumquist and from a Return on Investment Kit published by International Association of Refrigerated Warehouses.
18. From an article by C. Alan McCarrell, Ross Laboratories, in *Warehousing Forum,* Volume 8, Number 4, © Ackerman Company, Columbus, OH.
19. By Craig T. Hall, Lee-Shore Enterprises, from Volume 9, Number 11, *Warehousing Forum,* © Ackerman Company, Columbus, OH.
20. From an article by John A. Tetz, management consultant, Delaware, OH. Published in *Warehousing and Physical Distribution Productivity Report,* Volume 17, Number 11. © Alexander Research and Communications Inc., New York.
21. From a seminar presented for Warehousing Education and Research Council by Professor Joseph L. Cavinato of Pennsylvania State University and Professor Thomas W. Speh of Miami University.
22. Adapted from an article by Kenneth E. Novak, W. W. Grainger, Inc., *Warehousing Forum,* Volume 5, Number 11, © Ackerman Company, Columbus, OH.
23. From an article by Robert E. Ness, Ohio Distribution Ware-

house, *Warehousing Forum,* Volume 2, Number 5, © Ackerman Company, Columbus, OH.

24. Novak, op. cit.

25. From Volume 19, Number 9 of *Warehousing and Physical Distribution Productivity Report,* © Alexander Research and Communications, New York.

26. Alex Metz, Hunt Personnel, Ltd., in a newsletter published by Hunt Personnel.

27. Based on an article written for *Warehousing and Physical Distribution Productivity Report* by the late W. B. "Bud" Semco, President, Semco, Sweet & Mayers, Inc., Los Angeles. © Alexander Research and Communication Inc., New York.

28. Adapted from writing by W. J. Ransom, Ransom & Associates, *Warehousing Forum,* Volume 7, Number 7, © Ackerman Company, Columbus, OH.

29. From a technical paper by Leon Cohan management consultant, Gahanna, OH and published by Warehousing Education and Research Council.

30. By Jesse T. Westburgh of Citrus World, from *Warehousing Forum,* Volume 10, Number 9, © Ackerman Company, Columbus, OH.

31. From *Temperature Controlled Warehousing:* The Essential Differences, by Tom Ryan and Joel Weber, United Refrigeration Services, Inc. This article appeared in *Warehousing Forum,* Volume 3, Number 11, © Ackerman Company, Columbus, OH.

32. From *Operational Training Guide,* International Association of Refrigerated Warehouses, Bethesda, MD.

33. From Jesse Westburgh, Citrus World, Columbus, OH.

34. From an article by Lake Polan III of Allied Warehousing Services, Inc., *Warehousing Forum,* Volume 4, Number 4, © Ackerman Company, Columbus, OH.

35. From NFPA Flammable and Combustible Liquids Code, 1987 edition, pp. 30–35.

36. From presentations by Jeffrey A. Coopersmith of Columbus, OH and by James E. Dockter of P.B.D. Inc., Alpharetta, GA.

From Volume 2, Number 2, *Warehousing Forum,* © Ackerman Company, Columbus, OH.

37. From an article by J. T. Foley and F. S. Schorr, Volume 7, Number 11, *Warehousing Forum,* © Ackerman Company, Columbus, OH.

38. From an article by Eugene Gagnon, Gagnon & Associates, from Volume 11, Number 5, *Warehousing Forum,* © Ackerman Company, Columbus, OH.

39. From *Improving Existing Warehouse Space Utilization,* by David L. Schaefer. Published by W.E.R.C.

40. From Volume 17, Number 8 of *Warehousing and Physical Distribution Productivity Report,* by the late W. B. "Bud" Semco, © Alexander Research and Communications Inc., New York.

41. From an article by Bill Thomas, *Warehousing Forum,* Volume 9, Number 10, © Ackerman Company, Columbus, OH.

42. Significant contributions to the AGVS material came from Tom Ewers, Bob MacEwan, and Russ Gilmore. Portions appeared in Volume 20, Number 6, *Warehousing and Distribution Productivity Report,* © Alexander Research and Communications, New York.

43. From an article by Bill McDade of Terminal Corporation and Bill Blinn of Proficient Computing Solutions Corporation, Volume 8, Number 6, *Warehousing Forum,* © Ackerman Company, Columbus, OH.

44. By Morton T. Yeomans, from Volume 10, Number 2, *Warehousing Forum,* © Ackerman Company, Columbus, OH.

45. From an article by Russell A. Gilmore III, in Volume 6, Number 11, *Warehousing Forum,* © Ackerman Company, Columbus, OH.

About the author

Ken Ackerman has been active in logistics and warehousing management for his entire career.

Before entering the consulting field, he was chief executive of Distribution Centers, Inc., a public warehousing company that is now part of Exel Logistics USA. In 1980, Ackerman sold the company and joined the management consulting division of Coopers & Lybrand. In 1981, he formed the Ackerman Company, a management advisory service.

He is editor and publisher of *Warehousing Forum,* a monthly subscription newsletter. His three earlier warehousing books became recognized reference works on this subject. *Harvard Business Review* published "Making Warehousing More Efficient" that Ackerman co-authored with Bernard J. La Londe. He is the author of numerous other articles dealing with warehousing and management as well as a glossary called *Words of Warehousing.*

Ackerman holds a Bachelor of Arts degree from Princeton and a Master of Business Administration from Harvard. He is a past president of Council of Logistics Management and the 1977 winner of its Distinguished Service Award.

Ken Ackerman has provided management advisory services to companies throughout the United States, Canada, and Latin America. These clients include manufacturers, wholesale distributors, retailers, warehousing firms, and carriers. He has provided consulting support to several large consulting firms. In addition to advisory services, he conducts training seminars on warehousing. He has served as a speaker at conferences and conventions in North and South America as well as in Europe, Asia, and Australia. His fluency in Spanish enables him to lecture and consult in that language.

Ken Ackerman served as Columbus chapter chairman for Young President's Organization, and as a director of American Warehouse Association. He has been active in civic activities, serving as trustee and founding president of the Wellington School. He was a trustee of Columbus Association for the Performing Arts and a past president of Opera Columbus.

Index